NAVAL STRATEGY AND OPERATIONS
IN NARROW SEAS

CASS SERIES: NAVAL POLICY AND HISTORY
Series Editor: Holger Herwig
ISSN 1366-9478

This series consists primarily of original manuscripts by research scholars in the general area of naval policy and history, without national or chronological limitations. It will from time to time also include collections of important articles as well as reprints of classic works.

1. *Austro-Hungarian Naval Policy, 1904–1914*
   Milan N. Vego

2. *Far-Flung Lines: Studies in Imperial Defence in Honour of Donald Mackenzie Schurman*
   Edited by Keith Neilson and Greg Kennedy

3. *Maritime Strategy and Continental Wars*
   Rear Admiral Raja Menon

4. *The Royal Navy and German Naval Disarmament 1942–1947*
   Chris Madsen

5. *Naval Strategy and Operations in Narrow Seas*
   Milan N. Vego

6. *The Pen and Ink Sailor: Charles Middleton and the King's Navy, 1778–1813*
   John E. Talbott

7. *The Italian Navy and Fascist Expansionism, 1935–1940*
   Robert Mallett

8. *The Merchant Marine and International Affairs, 1850–1950*
   Edited by Greg Kennedy

9. *Naval Strategy in Northeast Asia: Geo-strategic Goals, Policies and Prospects*
   Duk-Ki Kim

10. *Naval Policy and Strategy in the Mediterranean Sea: Past, Present and Future*
    Edited by John B. Hattendorf

11. *Stalin's Ocean-going Fleet: Soviet Naval Strategy and Shipbuilding Programmes, 1935–1953*
    Jürgen Rohwer and Mikhail S. Monakov

12. *Imperial Defence, 1868–1887*
    Donald Mackenzie Schurman; edited by John Beeler

13. *Technology and Naval Combat in the Twentieth Century and Beyond*
    Edited by Phillips Payson O'Brien

14. *The Royal Navy and Nuclear Weapons*
    Richard Moore

15. *The Royal Navy and the Capital Ship in the Interwar Period: An Operational Perspective*
    Joseph Moretz

16. *Chinese Grand Strategy and Maritime Power*
    Thomas M. Kane

NAVAL STRATEGY AND OPERATIONS IN NARROW SEAS

MILAN N. VEGO

US Naval War College, Newport, Rhode Island

FRANK CASS
LONDON • PORTLAND, OR

First published in 1999
Second, revised edition published in 2003 in Great Britain by
FRANK CASS PUBLISHERS
Crown House, 47 Chase Side
Southgate, London N14 5BP

and in the United States of America by
FRANK CASS PUBLISHERS
c/o ISBS, 5824 N.E. Hassalo Street
Portland, Oregon 97213-3644

Website: www.frankcass.com

Copyright © 1999, 2003 Milan N. Vego

British Library Cataloguing in Publication Data

Vego Milan
Naval strategy and operations in narrow seas. – (Cass series. Naval policy and history; no. 5)
1. Naval battles – Planning – History 2. Naval strategy – History 3. Naval tactics – History
I. Title
359.4'23'09

ISBN 0-7146-5389-6 (cloth)
ISBN 0-7146-4425-0 (paper)
ISSN 1366-9478

Library of Congress Cataloging-in-Publication Data

Vego Milan
Naval strategy and operations in narrow seas / Milan N. Vego
   p. cm. – (Cass series–naval policy and history ; 5)
Includes bibliographical references and index.
ISBN 0-7146-5389-6 (cloth)/ISBN 0-7146-4425-0 (paper)
  1. Naval strategy 2. Naval tactics. 3. Sea control. 4. Coast defenses. I. Title. II. Series.
V163.V44 1999
359.4'2–dc21   99-18061
                                                                              CIP

All rights reserved. No part of this publication may be reproduced, stored in or introduced into a retrieval system, or transmitted, in any form or by any means, electronic, mechanical, photocopying, recording or otherwise without the prior written permission of the publisher of this book.

Typeset by Vitaset, Paddock Wood, Kent
Printed in Great Britain by
MPG Books Ltd, Bodmin, Cornwall

Contents

	List of Maps	vi
	List of Abbreviations	vii
	List of Place Names	xi
	Series Editor's Preface	xiii
	Preface	xv
1	Introduction	1
2	The Factor of Space	15
3	Positions	42
4	Bases	61
5	Theater Geometry	73
6	Strategic Objectives and Fleet Distribution	97
7	Sea Control and Sea Denial	110
8	Methods	129
9	Securing Control	147
10	Exercising Control	184
11	Disputing Control	206
12	Attack on Maritime Trade	225
13	Defense and Protection of Maritime Trade	254
14	Support of the Army Flank	267
	Conclusion	292
	Select Bibliography	301
	Index	311

Maps

1	The Mediterranean Sea	20
2	The North Sea	24
3	The Arabian (Persian) Gulf	25
4	The Korean Peninsula and Adjacent Seas	46
5	Western Mediterranean	49
6	The Caribbean Sea	52
7	Southeast Asia and Adjacent Sea Areas	58
8	The Baltic Sea	78
9	The Black Sea	79

Abbreviations

AA	Anti-Aircraft
ABDA	Australian–British–Dutch–American
AIP	Air-Independent Propulsion
AKA	Attack, Cargo
AOA	Amphibious Objective Area
APA	Attack, Transport
ASM	Air-to-Surface Missile
ASW	Anti-Submarine Warfare
ATAF	Allied Tactical Air Force
ATF	Amphibious Task Force
BB	Battleship
BDA	Battle Damage Assessment
BEF	British Expeditionary Force
C2	Command and Control
C2W	Command and Control Warfare
C3I	Command, Control, Communications, and Intelligence
C4I	Command, Control, Communications, Computers and Intelligence
CAS	Close Air Support
CINC	Commander-in-Chief
CODAG	Combined Diesel and Gas
CODOG	Combined Diesel or Gas
COMSTRIKEFORSOUTH	Commander, Allied Striking Force South
COOP	Craft of Opportunity
CTF	Commander, Task Force
CVBG	Carrier Battle Group
dB	Decibel
DWT	Dead Weight Ton
EC	European Community
EEZ	Exclusive Economic Zone

EU	European Union
EW	Electronic Warfare
GIUK	Greenland–Iceland–United Kingdom
Hz	Hertz
IFOR	Implementation Force
JTF	Joint Task Force
KTO	Kuwaiti Theater of Operations
LLOC	Land Line of Communications
LOC	Line of Communications
LOO	Line of Operations
LSMR	Landing Ship, Medium, Rocket
LSD	Landing Ship, Dock
LST	Landing Ship, Tank
MAG	Maritime Action Group
mbpd	million barrels per day
MCM	Mine Counter Measures
MEU	Marine Expeditionary Unit
MEZ	Maritime Exclusion Zone
MIF	Maritime Intercept Force
MGB	Motor Gunboat
MIO	Maritime Intercept Operations
MPA	Maritime Patrol Aircraft
MRL	Multiple Rocket Launcher
MTB	Motor Torpedo Boat
NATO	North Atlantic Treaty Organization
NBC	Nuclear, Biological, Chemical
NEI	Netherlands East Indies
NSFS	Naval Surface Fire Support
NFZ	Non-Flight-Zone
OOTW	Operations Other Than War
OPV	Offshore Patrol Vessel
PEO	Peacekeeping Enforcement Operations
PKO	Peacekeeping Operations
PO	Peace Operations
POA	Pacific Ocean Area
PRC	People's Republic of China
RAF	Royal Air Force
Ro/Ro	Roll-on/Roll-off
ROE	Rules of Engagement
ROK	Republic of Korea
SAG	Surface Action Group
SAM	Surface-to-Air Missile
S-BOAT	German Fast Attack Craft (*Schnellboote*)
SC	Submarine Chaser

ABBREVIATIONS

SEAD	Suppression of Enemy Air Defenses
SFOR	Stabilization Force
SHAPE	Supreme Headquarters, Allied Powers Europe
SLOC	Sea Lines of Communication
SOWESPAC	Southwest Pacific Area
SSBN	Nuclear-Powered Ballistic Missile Submarine
SSN	Nuclear-Powered Attack Submarine
STANAVFORLANT	Standing Naval Force Atlantic
STANAVFORMED	Standing Naval Force Mediterranean
TEZ	Total Exclusion Zone
TF	Task Force
TLAM	Tomahawk Land Attack Missile
UAE	United Arab Emirates
UAV	Unmanned Aerial Vehicle
UN	United Nations
UNPROFOR	UN Protection Force
UNSCR	United Nations Security Council Resolution
URG	Underway Replenishment Group
VDS	Variable Depth Sonar
VTOL	Vertical Takeoff and Landing
WEU	West European Union
WMD	Weapons of Mass Destruction

Place Names

Previous	Current
Batum	Batumi
Bay of Cattaro	Boka Kotorska
Bizerta	Bizerte
Bône	Annaba
Bougie	Bejaïa
Cape Planka	Cape Ploče
Cattaro	Kotor
Constantinople	Istanbul
Corfu	Kérkira
Curzola	Korčula
Dairen	Lüda
Danzig	Gdańsk
Durazzo	Durrës
Elbing	Elblag
Elliot Islands	Changshan Qundao
Fiume	Rijeka
Gotenhafen	Gdynia
Koenigsberg	Kaliningrad
Leningrad	St Petersburg
Liaotung Peninsula	Liaodong Bandao
Libau	Liepāya
Lesina	Hvar
Lissa	Vis
Lussin	Lošinj
Lussin Piccolo	Mali Lošinj

PLACE NAMES

Memel	Klaipeda
Nikolayev	Mykolayiv
Pernau	Pärnu
Phillippeville	Skikda
Pillau	Baltiysk
Plevna	Pleven
Pola	Pula
Port Arthur	Lüshun
Ragusa	Dubrovnik
Reval	Tallinn
San Giovanni di Medua	Šen Douani di Medua
Scarpanto	Kárpanthos
Sebenico	Šibenik
Sihanoukville	Kâmpóng Saôm
Songjin	Kimch'aek
Stettin	Szczecin
Swinemuende	Świnioujście
Tarent	Taranto
Valona	Vlorë
Wei-hai-wai	Weihai
Windau	Ventspils
Zealand	Sjælland

Series Editor's Preface

This book began as a research project in the Center for Naval Warfare Studies at the US Naval War College at Newport, Rhode Island, and was completed during Milan Vego's recent time there as Professor of Joint Operations. Thus it is not surprising that the book contains a balanced blend of theory and practice. It is designed both to stimulate 'open and vigorous debate' on the concept of naval strategy and operations in narrow seas among professionals, as well as to be of service to the current practitioners of the operational art of war in narrow seas.

The loose use of terms such as 'strategy' and 'operations' has caused (and still causes) confusion among historians, both civilian and military/naval. Relying on definitions developed by the German practitioners of armed violence, Vego establishes the theoretical basis of the book by introducing the reader to the terminology employed at the Naval War College. Vego defines *strategy* as 'the art and science of applying all sources of power in peacetime and in war to accomplish strategic objectives'. He then breaks strategy down into four sub-components: *national* or *national security strategy* ('the art and science of applying and coordinating all the elements of national power ... to achieve national objectives'); *coalition* or *alliance strategy* ('the theory and practice of applying and coordinating all the sources of power of a coalition or alliance for the achievement of coalition/alliance aims'); *military strategy* ('the art and science of applying the armed forces of a nation to accomplish national strategic objectives by the application of force or by the threat of force'); and *naval* or *maritime strategy* ('the art and the science of using sources of military power in a sea/ocean theater to accomplish naval elements of military strategy'). Finally, Vego defines the current US concept of *operational art* as 'the theory and practice of planning, preparing, conducting, and sustaining major operations and campaigns aimed to accomplish operational or strategic objectives in a given theater'. In all, a methodology worthy of the Prussian War Academy!

Having established the theoretical parameters of the book in Chapter 1,

Vego devotes the following 13 chapters to analyzing operations in narrow seas throughout history. Here, Vego paints with a broad bush, eschewing traditional narrowly focused portrayals in favour of broadly conceived 'space' depictions that include the elements of topography, oceanography/hydrography, and weather/climate in operations in narrow seas. Vego argues that the US Navy, despite having routinely operated in many enclosed or semi-enclosed seas, has failed to pay adequate attention to either the strategic or the operational problems of conducting war in such waters. Specifically, he suggests that the US Navy found itself unprepared in terms of force composition and training to fight in restricted waters during the Solomon Islands campaign of World War II, the Korean War, the Vietnam War, and in the Arabian Gulf during the Iran–Iraq War and the Gulf War of 1991. In a concluding chapter, the author pulls together the 'lessons learned' from his historical case studies and offers his insights into the kind of ships and combined forces required to conduct wars in narrow seas.

Taken as a complementary pair, Rear Admiral Raja Menon's *Maritime Strategy and Continental Wars* (Volume 3 of this series) and Vego's *Naval Strategy and Operations in Narrow Seas* offer refreshing and timely contributions to the ongoing global debate concerning the future of sea power in general, and the attempt of the US Navy to enunciate an American maritime strategy in particular.

Holger H. Herwig
Series Editor

Preface

War at sea has mostly been fought close to the shores of the world's continents or islands. Most merchant-ship sinkings by submarines, and submarine losses as well, have occurred near focal areas of maritime trade such as straits or approaches to major commercial ports. Almost all surface naval vessels of all belligerents in World War II were sunk or damaged in waters less than 600 feet deep. The reason for this is that, in all eras, the principal mission of navies has been either to provide protection for transport and to debark one's troops or to prevent enemy invasion from the sea. In the era of galley and sail, the limited speed, range and seaworthiness of ships, combined with a short range of visual observation and shipboard weapons, restricted combat employment of the fleet to the waters close to the shore. Yet, even in the era of steam, it was obviously better to prevent an enemy force from landing or to attack an enemy's trade close to shore than to search them out on the open ocean. Despite all the advances in ships' propulsion, sensors and weapons, most naval actions in the future will most likely take place in relative proximity to the shores of the world's continental landmass, in areas known as 'littoral waters', and part of a war in the littorals would take place in the waters of enclosed and semi-enclosed seas, the popularly called 'narrow seas'.

The problem of fighting wars in narrow seas has not been given the serious consideration it deserves by many blue-water navies. There is a widely held but erroneous view that a fleet capable of defeating an adversary on the open ocean could successfully operate in narrow seas. It is true that, until relatively recently, a blue-water navy did not face serious and diverse threats to operations by its large surface combatants and submarines in, for example, the Arabian (Persian) Gulf and the Red Sea and other narrow seas in Southeast Asia, but the situation is rapidly changing because many of the smaller navies operating in these waters have acquired the capabilities to challenge the unrestricted operations of a blue-water navy such as that of the US. A typical narrow sea also poses other challenges to the operations of large naval vessels because of the unique features of its physical environment, specifically, the small size of the area involved and the correspondingly short distances, the proximity of the land, the shallowness of the water and the presence of a large number of islands and islets. These features, in turn, generally limit maneuverability, speed and the use of weapons and sensors by one's surface ships and submarines, primarily designed for employment on the open ocean. The range of threats to the survivability of large surface combatants and submarines is greater in a typical narrow sea than on the open ocean because the weaker opponent can use land-based aircraft, small and fast surface combatants or conventional submarines, mines and even coastal anti-ship cruise missiles to contest control of a stronger navy. The small size of the theater and the proximity of the landing area in a typical narrow sea demand a much greater degree of cooperation among the services than in any war conducted predominantly on the open ocean.

The problem of employing naval forces in narrow seas has been, in general, insufficiently studied by most blue-water navies. This is perhaps hard to understand because in both the world conflicts and many regional conflicts fought since 1945 the blue-water navies have found themselves unprepared to employ their forces most effectively in narrow seas. No war has been lost because a blue-water navy was caught unprepared to carry out unforeseen tasks in narrow seas. Yet, this is not to say that the lack of awareness of the peculiarities of operating in such waters did not cause unnecessary losses of personnel and *matériel*, and, perhaps most importantly, of time. In each of these conflicts, the acquisition of knowledge and the understanding of the peculiar features of employing one's naval forces in narrow seas and other restricted waters was too slow, and the lessons were learned by trial and error. These and similar problems could have been overcome if sufficient attention had been given in peacetime to studying and understanding the problems of conducting naval warfare in narrow seas. Therefore, blue-water navies should formulate the necessary lessons of how wars in narrow seas were fought in the past, and how they differed from wars conducted on the open ocean. The strategic and operational concepts should be developed, and both operational and tactical doctrine should be written to take into full account the specifics of operating naval forces and aircraft in narrow seas. In contrast to blue-water navies, the navies of the countries that border narrow seas are obviously far more familiar with the effects the physical environment exercises on the employment of their maritime forces. This is, of course, reflected in their fleet composition which consists predominantly of smaller and less-capable ships but with a larger number of platforms. Yet, this does not necessarily mean that small coastal navies have developed a sound operational concept to employ their forces to best advantage when confronted with a stronger fleet. Hence, even these navies need from time to time to be reminded how important it is to continue to study all aspects of warfare in narrow seas.

The principal objective in writing this book was not to provide a historical overview of naval warfare in narrow seas. The focus instead is on the key aspects of naval (or maritime) strategy and combat employment of maritime forces in typical narrow seas. Whenever appropriate, theoretical aspects of warfare are illustrated by arbitrarily selected examples from distant and more recent wars in narrow seas. This, in turn, creates some problems, because the examples chosen by the author might not be the best to use to illustrate a theory or might be considered by some as completely inappropriate. A work such as this does not allow the author to describe selected historical examples in any great detail and, in turn, this requires a greater general knowledge of naval history on the part of the reader. Despite these and other disadvantages, it seemed that the best approach was to organize the book by topics. Hence, some of the fundamentals of naval strategy and operations are described and analyzed first, and this is followed by discussion and analysis of selected strategic and operational tasks that blue-water or coastal navies may be called upon to perform when operating in a narrow sea. Thus, Chapter 1 explains the meaning of the terms strategy, operational art, and tactics, and their relationships; explores the distinction between 'narrow sea', and the terms 'shallow waters' and 'restricted waters'; briefly surveys arbitrarily selected wars fought in narrow seas; and describes some key features of warfare in a typical narrow sea. Chapter 2 discusses the factor of 'space' – specifically, the size, shape, and configuration of a country bordering a narrow sea, and selected narrow seas in terms of their size and distances; and explains in generic terms the influence of geography – specifically topography, oceanography/hydrography, and elements of weather/climate – on the employment of one's ships and aircraft and their weapons.

The following three chapters describe and analyze selected elements of the factor of space. Chapter 3 deals with the geostrategic positions a country or one's fleet occupies or might occupy in respect to a given narrow sea. Chapter 4 describes and analyzes the naval and air bases required for a fleet operating in a narrow sea, while the next chapter focuses on the military organization of space and the various elements of a maritime theater, specifically the base of operations, physical objectives, 'decisive points', lines of operations, and sea lines of communication. Strategic objectives to be accomplished in peacetime and in time of war are discussed and analyzed in some detail in Chapter 6. This chapter also explains the relationships between one's naval objectives and fleet distribution and composition. The next chapter focuses on the question of sea control and sea denial, and their modern variants, and compares these two strategic concepts as practiced in a typical narrow sea and in a war on the open ocean. Chapter 8 sets the stage for further discussion by explaining the main methods of combat employment of naval forces operating in a narrow sea; it specifically discusses in generic terms the real meaning of naval tactical actions (including battles, engagements, strikes, and raids), major naval operations and campaigns. Chapter 9 describes and evaluates in some detail the actions to weaken, neutralize, destroy, or annihilate an enemy's naval forces and thereby obtain sea control. Methods of exercising sea control in a narrow sea in both war and peace is the topic of Chapter 10. This chapter specifically deals with methods of projecting power ashore by both blue-water navies and small navies in operations such as amphibious landings, attacks on enemy installations/facilities on the coast and peace-keeping and peace enforcement. The next chapter describes and analyzes the methods used by a weaker fleet to contest command of the sea by a stronger force. Specifically, it explains the active and passive 'fleet-in-being' concept, counter naval blockades, and defense against enemy invasion. The generic term 'trade warfare' refers to attack on or defense and protection of maritime trade and is discussed and analyzed in Chapters 12 and 13. These tasks are carried out regardless of whether a navy tries to obtain or already possesses sea control in a given theater or sea area, or contests sea control by a stronger navy. Offensive and defensive aspects of mine warfare are discussed in some detail in the chapters dealing with the operations of both the stronger and the weaker sides in typical narrow seas. A unique feature of war in a typical narrow sea is the close cooperation between navy and ground forces in both offensive and defensive operations on the coast and is explained and analyzed in some detail in Chapter 14. The last chapter highlights the key aspects of war in narrow seas, draws some conclusions and discusses the types of ships most suitable for employment in such waters.

This book started as a project for the Center for Naval Warfare Studies (CNWS), US Naval War College, Newport, Rhode Island, USA. My thanks and gratitude for endorsing my proposal go to Dr Thomas Etzold, former Head, Advanced Research Projects at the CNWS. I also want to thank Professor Frank Uhlig, former editor of *Naval War College Review*, for his enduring support, frequent sound advice, and great patience, especially in the early days of the project. I hope that this work will not only lead to an open and vigorous debate on this important subject among professionals, but actually be useful to theoreticians and practitioners in developing sound operational concepts for employment by the US and other friendly navies in narrow seas – the waters that present both the opportunities and the challenges. I also want to thank Patrick Rossoni and Joseph Nunes of the Graphics Division, US Naval War College, for their highly professional work in preparing the maps. Finally, I would like to express my gratitude and thanks to Andrew Humphrys, Senior Editor at Frank Cass Publishers, for his work and great patience in preparing this material for publication.

*To all sailors who fought
and died in narrow seas*

1

Introduction

The terms 'strategy', 'policy', and 'operations' are frequently used interchangeably as if they meant the same thing. However, there is significant difference in the real meaning of each of these terms. In generic terms, *strategy* is the art and science of applying all sources of power in peacetime and in war to accomplish strategic objectives. National, coalition/alliance, and military strategy are differentiated. *National* (or *national security*) *strategy* deals with the art and science of applying and coordinating all the elements of national power (military, economic, financial, diplomatic, psychological, technological, and others) to achieve national objectives in peacetime and in war, to secure national objectives.[1] *Coalition* or *alliance strategy* deals with both the theory and practice of applying and coordinating all the sources of power for the attainment of coalition/alliance aims. It is formulated by heads of state and their principal civilian and military advisors.

Military strategy is the art and science of employing the armed forces of a nation to accomplish national strategy objectives by the application of force or by the threat of force in the support of national strategy.[2] It must be coordinated with the use of other elements of national power. National policy makers, their military advisors, and the nation's senior military leadership are principally responsible for determining military strategy both in peacetime and in war. They decide the war aims and assign specific military strategic objectives to individual theaters of war.[3] *Theater strategy* is a subset of military strategy and is applicable to powers facing a potential conflict which may require the threat or use of military force in multiple theaters. Normally, for medium and small powers, there is no intermediate step between national military strategy and the practical application of operational art.

In the strict meaning of the term, military strategy does not involve the actual application of military force, but is principally focused on accomplishing national military and/or theater-strategic objectives through national or coalition/alliance strategy, determining principal and secondary theaters of war and overall military posture (offensive or defensive), deciding on the distribution of armed forces to various theaters, and conflict/war termination. Depending on the medium where sources of military power are to be applied, military strategy encompasses naval, airspace, and land aspects. Many theoreticians differentiate naval from airspace strategy, while the term 'military strategy' is commonly used when referring to land warfare.

In a narrow definition of the term, *naval* (or *maritime*) *strategy* encompasses both the art and the science of using sources of military power in a sea/ocean theater to accomplish naval (maritime) aspects of military strategy.[4] Among other things, naval strategy determines strategic objectives to be accomplished at sea based on military strategy (or theater strategy): selecting naval and air basing areas and acquiring new ones by diplomatic or military means, developing the overall concept of war at sea, deciding which theater will be principal and which secondary, and determining maritime strength in various theaters.

A naval strategy is an integral part of a national security and national military strategy. Therefore, in peacetime and in a crisis, the diplomatic, political, economic, social, and psychological sources of national power backed by military strength are used primarily to accomplish national strategic objectives. In a war, these aspects of power recede into the background, and military force is employed to accomplish the objectives of national strategy.

All too often, naval strategy is confused with naval policy and vice versa. However, there is a considerable difference between these two. In the strict definition of the term, a *naval policy* constitutes the sum of all political, diplomatic, financial, social, and purely military decisions taken by the country's highest political and naval leadership that affect the country's maritime situation in general, the size and composition of the navy, number of personnel, and organization of maritime forces. Naval policy, in contrast to naval strategy, is also heavily influenced by domestic politics. Foreign policy and domestic policy are mutually dependent. Foreign policy should define which countries or group of countries will be regarded as friends, neutral, or potentially hostile. The size and the composition of the navy depend largely on the country's industrial capacity, financial strength, and the willingness of the country's political leadership to allocate resources for expansion and modernization of the fleet. Therefore, foreign policy serves as a framework, while domestic policy is the principal factor in the realization of naval policy objectives.

Naval strategy is all too often confused with what is properly understood as 'operational warfare' (or 'operations') or, in US military terms, 'operational art'. The highly respected naval strategist Rear Admiral Alfred T. Mahan clearly did not differentiate between strategy and what is known today as 'operational art'. One of his most important works, *Naval*

INTRODUCTION

Strategy (published in 1911), did not deal with naval strategy *per se*, but rather with various elements of 'operational art'. Likewise, the US Navy in the 1920s and 1930s believed that strategy was concerned with campaigns and that tactics applied to battles or combat.[5] This confusion is regrettably common today as well, because many naval officers in discussing employment of naval forces do not differentiate between what pertains to naval strategy and what to operational art. To make things more complicated, the term 'operations' is used by the US/Western military so imprecisely and loosely that its real meaning is obscured or completely lost.[6]

In contrast to strategy, *operational art* deals with the actual employment of one's combat forces. Operational art is a component of military art concerned with both the theory and the practice of planning, preparing, conducting, and sustaining major operations and campaigns aimed to accomplish operational or strategic objectives in a theater. It is applied across the entire spectrum of warfare. However, its full potential is possible only in a high-intensity conventional conflict, not a nuclear war or low-intensity conflict.[7]

There are many commonalities in the combat employment of various services when employed to accomplish strategic or operational objectives in a theater. However, there are also significant differences in the methods and the techniques in which the forces of various services are employed in combat. These distinctions are largely due to the diverse characteristics of the physical 'medium' (land, air and sea) in which each service operates, and of their unique weapons platforms. Hence, some theoreticians argue that each service has its own operational art.[8] They have a point, because operational and partial strategic objectives can often be accomplished by forces exclusively or predominantly of a single service. For example, a major operation aimed to cut off the enemy's maritime trade or defend one's own trade is usually conducted by naval forces and aviation. The destruction or neutralization of a major part of the enemy's fleet would require the planning and execution of one or several major operations in which naval forces would play a dominant role. Likewise, an offensive or defensive counterair operation would be planned and conducted predominantly by air forces.

Operational art is greatly influenced not only by strategic, but also by political, economic, social, and other aspects of the situation in a given theater. The opposite is also true, however. The outcome of major operations or campaigns considerably affects the political, economic, social, or even environmental situation in a given theater. In contrast to tactics, the deployment of one's forces is the very heart of a major operation or campaign plan. Mistakes made in operational or strategic deployment of the combat forces can only be corrected with difficulty, if at all. Also, logistical support and sustainment are much more critical to the successful conduct and outcome of major operations or campaigns than they are in fighting battles and engagements.[9]

Naval strategy guides operational art by determining objectives, allocating forces, and imposing conditions on tactical combat. At the same time, naval strategy should take into account the existing and projected *operational realities* of the situation. This means, among other things, that strategic or operational objectives must be based on realistic assessment and balancing of the factors of space, time and force. A serious mismatch or a disconnect between the means and ends is invariably fatal.

Operational art translates the intent and conditions of strategy into a plan (or plans) for the defeat of a mjor part of the enemy force. It determines who is employed where and when to achieve tactical victories in support of strategic or operational objectives.[10] Brilliance at the operational level may only delay not prevent an ultimate defeat due to a seriously flawed strategy. Also, a sound and coherent strategy in itself cannot secure victory in a war, because a high level of operational competence is required to accomplish strategic objectives through the actual employment of one's combat forces.[11]

The practical application of operational art requires that all the actions be sequenced and synchronized in such a way as to contribute most directly to the accomplishment of the assigned strategic or operational objectives. Therefore, there should always be a clear appreciation not just of what is *operationally desirable* but of what is *strategically possible*. In practicing operational art, everything must be subordinated to a simple operational scheme or idea that accomplishes an operational or strategic objective. A properly determined military objective should normally be focused on the enemy forces that must be either destroyed or maneuvered into a situation that makes it impossible for them to accomplish their assigned objectives. However, the politico-strategic purpose should never be forgotten. Serious errors could be made by unrealistic and over-costly demands on tactics.[12]

Tactics encompass both the art and the science of planning, the preparation and conduct of battles, engagements, strikes, and attacks aimed to accomplish tactical (and sometimes operational) objectives in a given combat sector or zone. They must ensure that results are in harmony with both operational and strategic objectives. Poor application of operational art can lead to tactical defeats, which may in turn often have strategic consequences. The accomplishment of strategic objectives depends upon the results gained by tactics. Strategy must ensure that tactical combat is conducted under conditions favorable to accomplishing operational or strategic objectives. It must also always consider the limitations imposed by tactics. The results of tactical actions are useful *only* when linked together as a part of some larger design framed by military or theater strategy. Bad tactics may invalidate a good strategy.

INTRODUCTION

Therefore a sufficient level of tactical competence is absolutely necessary to accomplish strategic or operational objectives.

One of the principal aims of operational art is to bring about a battle under the most favorable terms possible, to make the enemy fight at a disadvantage. However, tactical brilliance can rarely overcome inadequate operational performance. Also, the critical importance of tactics is not reduced with the application of operational art. It is a truism that a lost battle cancels all the advantages obtained by a good (major) operation.[13]

Military art guides the planning, preparation, and employment of one's armed forces in time of peace and war. Each service of the armed forces operates in an environment that is unique in terms of its physical size and characteristics, and the presence or lack of human settlements or man-made structures.

Navies operate in the open ocean and in what are euphemistically called 'narrow seas'. However, the latter term is not always properly defined and understood. It is often used interchangeably with the terms 'coastal waters', 'shallow waters', or 'confined waters'. Yet, despite some apparent similarities, each of these terms has a different meaning. A body of water can be both narrow and deep throughout, as are the Mediterranean Sea and the Caribbean Sea, or both narrow and predominantly shallow, as are the North Sea and the Baltic Sea. Seas are classified as either enclosed or partly enclosed seas. Enclosed seas penetrate deeply into the continent and are connected with the open ocean by a narrow passage. Because of their restricted communication with the open ocean, they are characteristically tideless or have a small tidal range. Enclosed seas are also called 'continental seas' because of their restricted connections with the open ocean. These can be shallow if they rest on shallow depressions, as are the Azov Sea, Hudson Bay, and the Baltic Sea; or they can be epeiric or deep continental seas and intercontinental seas, and their depths may exceed 2,500 fathoms. The term 'mediterranean' or 'middle-earth' sea can be used instead of intercontinental sea but this is usually reserved for the almost landlocked basins of the Arctic Sea, the Mediterranean Sea, the Central American Sea and the Indonesian Sea.[14]

In the *geographic* definition of the term, narrow sea encompasses enclosed and semi-enclosed seas. In addition, large inland bodies of water such as the Caspian Sea, Lake Erie and Lake Ontario can also be considered as narrow seas. An *enclosed sea* lies wholly within the continental shelf, and is surrounded by a landmass, except for a strait or several straits that connect it to an ocean or another enclosed or semi-enclosed sea. *Semi-enclosed seas* may be linked with the open ocean by one or more straits or narrows (as is the North Sea), or by several passages between islands. They are characterized by large tidal ranges and are more oceanic in character than enclosed seas. Because they are open to the ocean, semi-enclosed seas may be referred to as 'pelagic' seas. Shallow pelagic seas lie wholly upon the continental shelf (they are sometimes called shelf seas) and their depths rarely exceed 100 fathoms. Deep pelagic seas are those separated from the open ocean by island arcs or shallow submarine ridges and their depths often exceed 1,000 fathoms. Such seas are typical of western Pacific and Central American waters.[15] A *marginal* or *peripheral sea* is part of an ocean bordering a continental landmass or an island archipelago. The outer boundary of such a sea is widely open to if not an integral part of the respective ocean. A marginal sea does not lie beyond the continental shelf but on a downward portion of it. It is militarily regarded as a narrow sea when surrounded on all sides by large islands or by an island festoon, even if situated in the midst of an ocean, as are the Coral Sea and the Solomon Sea. Some geographers consider the Mediterranean as one of the marginal seas of the Atlantic Ocean. In the proper understanding of the term, however, the Mediterranean should be considered as a large enclosed sea.

The terms 'shallow', 'coastal', 'confined', 'inshore', and 'restricted' waters themselves have different meanings. *Shallow waters* are those waters wherein wind-generated waves travel in less than one half of their length, and are defined as the marginal or inland extension of ocean having prevailing depths of under 600 feet.[16] In the hypsometric definition of the term shallow waters encompass coastal and inshore waters less than 100 fathoms deep. In acoustic terms shallow water exists whenever the propagation of underwater sound is characterized by numerous encounters with both the sea surface and the sea bottom. In some cases a body of water could be considered hypsometrically shallow but 'deep' in respect of the propagation of underwater sound.[17]

Coastal waters are the waters on the continental shelf and adjacent semi-enclosed seas. The continental shelf encompasses the seabed and subsoil adjacent to the coast to a depth of about 660 feet or beyond that limit where the depth of the adjacent waters allows exploitation of any natural resources. Depth of the water on the continental shelf worldwide varies from 65 to 200 fathoms. In some places the continental shelf extends for several hundred miles, while in other places it is much narrower or even completely absent. The greatest width of the continental shelf is in the Barents Sea (about 750 miles); off western Europe it is approximately 200 miles. The smallest width of continental shelf exists along the edge of the eastern part of the Pacific, specifically off the US West Coast, where it extends on average to about 20 miles. *Inshore waters* encompass the waters of gulfs, bays, deltas and estuaries.[18]

Restricted waters are the waters wherein the movement of surface ships and submarines and the use of shipboard sensors and weapons are limited by the small surface area and the depth of water. They encompass the coastal waters between a mainland coast and offshore islands, all inshore waters, shallow waters of archipelagoes, and the waters of straits or narrows, channels, canals and navigable river systems. However, the term restricted waters should not be applied to narrow seas and coastal waters facing the open shore of archipelagoes, straits, narrows and channels.

In the military meaning of the term, a narrow sea is a body of water that can be controlled from both of its sides and some sea areas, notably the English Channel (La Manche), the Danish Straits (Skagerrak and Kattegat), the Turkish Straits (Dardanelles and Bosporus), the Strait of Gibraltar and the Sicilian Narrows, have been regarded as narrow seas throughout the ages. The term 'narrow sea' was reputedly first used in reference to the rise of England's sea power in the late sixteenth century. In its many wars with Spain, the Dutch Republic, and France in the sixteenth and seventeenth centuries, England essentially commanded only the Strait of Dover. By the end of the nineteenth century, however, the Royal Navy controlled a number of international waterways and strategic positions, specifically the Strait of Gibraltar, Malta, the Suez Canal and Aden. Britain therefore held the keys to four internationally important narrow seas: the English Channel, the North Sea, the Mediterranean, and the Red Sea. Moreover, by controlling the choke points through which seaborne commerce had to pass to reach the great centers of world trade, Britain dominated large areas of the world's oceans.

The term *narrow seas* became extensively used during World War II in reference to the numerous clashes between small surface combatants in the waters surrounding the European landmass, especially the English Channel and the North Sea. With the steady extension of the range and endurance of surface ships and the increase in the effective range of weapons, especially after the advent of aircraft and cruise missiles, ever-larger parts of the ocean bordering the world's continents and large enclosed seas such as the Caribbean and the Mediterranean became in fact narrow seas. A large force of surface ships even in large enclosed seas cannot hide from an adversary and, today, the ever-present threat from the air considerably restricts high-seas operations in even the largest narrow seas.

The terms 'narrow sea' and 'littoral waters' are often used as though they are identical in meaning. This is not the case. In its precise meaning, 'littoral' pertains only to a coastline, encompassing both land and near-shore waters, particularly that part between high and low tide. In a broader definition it means the 'coastal region' or the area 'pertaining to a shore'.[19] In military terms, the extent of the 'littoral' depends on one's ability to project power effectively from the shore into the enemy's interior. The term, 'littoral waters' encompasses not only the coastal waters of a narrow sea but those of an ocean as well. Moreover, 'littoral waters' are smaller in their extent than a narrow sea. Officially, the US Navy divides littoral into two parts: seaward – the area from the open ocean to the shore, which must be controlled to support operations ashore, and landward –

the area inland from shore that can be supported and defended directly from the sea. Hence, littoral warfare is conducted not only in semi-enclosed and enclosed seas but also in the waters of an open ocean bordering a continental land mass.[20] Littoral waters are not necessarily shallow but can be very deep. The biggest differences between littoral waters and deep ocean are in tides, tide-generated currents and the topography of the seabed.

Throughout recorded history control or lack of control in a narrow sea has had a significant, and at times a key role in a war on land. It also had a large influence on the diplomacy of the great powers of the day. Control of their adjacent narrow seas was a critical factor for the development of many great sea powers, as the example of England illustrates. Because of its size, France in the seventeenth century was largely invulnerable to blockade by the British Navy. However, the French, because of their military power, posed a real invasion threat to England whenever they obtained control of the adjacent narrow seas. France threatened the balance of power in western Europe by its attempts to gain control of the Low Countries, and by its persistent efforts to obtain a dominant position on the Iberian Peninsula.[21] The War of 1689 was a consequence of French naval superiority. James II landed with about 5,000 men in Ireland in March 1689 to raise that country against the Protestant Succession. The Royal Navy's attempt to close the Jacobite supply route to France led to an inconclusive battle in Bantry Bay in May 1689.[22] This threat to the British Isles forced King William to divide his armies between Ireland and the Netherlands and, for the next two years, he had to devote the greater part of his time and energy to home instead of continental defense.[23]

In the early eighteenth century, by combining diplomacy and naval strength, Britain secured its interests in the Baltic. This primarily concerned the timber trade and the preservation of the balance of power between the northern Baltic states. Between 1715 and 1718 the British sent strong fleets into the Baltic to oppose attempts by Sweden to interrupt British maritime commerce, to threaten Norway, and to support the Jacobites. Yet, in 1719–21 and again in 1725–27, the Royal Navy sailed into the Baltic to support the Swedes against the rising power of Russia under Peter the Great. However, despite all its efforts, Britain was unable to check the steady weakening of Sweden's influence and the rising power of Russia by deploying its navy in the Baltic. This would have been possible only if Austria, Prussia, and France had acted as Britain's continental surrogate – and none of them wanted to play that role.[24]

The Baltic became a critical theater in the struggle between the European powers in 1807, in the aftermath of the Peace of Tilsit. Russia declared war on Great Britain, while Prussia acquiesced to the new situation. The Russians acquired Finland in February 1808 after their army invaded that country. Denmark hoped to expand at Sweden's expense, and prepared to invade Sweden across the Sound.

The Baltic and the Black Sea were the scenes of many naval actions during the Crimean War (1853–56). The undisputed control of these seas allowed the British and French fleet to besiege the naval base of Sevastopol' and to bombard the Russian coast in the Baltic. The Adriatic Sea was the scene of conflict between the Austrian Empire and the Kingdom of Sardinia in 1848–49, when the latter came to the aid of Republic of Veneto.

During the war of Austria and Prussia against Denmark in 1864, the North Sea became the principal maritime theater. The combined Austro-Prussian squadron led by the Austrian Commodore Wilhelm von Tegetthoff inflicted a decisive defeat on the Danish fleet at Heligoland in May 1864. This action ended the Danish blockade of the northern German coast and opened Hamburg to commercial traffic. By the end of June 1864 the situation had changed. The entire Danish army was concentrated in Alsen and Funen, and, although the Danish fleet held command of the Baltic, the situation in the North Sea and Kattegat was by no means unfavorable for the Austrians and Prussians, whose troops crossed the narrow strait and landed on Alsen on 29 June. Subsequently, the Prussian Chief of General Staff, General Helmuth von Moltke, Sr, drew up a plan to transfer actions to Sjælland, but the treaty of Vienna, signed on 30 October 1864, led to the cancellation of the plan before it was carried out.[25]

The Black Sea was the scene of naval actions during the Russo-Turkish War of 1877–78. The Russians did not have a fleet worthy of the name, and hence the Turks held undisputed command of the sea. This meant that the Russian attacks against Constantinople and in the Caucasus had to be made overland. The 300-mile-long Russian overland advance towards Constantinople was characterized by heavy fighting at Plevna and elsewhere. Despite their weakness at sea, the Russians, in a series of joint army and navy operations, quickly seized control of the Danube estuary. The Danube was also the scene of the first torpedo attacks by Russians against the Turkish ironclads and monitors.

In the Far East, the Yellow Sea was the scene of naval fighting in the Sino-Japanese War of 1894. Both the Yellow Sea and the Sea of Japan saw almost continuous fighting during the war between Russia and Japan in 1904–5. Several major battles, notably those of the Yellow Sea and Tsushima, played vital roles in the outcome of the war on land.

Most naval actions in World War I were fought in narrow seas: specifically, in the North Sea, the English Channel, the Mediterranean Sea, the Black Sea, and the Baltic Sea. Control of narrow seas was critically important to the Entente and its associated Powers and to the Central Powers. It was absolutely essential for Germany to control the central and western Baltic because it received most of the iron ore necessary for its war effort from Sweden. The coal from the Ruhr was carried by sea, and the supplies for the army fighting in Russia went by sea as well.[26] During World War I, even the distant Persian (Arabian) Gulf emerged as an important theater of operations. The British objective in their ill-fated Mesopotamian campaign was to secure the Shatt el-Arab estuary, since the Anglo-Persian oil pipeline ran down to the river about 30 miles above its mouth. The British had secured a controlling interest in the company in 1914, just before the war had started.

In World War II, most of the world's narrow seas, including all the seas adjacent to Europe, the Caribbean and the southern Pacific were the scenes of numerous naval encounters. However, not all armies had realized before the outbreak of war the value of the control of narrow seas in their land campaigns. The German Supreme High Command showed almost complete neglect of the value of controlling narrow seas in its plan for their invasion of the Soviet Union (Plan *Barbarossa*) in June 1941. The German Navy's task (besides defending the country's coast) was to prevent any attempt by the Soviet Navy to break out and operate in the Baltic. No major naval actions were planned, however, because the Army would seize Leningrad and thereby eliminate the Soviet Baltic fleet. Once that objective had been accomplished, the German *Kriegsmarine* would resume transporting supplies to Army Group North. This also included mine clearance along the shipping routes. Perhaps surprisingly, the Germans did not take into their plans the use of direct sea communications from Stettin, Luebeck and Koenigsberg and further northwards along the eastern coast of the Baltic.[27]

Several marginal seas and enclosed seas in the Pacific became the setting for numerous naval actions in World War II. The Japanese invasions of Malaya and the Netherlands East Indies (NEI) encompassed all enclosed and semi-enclosed seas in the area. Neither the Japanese nor the United States had anticipated in their plans that the Coral Sea and the waters adjacent to the Solomons Archipelago would become the areas of a protracted naval struggle.

Two narrow seas, the Sea of Japan and the Yellow Sea, were the scene of naval actions during the Korean War of 1950–53. This conflict was essentially waged between purely land powers (North Korea and the People's Republic of China) and a coalition of maritime powers (the United States and its Western allies). The North Koreans were forced to transport troops and supplies entirely overland, because the UN held undisputed control of the sea. Their land supply routes were almost continuously attacked from the air, but were never completely cut off. Although the Korean War ended in a stalemate, a sea power had demonstrated its ability to maintain the *status quo* in the face of communist aggression.[28]

Naval forces and aircraft have been employed in narrow seas on numerous occasions since 1953, most notably during the Cuban Missile

Crisis (1962), the Vietnam War (1965–75), three Arab–Israeli conflicts (1956, 1967 and 1973), the Iran–Iraq conflict (1980–88), the Gulf War (1990–91), the conflict in the former Yugoslavia (1991–95), and the crisis in the Gulf (February–March and November–December 1998). There is no doubt, that in any future conflict, regardless of its scope, narrow seas surrounding the Eurasian landmass will become the operating area of naval action.

Narrow seas can become the scene of naval actions in a *general* or *global* conflict. They can also be an integral part of a *limited* war or *regional* conflict, a war restricted to a specific geographic area and fought either between a major and one or more minor powers, or a war between two major powers to accomplish limited strategic objectives. Possible scenarios for limited conflict at sea include: a limited war of intervention; a limited war between major powers evolving into an entirely naval conflict; harassment of enemy and/or neutral shipping in international straits or narrows; a dispute over an *economic exclusive zone* (EEZ); and a *local conflict* in a typical narrow sea fought between two or more minor powers while major powers remain strictly neutral. In practice, a major power may provide political, diplomatic, and logistical support to one or more of its clients.

War in a narrow sea can be fought between two or more riparian states. It also can involve a major out-of-area naval power, a single navy or combination of small navies. A major power would usually have limited objectives, such as opening the sea's exits for the free flow of maritime trade or protecting its own shipping within and at the approaches of a given narrow sea. For a small navy facing a stronger adversary at sea, the objectives would be less limited in terms of scope and duration.

A conflict between two countries bordering a narrow sea would rarely be limited to hostilities at sea or in the air, but would encompass the land area as well. The land front would be a principal theater of operations, while combat at sea and in the air would remain of secondary importance. Outcome of the war as a whole is decisively influenced by the situation on land, but the actions at sea and in the air can significically contribute to the success of operations on land.

War in a typical narrow sea differs considerably from that on the open ocean, primarily because of the lack of physical space and the proximity of the continental landmass. No maritime theater is more directly affected by the geomorphological and hydrographic (or oceanographic) features of the environment than a narrow sea. In a typical narrow sea, highly indented coasts are endowed with numerous islands and islets, which often restrict the maneuverability of surface ships, especially major surface combatants and submarines. In shallow waters, large surface ships must greatly reduce their speed. Narrow seas also often have numerous shoals, reefs, strong tides and currents, which make safe navigation difficult.

Combat actions normally encompass the entire sea area, and are now more decisive than those in the past because of the massive use of missiles and other high-precision weapons. A major part or even an entire ship formation can be destroyed by a mass missile strike. Given the high mobility of naval forces, the long range and high destructiveness of modern weapons, combat in a typical narrow sea would be fought at high intensity from the outset, because each side would try to seize the initiative and pre-empt the strikes of the other. If the balance of forces is roughly equal, the side that acts more quickly and with greater determination has an inherent advantage over its adversary.

One of the main features of a modern naval combat in a narrow sea is a frequent and drastic change in the tactical and operational situation. The extensive use of electronic warfare can make difficult – if not impossible – the effective use of sensors and guided weapons. The high speed of modern ships and aircraft and their ability to combine maneuver and firepower may allow one side to achieve surprise as well as to obtain superiority in place and time. The high intensity of naval actions will result in a rapid change in a situation, which in turn may often lead to a sudden shift from offense to defense or vice versa. Because of the extended range of modern weapons, their maneuverability, large radius of action, and the endurance of naval platforms, combat actions in a typical narrow sea can involve a large area. The large area of involvement is further caused by the participation of diverse combat arms operating on the surface, sub-surface, and, above all, in the air and on the coast.

Naval actions in a narrow sea are conducted largely during the night or in bad visibility. Because of limited visibility, night combat is fought at close ranges, thereby making the deployment/redeployment and maneuver of one's tactical-sized forces more difficult.

Air power in a typical narrow sea will have a decisive role in the outcome of war at sea when focused on the destruction of the right set of targets and applied massively and in a sustained manner. The enemy should not be given sufficient time to recover after each new round of strikes and attacks. Land-based air power is one of the most effective means to carry out attacks against enemy shipping and shipping-related facilities on the coast in an enclosed sea theater. Their high degree of readiness and maneuverability enable aircraft to concentrate their strikes against transports, warships or aircraft covering enemy ships at sea.

NOTES

1 Department of Strategy and Tactics, 'Strategic Studies, Resumes of Committee Discussion, 14–25 March 1952', classes of June 1952 (Newport, RI: The United States Naval War College), p. 1.
2 Ibid.

INTRODUCTION

3   Stephen E. Runals, 'A Different Approach', *Military Review*, 10 (October 1987), p. 47.
4   In another definition, naval strategy is described as the art and science of employing the navy of a nation to secure the objectives of national policy by the application of force, or the threat of force, in the support of military strategy; Department of Strategy and Tactics, 'Strategic Studies, Resumes of Committee Discussion, 14–25 March 1952', p. 1.
5   M. G. Cook, 'Naval Strategy' 2 March 1931, Air Corps Tactical School, Langley Field, VA, 1930–31, Strategic Plans Division Records, Series, Box 003 (Washington, DC: Naval Operational Archives), p. 2.
6   The term 'operation' in US military parlance refers to actions ranging from a sortie to the actual employment of combat forces. The US joint community defines operation as 'a military action or the carrying out of a strategic, tactical, Service, training, or administrative military mission; the process of carrying on combat, including movement, supply, attack, defense, and maneuvers needed to gain the objectives of any battle or campaign', Department of Defense, *Dictionary of Military and Associated Terms and Abbreviations* (Washington, DC: Government Printing Office, 1994), p. 274.
7   This term more accurately reflects the nature of military operations short of major conflict and hence will be used throughout the text. The US term is 'operations other than war' (OOTW).
8   The Soviets and the Russians today differentiate among operational art of ground forces, navy, and air forces.
9   Ludwig Beck, *Studien*, ed. Hans Speidel (Stuttgart: K. F. Koehlers Verlag, 1955), p. 79; Arbeitspapier, 'Operative Fuehrung', Hamburg: Fuehrungsakademie der Bundeswehr, August 1992, p. 15.
10 Leon H. Rios, *The Linkage of the Strategic and Operational Levels of War* (Fort Leavenworth, KS: School of Advanced Military Studies, US Army Command and General Staff College, 12 May 1986), pp. 19–20.
11 Meinhard Glanz, *Rommels Entschluss zum Angriff auf Alexandria* (Hamburg: Fuehrungsakademie der Bundeswehr, February 1963), p. 5.
12 Ibid., pp. 15–16.
13 L. D. Holder, 'A New Day for Operational Art', *Army*, 3 (March 1985), pp. 4–7.
14 Charles H. Cotter, *The Physical Geography of the Oceans* (New York: Elsevier, 1966), pp. 71, 72, 82.
15 Ibid., p. 72.
16 David F. Tver, *Ocean and Marine Dictionary* (Centreville, MD: Cornell Maritime Press, 1979), p. 280.
17 Robert J. Urick, *Sound Propagation in the Sea* (Washington, DC: Defense Advanced Projects Agency, 1979), p. 8-1.
18 Although terms *gulf* and *bay* are often used as synonyms they do not mean the same thing. A *gulf* is a sheet of water whose length is greater than its width; Cotter, *The Physical Geography of the Oceans*, p. 70. A *bay* is a recess in a shore or inlet of sea between two widely spaced capes or headlands. It is not as large as a gulf but is larger than a cove; David F. Tver, *Ocean and Marine Dictionary*, p. 29; Cotter, *The Physical Geography of the Oceans*, p. 71. *Confined waters* are those bodies of water, which, owing to their width and depth, constrict the maneuvering of surface ships and submersibles and in some cases also limit the employment of shipboard sensors and weapons. In general, they encompass the waters of archipelagoes, straits, channels, narrows, sounds, and artificial canals. A *strait*, in the geographic definition of the term, is a relatively narrow channel connecting two water bodies separated by islands, or by an island and the mainland. In another

definition, a strait pertains to breaks in a chain of islands or breaks in former isthmuses, such as the Dover Strait, the Bosporus and the Dardanelles; Cotter, *The Physical Geography of the Oceans*, p. 71; Tver, *Ocean and Marine Dictionary*, p. 307. A strait can separate two or more islands, two continents, a continental landmass and an oceanic island, a continent and a continental island, two peninsulas, and a mainland coast and an offshore island. Some straits represent the most practical or sometimes even the only passage between two commercially or militarily important bodies of water. Therefore, they are critically important for the riparian states. Some 35 straits worldwide are considered to have significance for international traffic. A unique case of a strait in all but name is the sea route around a southern tip of Africa, where normal navigation is restricted to between six and 30 nautical miles outside the ten-fathom line in order to avoid the influence of the strong Agulhas current. Constricted sections of a strait are usually called *narrows*. A long passage of water connecting two large bodies of water but too wide and extensive to be defined as a strait is called a *sound*. It usually runs parallel with, and may separate an island from, the mainland coast, or connect the ocean with a sea or lake. A *channel* is a comparatively narrow stretch of water but wider than a strait separating two land areas and linking two more extensive seas. The term channel also refers to the deep navigable part of a bay or estuary, or to the navigable route between the shoals which affords the best and the safest passage for ships. The term *canal* pertains to an artificial waterway cut for the purpose of navigation or irrigation. The canals linking an ocean and an extensive sea or two extensive seas are international waterways and have great importance for both the movement of cargo and the transit of naval vessels. This term is also used in referring to a long but narrow stretch of water between the mainland coast and offshore islands or within a group of islands.

19 Tver, *Ocean and Marine Dictionary*, p. 182.
20 US Department of the Navy and US Marine Corps, ... *From the Sea: Preparing the Naval Service for the 21st Century* (Washington, DC: September 1992), p. 5.
21 Paul Kennedy, *The Rise and Fall of British Naval Mastery* (London: Macmillan, 1983, reprinted 1985), p. 74.
22 G. J. Marcus, *A Naval History of England*, Vol. I: *The Formative Years* (Boston, MA/Toronto: Little, Brown, 1961), p. 199.
23 Kennedy, *The Rise and Fall of British Naval Mastery*, pp. 76–7.
24 Ibid., p. 89.
25 Frederick E. Whetton, *Moltke* (London: Constable, 1921), p. 83.
26 John Creswell, *Naval Warfare: An Introductory Study* (London: Sampson, Low, Marston, 1936), p. 104.
27 Friedrich Ruge, *Seekrieg: The German Navy's Story 1939–1945*, trans. M. G. Saunders (Annapolis, MD: Naval Institute Press, 1957), pp. 197–8.
28 Donald Macintyre, *Sea Power in the Pacific: A History from the 16th Century to the Present Day* (London: Military Book Society, 1972), p. 256.

2

The Factor of Space

Naval strategy in narrow seas has much in common with that required on the open ocean, but there are also considerable differences because of the much smaller size of typical narrow seas and their proximity to the continental landmass; hence, their greater effect on other sources of national strength and in the employment of fleet forces and aviation of both sides both in peacetime and in war. *Space* is an essential factor for the successful employment of the forces of both sides in a conflict. The higher the level of war, the more important this factor is. In general, the factor of space requires proper consideration of many elements, such as the country's maritime position and its shape, the sea area, its distances, shape and configuration; the physical characteristics of the coastline and the coastal area in general, and its hydrography/oceanography. These factors are far more critical in narrow seas than they are for a fleet employed on the open ocean.

Country's Position

In geographic terms, a country or a territory may occupy, in relation to the adjacent sea and land area, a central, a semi-central, a peninsular, or an insular position. A country occupies a *central position* when none of its land areas borders a sea or an ocean. If a country is located on the rim of a continental land-mass, but borders a sea or an ocean, it is said to occupy a *semi-central* position. This position has not always been favorable to the development of sea power. This was especially true if a country was rich in natural resources, because the ambitions and energies of the entire nation or community were not necessarily directed towards the use of maritime resources, as the examples of France and Tsarist Russia show. Each country was also tempted to maintain a large standing army to seek an outlet for its energies for conquest on the land.

A territory occupying a semi-central position, but lacking natural resources, usually drew its population to the sea, as the example of Holland in the medieval era illustrates. However, the resources and energies of Holland were exhausted early on, because of the need to maintain a large army and carry out costly wars to preserve its independence against its adversaries, Spain and France.

A territory having a longer frontier with the sea than with the land is said to occupy a *semi-isolated* or *peninsular* position. Although generally favorable for the development of sea power, a peninsular position may require spending most of the country's scarce resources for defense against an invasion across its border on land.

A country situated on a large island or several large islands occupies an *insular* position. The examples of the ancient Crete and Cyprus and medieval England illustrate that the population of an insular country or community lacking natural resources is forced to seek its fortune at sea. However, there is usually no advantage for a country in occupying an insular position unless it controls the surrounding seas.

An insular position is usually not detrimental to the development of sea power even in the case where a country is spread over several large islands, because of the high degree of physical concentration. If a country is spread over a large number of islands encompassing a rather large sea area, as with Indonesia and the Philippines, it lacks the necessary physical concentration and this, in turn, often has adverse effects on the development of sea power.

England probably provides the best example of the advantage accruing from occupying an insular position. The British Isles form a 700-mile-long barrier lying off the coast of Western Europe. Their coasts border the open waters of the Northern Atlantic and three narrow seas: the North Sea, the English Channel and the Irish Sea. By having its entire frontier with the sea, England had little or no reason to fear a surprise attack by its adversaries so long as the Royal Navy controlled the adjacent narrow seas. Also, the British, with no land frontiers to raise and maintain a large standing army, concentrated their ambitions and energies for the most part of their history towards the development of seaborne trade and naval strength. This stood the British in good stead against the maritime powers of France and Holland, their main rivals and adversaries on the European continent.

Japan occupies a similar position in respect to the adjacent Asian landmass as England has to Europe. Japan's mainland stretches over four large islands whose shores are washed by the waters of the Pacific and two narrow seas: the Sea of Japan and the Inland Sea. Japan was secure from a surprise attack by its hostile neighbors on Asia's mainland so long as its navy controlled the surrounding seas. Because Japan had at times ambitions to conquer the territories on Asia's mainland it was forced to maintain also, besides a large navy, a large standing army.

A country can border one or several narrow seas, a narrow sea and an ocean, or two or more narrow seas. A country occupying a semi-central position but fronting an enclosed sea usually does not enjoy a favorable position for the development of its sea power, as demonstrated by medieval Poland and the Austrian Empire in the eighteenth and nineteenth centuries.

Wilhelmine Germany bordered two narrow seas: the North Sea and the Baltic Sea. Because its seaboard was divided by the Jutland Peninsula, Germany was unable to ensure the cooperation of its two fleets in wartime unless Denmark allowed unhindered transit of the German ships to and from the Baltic. The problem was alleviated to some extent in 1898 when the Kiel Canal (Nordsee–Ostsee Kanal) was completed. In contrast to Germany, Denmark occupies a more favorable position at sea because the two narrow seas washing its shores are adjacent to each other.

Russia stretches for about 8,000 miles across two continents and its seaboard encompasses four maritime theaters, each widely separated and requiring the establishment and maintenance of four fleets. Tsarist Russia and its successor the Soviet Union often tried to overcome natural disadvantages in their maritime position by obtaining access to the warm waters of the Mediterranean and the Indian Ocean.

An insular country may border a narrow sea and an open ocean, or it may front two or more narrow seas and an open ocean. Such a position favors the development of sea power, because the country's seaboard is continuous and the access to the high seas is unhindered. In contrast, a country occupying a semi-central position and facing one or more narrow seas and an open ocean, separated from each other by intervening land, has its strength at sea fragmented.

The development of sea power was often greatly affected by a country's position within a given narrow sea. A country can be so situated either to control wholly or partially a sea's only exit, or to occupy a position at some distance from a sea's exit. The riparian states guarding both shores of the sea's only exit enjoy the most favorable position in a narrow sea. Byzantium and its successors, the Mussulman Turks and modern Turkey, held undisputed control of the Straits, the only exit from and to the Black Sea. Therefore, they dominated not only all seaborne trade to and from the Black Sea, but were able to prevent naval incursions into the area. Likewise, Denmark's maritime position has always been strong, owing to its control of the Skaggerak and Kattegat. France and England share control of the English Channel and the Strait of Dover. However, England's position in the North Sea was much stronger than that of

France because it controlled the approaches to the northern exit between the Shetlands and Norway. And whoever controlled the Strait of Gibraltar – as the Carthaginians, Romans, Arabs and Spaniards, and England have since 1701 – also controlled seaborne trade to and from the open waters of the Mediterranean.

Powers situated at some distance from the sea's exit generally hold a weak strategic position, because their links with overseas and their naval movements to and from the particular enclosed sea are controlled by potentially hostile powers. Moreover, the powers controlling the sea's only exit are in a strong position to carry out naval incursions into the waters of other riparian states. In contrast, Austria-Hungary and the former Yugoslavia in the Adriatic, and Iraq in the Arabian Gulf have their maritime positions greatly weakened because the sea's only exit is controlled by other, potentially hostile states. The most unfavorable position among sea powers has been that of Tsarist Russia and its successor states. The exits from the Black Sea, the Baltic Sea, and the Sea of Japan (prior to 1945 also from the Sea of Okhotsk) to the warmer waters of the world's oceans were, and still are, controlled by the country's traditional adversaries. The Russian Federation currently enjoys favorable access to the open ocean only in the Barents Sea and the Sea of Kamchatka. The access of the Russian home-based fleets to the waters of the Northern Atlantic from the Barents Sea in wartime is difficult, because potentially hostile powers control the stretch of water from Greenland to Norway. By contrast, Russia's access to the open ocean in the Far Eastern theater is relatively more favorable, because the Sea of Kamchatka is a marginal sea of the Northern Pacific. However, the Kamchatka Peninsula, on which Russian major naval bases are located, lacks well-developed road and railroad links with the mainland and is, for all practical purposes, an island with all the inherent advantages and disadvantages accruing from such a position.

A maritime country can occupy a position astride or flanking the routes of potential adversaries at sea. England's rise as the world's foremost sea power was largely because of its favorable position in respect to the sea routes of its main potential opponents at sea. England was astride the principal Dutch sea routes and also flanked the routes to and from France's Channel ports. The sailing ships of the riparian states in the Baltic hugged England's coast on their way to and from the English Channel and the North Sea. Likewise, France's coast flanked the English sea routes in the Channel. The 300-mile-long island of Ireland enjoys a very favorable position, because it lies astride England's routes to and from the Eastern Atlantic and could serve as a stepping stone for an invasion of England.

THE FACTOR OF SPACE

Within the Mediterranean, Italy enjoys a commanding geostrategic position in the Central Mediterranean. One leg of the Apennine Peninsula faces the Strait of Messina, while the other borders the Strait of Otranto. Whoever controlled the southern part of the Peninsula held the keys to Sicily, and guarded the exit to and from the Adriatic Sea. The power that controlled Sicily also enjoyed an excellent position to dominate the approaches to and from the Central Mediterranean. The security of Italy's communications between ports on its western coast and those elsewhere in the Mediterranean, however, have been perennially threatened by the powers that controlled the islands of Corsica and Malta. These islands lay both astride and flanked Italy's sea routes in the western and central part of the Mediterranean.

A power that controls both shores of an international strait, as Turkey does, lies both astride and flanks sea lines of communications to the respective enclosed or semi-enclosed sea. A power controlling only one shore of a strait important for international trade – in the case of Oman or Iran with respect to the Strait of Hormuz or Malaysia with regard to the Malacca Strait – also flanks the routes of its potential adversaries at sea. However, such a power does not possess absolute local command, because another power controls the other shore of the respective strait. In the same way, any power that controlled Denmark, as Germany did in World War II, also controlled the eastern and western approaches of the Baltic Sea. Moreover, such a power was in a situation to dispute the command of the northern part of the North Sea and, by extension, the eastern part of the Northern Atlantic.

The distance of a country or territory from the most probable theater of war at sea is a significant factor that affected, and still retains its value in, the development of that country's sea power. England enjoyed an excellent geographic position from which to attack many trade routes and to conduct naval actions against its main adversaries at sea: France, Holland, and Germany and also its commercial rivals in the Baltic. The 120-mile-long English Channel resembles a huge funnel, only 21 miles wide at the narrowest point. Because of the short distances separating various points in the Channel, French *corsairs* used bases between Brest and Calais to threaten the safety of English and Dutch shipping. Likewise, Ireland and the British Isles are separated by the Irish Sea, which has the 100-mile-long North Channel only 11 miles wide at one end, while at the other is the St George's Channel varying in width from 60 to 140 miles. England repeatedly prevented any hostile power from obtaining a foothold in Ireland from which it would have been within striking distance to threaten the ports on England's western coast.

Map 1. THE MEDITERRANEAN SEA

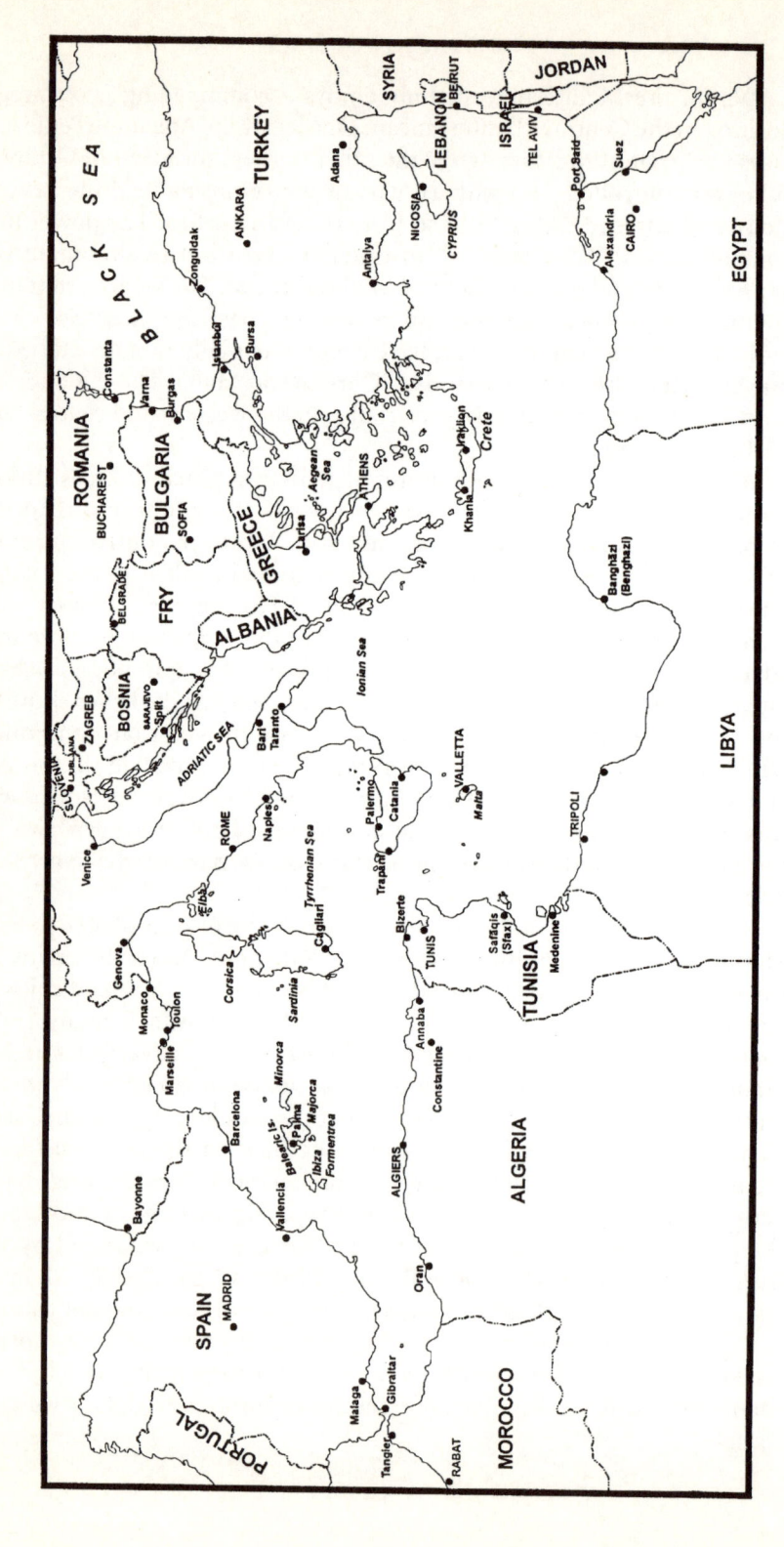

Shape

The country's shape may considerably affect a country's maritime position when other factors are taken into account. The compact, elongated, and prorupt shape of a country may be distinguished. A country is said to have a *compact shape* when all the points of its boundary lie roughly at the same distance from the country's geometrical center. Germany, the People's Republic of China (PRC), Belgium, and Poland all seem to have nearly round or rectangular boundaries. A country's compact shape in general offers relatively easy integration of its industrial base, transportation network and administration. It offers the shortest possible boundary relative to the area enclosed. Because such a country does not have peninsulas, large islands, or other protruding parts, the establishment of the communications network is generally easier.

A country is said to have an *elongated* (or *attenuated*) shape when its length is at least six times greater than its width. The land frontiers of a country with an elongated shape in combination with narrowness are difficult to defend and usually require the maintenance of large standing armies. When a country – Thailand for example – is nearly compact in shape but also has part of its territory in the form of a 'corridor' leading away from its main body, it is said to have a *prorupt* shape. Such a country usually has a long seaboard and narrow width of coastal area. Thus, it is highly vulnerable to an attack from across the sea and on its land borders.

Whenever a maritime country stretches across a land area, either on the rim of a continent or an island, it is said to be *physically concentrated*. Such a position is normally favorable for defense against external attack. A country is physically concentrated even if it extends across two or even more islands, provided that each is separated from the other by a relatively narrow channel or strait, as is the case of the British Isles. Denmark enjoys a high physical concentration, because almost all its 600-odd islands are situated close to the 210-mile-long Jutland Peninsula and adjacent to each other. If a country or territory spreads over too many islands widely separated, as is the case with Greece, Indonesia, and the Philippines, such a country is said to be *physically fragmented*. The Philippines extend for about 1,150 miles from north to south, and for about 700 miles from east to west. The Philippine archipelago encompasses about 7,100 islands of which only 460 are larger than one square mile. Indonesia is even more physically fragmented than the Philippines. The country consists of about 13,700 islands spread over some 3,200 miles of sea area.

When a narrow sea not only borders or surrounds, but also separates, a country into two or more parts then control of the surrounding seas becomes not only desirable but an absolute necessity in wartime. Control of Ireland by a hostile power always threatened the security of Britain, for example, because the hostile army could cross the Irish Sea and land on the British coast. However, at the same time, the very narrowness of the Irish Sea made it easier for the Royal Navy to control. The French in their repeated attempts to conquer England in the seventeenth and eighteenth centuries realized this, and did not dare to send ships-of-the-line into the Irish Sea. The French expeditions to land troops on England's western shores were directed instead towards Ireland's western and southern shores.

Sea Area

One of the main tasks of naval strategy in peacetime is to obtain sufficient space for the operations of one's naval forces and aircraft in time of war. The larger the size of a sea area, the larger the space for the movement of ships, submarines and aircraft. In contrast to the vast expanses of an ocean, a typical narrow sea presents a much smaller area to be controlled or defended and protected. The Baltic Sea extends along its main north–south axis for about 920 nautical miles and has an average width of a little over 105 nautical miles and covers an area of sea surface of 163,000 square miles. By contrast, the North Sea extends over 700 miles, from the Straits of Dover in the south to the Shetlands in the north and its maximum width is about 420 miles. The Arabian Gulf encompasses some 90,000 square miles and extends about 600 miles from the estuary of the Shatt el-Arab to the Strait of Hormuz, while its width varies from 35 to 210 miles.

The Mediterranean Sea is a typically large narrow sea, covering an area of about 970,000 square miles. It extends from west to east more than 2,400 miles, while its maximum width is about 1,000 miles. The Mediterranean encompasses a number of smaller seas, specifically the Tyrrhenian, the Ionian, the Adriatic, and the Aegean. The Adriatic encompasses only about 60,000 square miles and is, in all but name, a large gulf. The 400-mile-long and 180-mile-wide Aegean is a semi-enclosed sea within the Mediterranean. Another narrow sea adjacent to the Mediterranean is the Black Sea. This body of water stretches for about 750 miles from east to west and its maximum width is about 380 miles and encompasses an area of about 168,500 square miles.

A narrow sea can be surrounded on all sides by a continental landmass, while having a single exit to the open waters of an ocean or another narrow sea. The strategic importance of a sea exit is clearly greater, if it constitutes the single link between the riparian states and the open ocean or with some other larger enclosed or semi-enclosed sea. A sea exit can also represent the only link between the riparian states of a given narrow sea and a larger narrow sea, as do the Turkish Straits. The Baltic is surrounded on all sides by land, and its only exit with the other narrow sea – the North Sea – is through the Danish Straits. Similarly, the Strait of Otranto is the only link between the Adriatic Sea and a larger narrow sea, the Mediterranean.

Employment Area

In general, distances between various points within a given employment area involve both advantages and disadvantages. Short distances imply a small employment area and hence pose difficulties for the movements of a large attacking force, while a defender is in danger of having his supply lines cut off or his forces enveloped. The distances between various points in a typical narrow sea are relatively short. This considerably affects all aspects of the employment of one's naval forces and aircraft. In the Baltic, the distance between Kiel and Helsinki is about 625 nautical miles, the port of Tallinn lies only about 220 nautical miles away from Stockholm. Some 230 nautical miles separates Copenhagen and Rostock. The distances from Skaw in Denmark to Helsinki and to the Kiel Canal are about 840 and 250 miles respectively. Copenhagen lies only some 150 miles from Skaw. Luleå in the Gulf of Bothnia lies about 1,100 miles from Skaw, while Liepāya is about 620 miles away.

In the North Sea, the British port of Hull is only about 280 nautical miles away from the German port of Emden, while the Belgian ports of Ostend and Zeebrugge lie 207 and 205 nautical miles respectively from Hull. From Heligoland to Hull there are about 300 nautical miles of sea surface to cross. The German base at Cuxhaven lies about 420 nautical miles from the Firth of Forth. Skaw and Cuxhaven are 350 and 475 nautical miles, respectively, distant from Scapa Flow in the Orkneys, while only 40 nautical miles separates the Orkneys from the southern tip of the Shetlands, and from the latter to the Norwegian port of Haugesund there are only about 200 nautical miles to traverse.

In the Arabian (Persian) Gulf, the main Iraqi naval base of Basra (Al Başrah) lies about 154 nautical miles from the Iranian port of Bandar Khomeini (Bandar-e Khomeyni) and about 570 nautical miles from Bandar Abbas (Bandar-e 'Abbas) on the Strait of Hormuz. In the Sea of Japan, the main Russian naval base and its largest commercial port of Vladivostok lies about 325 nautical miles from the North Korean port of Wonsan and about 515 nautical miles from the South Korean port of Pusan.

Map 2. THE NORTH SEA

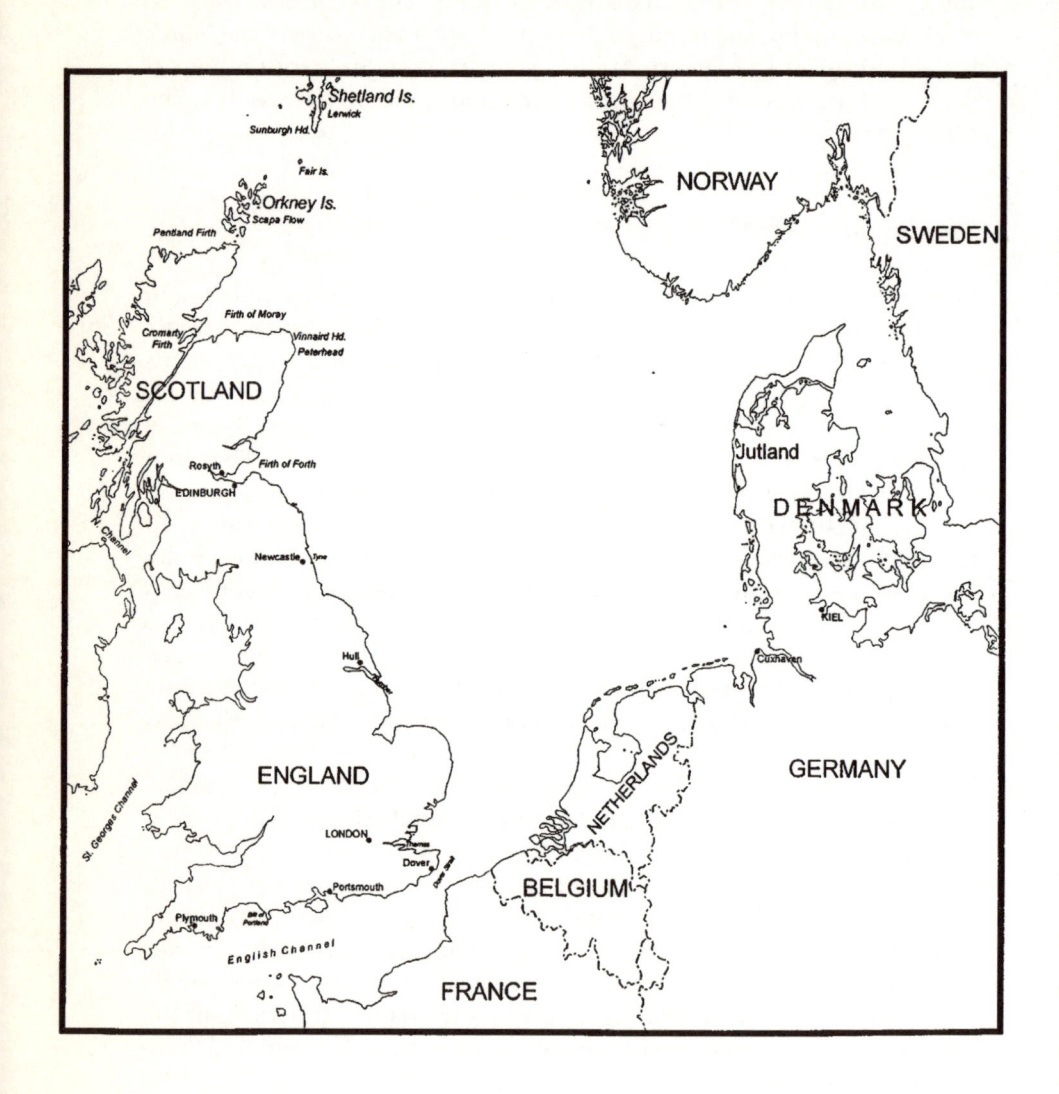

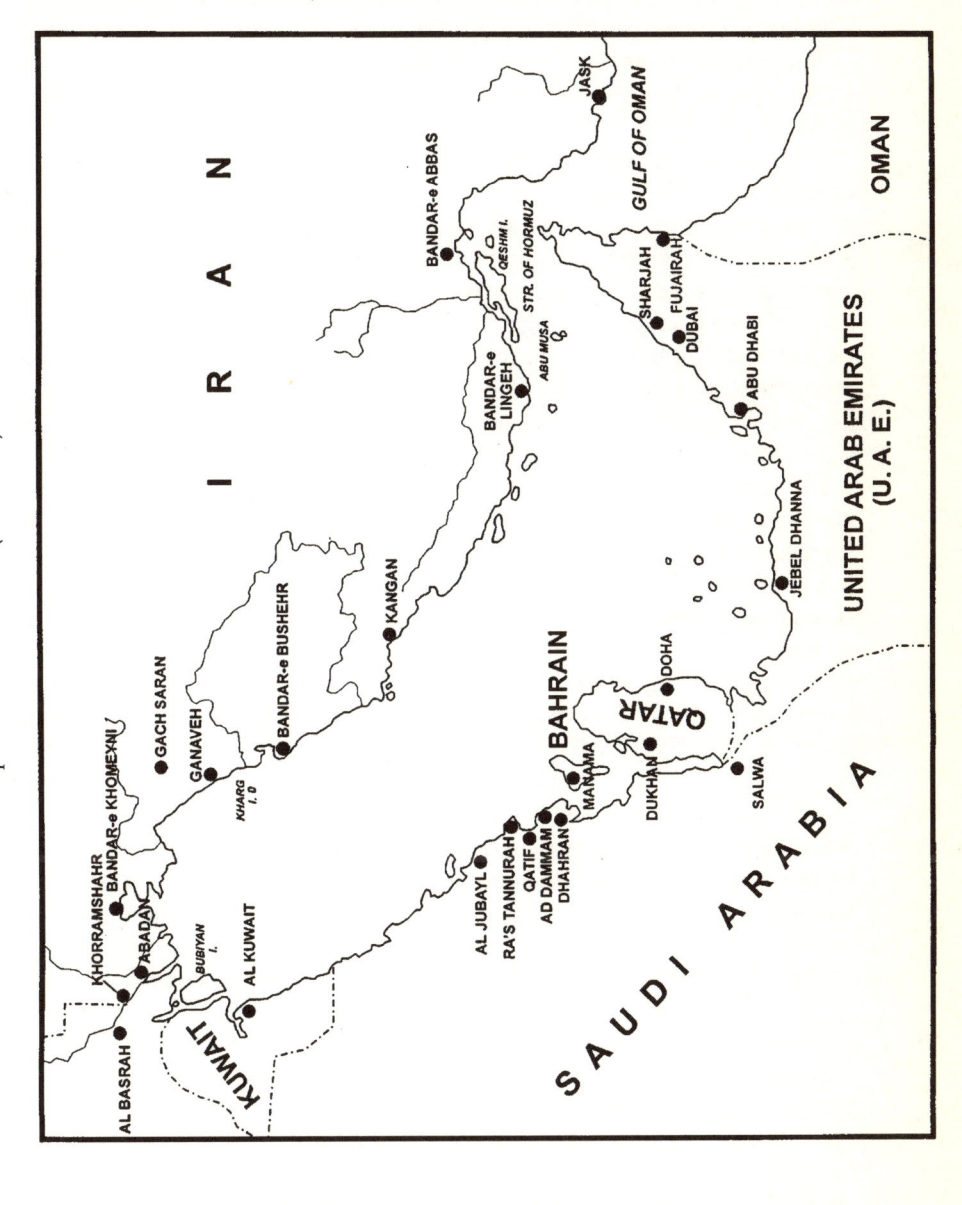

Map 3. THE ARABIAN (PERSIAN) GULF

In the Mediterranean, the distance between Gibraltar and Beirut is about 2,015 nautical miles, while La Valletta on Malta lies 1,005 nautical miles away from each of them. From the Strait of Gibraltar to Marseilles and Naples are distances of about 710 and 980 nautical miles respectively. The sea distance from Marseilles to Tunis or Algiers is some 410 nautical miles, while Ajaccio on Corsica is about 190 nautical miles away. The farthest point from shore that one may be anywhere in the Mediterranean is about 230 miles.

Because of the short air distances, all types of fixed-wing aircraft and helicopters can be used in a narrow sea. The short flying times related to a narrow sea allow more aircraft sorties and longer periods over the target area, whereby a larger area can be reconnoitered within a given time. The probability of achieving a surprise is greatly enhanced, especially if the aircraft approaches the target area at a low altitude. Also, aircraft, if damaged in combat, have better chances of reaching a friendly base than when they operate over the open ocean.

The short distances involved in a typical narrow sea allow the side stronger in the air to dominate the theater to a far greater degree than is the case with a similar ratio of forces in a war on the open ocean. The air threat alone severely restricts, or can even preclude, the use of major surface combatants such as cruisers and destroyers in a typical narrow sea, unless they operate under a strong air cover, though even then they should not operate too close to the enemy coast and mine-infested waters.

In a typical narrow sea, the side inferior in the air is forced to operate mostly during the hours of darkness or in bad weather, if then. However, the small size of a typical narrow sea means that even the weaker side can carry out surprise strikes against enemy targets on the sea and in the coastal area with a high probability of success. The short distances also allow one's ships and aircraft to change their respective areas of deployment within hours. The strikes may be carried out in rapid succession, thereby making it very difficult for the defender to recover from blows suffered in previous strikes.

The longer the distance from a home base to the prospective operating area, the more complicated and time-consuming it is for a blue-water navy to project its power ashore. In World War II, Axis control of both shores of the Central Mediterranean forced the British to use a much longer sea route, around the Cape of Good Hope, to supply their forces in North Africa – some 12,000 miles, compared with 2,400 miles via Gibraltar. By 1943 the Axis position in the Mediterranean had collapsed, and this allowed the British to save about five million tons of shipping space by shortening the distance to the Middle East.[1] The Korean War, the Vietnam War, and the Gulf War of 1990–91 all posed considerable problems for the employment and sustainment of US forces, because of long distances between the sources of power and the combat employment area. The sea distance from Norfolk, VA, via the Cape of Good Hope to Kuwait is about 11,900 nautical miles, while it is about 8,600 nautical miles via the Strait of Gibraltar and the Suez Canal.

Gaining/Losing Space

Space is a highly dynamic factor once hostilities at sea start. Additional space can be gained by offensive actions against the enemy-controlled sea area or by deploying and massing one's combat forces in greater depth. In peacetime, additional space for potential employment of one's forces can be obtained by diplomatic means or by creating alliances or coalitions. The US position on the European continent and over the adjacent narrow seas has been greatly improved, since the establishment of NATO in 1949. Likewise, the US bilateral defense treaty with Japan and the Republic of Korea has greatly facilitated the forward presence of US military forces in the area. The geostrategic position of the former Soviet Union in Europe was immeasurably strengthened between 1945 and 1991 because of the country's control of a string of communist-ruled East European countries. A country's geostrategic position can also be enhanced by obtaining additional territory through a diplomatically negotiated settlement. However, this is not often possible. The Germans failed to reach political agreement with Denmark in the early days of World War I to secure the Baltic exit and thereby to improve considerably their strategic position.[2]

An example of the acquisition of a much better strategic position through conquest was Germany's conquest of Norway in April–June 1940. The 2,125-mile-long Norwegian mainland coast was then more than twice as long as Germany's coast in the North Sea and the Baltic. Moreover, Norway possessed a large number of good natural harbors, and the adjacent waters were deep and mostly unsuitable for enemy mining. The nearest British base, in the Shetlands, was 170 miles away from Bergen, while Scapa Flow was 240 miles from Stavanger. The route through and along the Norwegian Leads was protected by German air power, coastal guns and naval forces. At the same time, German opportunities to conduct raids against Allied shipping in the northern Atlantic and the Barents Sea were greatly enhanced.[3]

One could obtain larger space because of the collapse of an opposing alliance or coalition, as happened after the breakup of the former Soviet Union in 1991, when the Russians lost their access to the Baltic, except for the enclave of Kaliningrad and the eastern part of the Gulf of Finland.

Space for the employment of naval forces of both sides is reduced by the existence of formally neutral countries or territories, whose legal status precludes the use of their territory, waters, or airspace by all belligerents. Also, the sides in conflict are not allowed to move troops or war material or supplies across neutral territory. Troops that enter neutral territory must be disarmed and interned until the end of the conflict. On a non-discriminatory basis, neutral nations can (but are not obliged to) close their ports and anchorages to belligerents.[4] Naval vessels of the belligerents are limited in respect of the time they can stay in a neutral port or anchorage.

The most elementary factor in the stability of a neutralized zone in wartime is the symmetry of the advantages both sides draw from its existence. Normally, one of the sides in a conflict will respect the neutralized zone, as long as the advantages of its use are at least as great as its disadvantages. In World War I both Britain and Germany had equal interest in maintaining Norwegian neutrality. The British used coastal waters off northern Norway to supply its Russian ally, while the Germans used the central and southern Norwegian coastal waters for the transport of Swedish iron ore. This situation changed in the initial phase of World War II, when Russia was not a belligerent and Germany had taken advantage of Norway's neutrality by using its coastal waters for transporting Swedish iron ore.[5] Sweden's neutrality in World War II allowed Germany to use the much safer route from Luleå to German Baltic ports for the import of iron ore. Denmark's neutrality in World War I prevented possible British naval incursion into the Baltic and establishment of a link with Russia. At the same time, Denmark's neutrality prevented the Germans from employing their naval forces based in the Baltic on operations in the North Sea.

A conflict between two countries or groups of countries can indirectly involve third countries, which may essentially be uninterested in the outcome of the conflict but want to remain on friendly terms with both sides.[6] These third states or 'neutrals' may reduce the space for the attacker and thereby benefit the side on the defensive. Space for military action can also be reduced by the existence of the countries or territories in a state of nonbelligerency. Another option for neutrals is to avoid involvement in a conflict by declaring a neutrality zone. Such a zone can be effective if the diplomatic situation demands that each side in a conflict feel bound to placate neutrals.

Blockade and a counter-blockade can also be used to deny space, at least in legal terms, to an opponent at sea. A blockade is a hostile act intended to prevent ships or aircraft of any nation, enemy or neutral, from entering or exiting specified ports, airfields, or coastal areas belonging to, occupied by, or under the control of the enemy nation. The objective is to deny the enemy the space in which to use his or neutral vessels or aircraft to transport personnel and goods to or from enemy territory. The Entente Powers conducted a naval blockade of the Central Powers in World War I, while the Central Powers conducted a counterblockade. Great Britain declared a blockade of Germany on 3 September 1939, and on 27 November of that year it reasserted the principle of a distant blockade. The neutral Scandinavian countries protested at the British action, but their protests became muted after Germany invaded Norway in April 1940.[7] In June 1940, the British blockade was extended to Italy and its possessions.

Since 1945 the most prevalent method of preserving space for one's forces and denying space to the opponent at sea has been the declaration of various *exclusion zones*. The effective enforcement of the exclusion zone requires the proper balancing of forces in terms of space and time. Otherwise, the side that declared the exclusion zone is forced to conduct random searches or attacks in the hope of deterring the opponent's ships or aircraft from entering and operating within the zone. In the Iran–Iraq War (1980–88), both sides declared *exclusion zones* in the Arabian Gulf shortly after the outbreak of hostilities in September 1980.[8]

The factor of space can be temporarily reduced or enlarged according to seasonal climatic changes in a given sea area. The changes in the ice limit in the Barents Sea considerably changes the area of open sea for the free movement of surface ships. By September the pack ice line runs north of the Svalbards (Spitsbergen), while by March its southern limits include the island of Jan Mayen and Bear Island. Likewise, the larger part of the Gulf of Bothnia and the Gulf of Finland in the Baltic are frozen during winter months, as is the major part of the Sea of Okhotsk.

Configuration of the Coast

The development of a sea power in a typical narrow sea has always been affected by the *configuration* of the country's coast, the number and quality of natural harbors, the character of terrain in the littoral area, the abundance or scarcity of natural resources, and inland communications. The physical character of a coastal area and its *hinterland* has sometimes considerably affected the development of sea power. A flat and low-lying coast at medium geographic latitudes is generally more fertile and suitable for agriculture than a coast with similar features at higher geographic latitudes. However, such a coast did not necessarily draw the majority of its inhabitants towards the sea, as exemplified by France's coast along the English Channel. In the past, if the coast was below sea level, as the

Dutch coast or the lagoon coast around Venice, then the scarcity of natural resources usually drove the populace to seek its fortunes at sea, because the available arable land area could not provide a livelihood for the entire populace. The low-lying coast of Germany and northern Italy facilitates communications with the country's interior. At the same time, these coasts are also much more vulnerable to attack from across the sea than a high mountainous coast, such as the eastern coast of the Adriatic or the Iranian coast of the Gulf.

A space that allows the easiest approach into space controlled by hostile forces is a definite advantage in a war. In defense, the commander often cannot easily determine the place of the enemy attack, but he always has a chance to increase the combat power of his own forces through skillful use of the advantages offered by terrain.[9]

Physical features of the theater significantly affect the employment of both the enemy's and one's own forces. The advantages and disadvantages of various physical features in a narrow sea should be evaluated while considering the entire maritime theater not just the area or zone in which tactical combat is to take place. This means that, in considering and analyzing the physical environment, an *operational* rather than a tactical focus must be in place in order to employ successfully one's forces to accomplish operational or strategic objectives.

Character of the Coast

The coastline's length and the mutual position of the opponents' shores also play significant roles in determining the conduct of war in a typical narrow sea. Sweden's coast extends for about 2,000 nautical miles. It runs roughly in a straight line from the northern-most corner of the Gulf of Bothnia to Karlskrona in the south. The southern part of Sweden's coast (Skåne) is sharply triangular, while, in contrast, the eastern shore of the Baltic extends in the form of a semi-circle. Sweden's coast allows a rapid shifting of forces along its longitudinal axis, while its southern coast curves convexly toward the opposing shore.

The character and configuration of the coast have a direct influence upon the size, complexity and degree of vulnerability of one's base of operation, its coastal surveillance and defense systems, and the degree of threat from an attack from across the sea. The fiord-skerry coasts of Sweden and Finland are highly indented and the high terrain backing the shore allows construction of both conventional naval bases and underground shelters in contrast to the southern and eastern shores of the Baltic. In general, the high, steep, and rocky coast along both the northern and the southern part of Croatia's coast makes it difficult to make timely observation of movements of small surface ships if they sail close to the coastline.

The existence (or absence) of offshore islands greatly influences the conduct of war in narrow seas. In general, a multitude of islands and islets enhances the defensive value of the coast. They provide refuge for ships from bad weather. Large numbers of offshore islands, as, for example, along Sweden's and Finland's coast in the Baltic and Dalmatia's coast in the Adriatic, allow a greater depth of defenses against attacks from across the sea or from the air. The more numerous the islands the more difficult becomes the detection of small surface combatants from the air. The presence of a large number of islands also greatly eases the problem of basing one's naval forces. The multitude of protected bays or channels, as they exist off the central part of Sweden, offers a refuge to ships. Offshore islands also facilitate greater depth in defense against attack, both from across the sea and from the air.

Islands situated relatively far off the mainland coast can provide early warning of an impending attack, especially from the air. Whenever islands run parallel to the mainland coast in several rows, they facilitate surveillance of the adjacent coastal waters by one's forces. They also simplify control and protection of one's shipping. If the islands extend transversely to the coast, as they do in central Dalmatia, the channels separating the adjacent islands are often wider and deeper, thereby allowing quick, concealed, and relatively easy deployment and redeployment of one's ships. At the same time, the attacker has a greater chance of penetrating the outer defenses to strike at targets along the mainland coast.

The archipelago-type of coast, such as in the Aegean or the Baltic, allows greater flexibility in the selection of lines of operations and easy and secure '*castling*' (leap-frogging) of naval forces. It also provides excellent chances to use mines to protect one's own naval base, commercial ports, and sea traffic.

A long coast without offshore islands, as the Iranian coast of the Arabian Gulf, is highly vulnerable to enemy attack from across the sea. In contrast, a coast fronted with numerous islands is easier to defend. A larger number of offshore islands facilitate basing. Surface warships and especially combat craft can change their bases or anchorages quickly, in a matter of hours. The more offshore islands that exist, the more difficult it is to detect combat craft from the air. Normally, the aircraft must fly more sorties to sweep effectively along an open coast in a narrow sea than they do on the open ocean.

Landform

Landform and *relief* affect both the tactical and the operational employment of one's forces. The character of landform in the coastal area always has great influence on the extent and quality of the transportation network. A low-lying coast is usually favorable for the development of both longitudinal and lateral networks of roads and railroads. This in turn makes it easy to transport troops and *matériel* and generally reduces the need for local coastwise traffic. A coast with a well-developed road and railway communications makes it difficult for the opponent to interrupt one's traffic for any extended period of time. At the same time, such a coast facilitates the speedy advance of the enemy army into the country's interior. In contrast, a coast with poor land communications requires a greater reliance on coastwise traffic to transport troops and *matériel*. This land traffic can easily be interrupted for long periods of time especially if the principal roads or railways run close and parallel to a coast backed by steep high mountains. In general, a coast without any, or only a few, lateral communications favors the defense, because it offers few access routes into the country's interior for an army, which has been successfully landed on the coast. A highly elevated or mountainous coast often lacks entirely or has great scarcity of communications. If a mountain chain runs close and parallel to the coast, then the roads and railroads usually run in the same direction.

A flat coast with few or without offshore islands usually complicates the defender's problem in repulsing enemy amphibious assaults. The high and rocky coast offers, in contrast, few if any places for enemy landings. Coral reefs and shallow waters favor defense against conventional amphibious landings. Swamps and marshes in the coastal area can significantly inhibit vehicular traffic especially the movement of heavy armor. A steep, rocky and highly indented coast, or with fiords separated by rocky headlands and numerous rivers, makes longitudinal communications difficult, while the lack of beaches makes it difficult to carry out conventional large-scale amphibious landings. The Jutland Peninsula's northwestern coast consists of long wide beaches of packed sand, while the southeastern coast is suitable for only small landings during good weather. The eastern coast of Jutland, as the coast of Sjaelland and other smaller islands, is suitable for landings because of its continuous beaches. However, the seaward approaches are shallow and landings and naval surface fire support (NSFS) are rendered difficult. The 1,000-mile-long coast of the White Sea is characterized by high and precipitous cliffs, a barren interior and great scarcity of beaches and these, combined with severe climate, make any large-scale, conventional amphibious landing difficult.

Slopes on certain landforms prohibit or make movement very difficult for some types of vehicles, while certain slopes may form natural obstacles to all forms of movement.[10] In preparation for the Normandy invasion (Operation *Neptune*), the Allied planners properly analyzed the tactical characteristics of the terrain in the amphibious objective area, but failed correctly to evaluate the operational effect of the hedgerow (or *bocage*) terrain and the existence of large marshes on the Carentan Plain.[11] The hedgerows and the inundated terrain in the western part of the peninsula considerably restricted the mobility of the US forces and greatly favored the German defenders. Further east, the combination of the hedgerows and shattered villages greatly hindered General Bernard Montgomery's efforts to break out of the beachhead. In contrast, the German use of the hedgerow to conceal their defensive positions significantly reduced the effectiveness of Allied air attacks.[12] The defensive features of the hedgerow could have been overcome by allying the modification of existing equipment to a much closer combined arms cooperation. The end result of this failure was that the war in the West was unnecessarily prolonged.[13]

Rivers

Abundance or scarcity of the rivers in the littoral area can significantly influence the employment of naval forces in a narrow sea. A coast endowed with a larger number of navigable rivers has inherently greater value than a coast without them. Rivers do not separate but provide means of communication, because they facilitate the concentration of the country's internal trade and act as outlets for overseas trade. For example, the River Po and its tributaries in Italy and the Rhône River in France have had such a role in the past. Every river is a military obstacle that exercises some degree of influence on offensive as well as defensive operations. Wide, deep rivers whose valleys offer concealment and may provide a satisfactory defense, being an effective barrier to offensive movement. Marsh and bog lands generally favor defense but are unsuitable for cross-country movement.

Oceanography

The most significant factors directly influencing the employment of one's surface ships, submarines and their weapons in narrow seas are the water's depth, the character of the seabed, the tides, and the water transparency. Most small enclosed and semi-enclosed seas are characterized by the prevalence of shallow waters. About 60 per cent of the Baltic Sea is less than 150 feet deep. The deepest water, at about 1,400 feet, is found between the island of Gotland and port of Nyköping. The depths in the western part of the Baltic far offshore range between 55 and 140 feet, and in the central basin from 165 to 550 feet. The depths of water in the Gulf of Finland vary from 110 to over 300 feet. The average depth within 15–20 nautical miles off the eastern coast is only 55–110 feet.

The average depth of the North Sea is about 300 feet, but there are great variations of depth between its northern part, which is generally deeper (760 to 790 feet), and its southern part. In the south and east there is a broad coastal strip of sea where the depth nowhere exceeds 120 feet. The depths in the English Channel range between 90 and 180 feet.

In the Arabian Gulf, the mean depth is about 80 feet and the water is rarely deeper than 300 feet. The deepest water is found off the Iranian coast, while the depths off the Saudi Arabian coast averages a mere 110 feet. By contrast, the depth of water in several smaller and larger narrow seas is considerably deeper.

The average depth of water in the Mediterranean is about 4,900 feet, but its maximum depth (off the southern coast of Greece) is about 16,900 feet. The Caribbean Sea is even deeper, with a maximum depth of about 21,600 feet and an average depth of about 13,400 feet.

The Black Sea is also much deeper than most other smaller narrow seas. The depth in its central basin averages 6,000 feet and its maximum depth is about 7,000 feet. However, the waters along its eastern and southern coasts are shallow. The shallowest part is the Sea of Azov with depths ranging from only three to about 50 feet. This body of water is linked with the Black Sea proper by the only 15-feet-deep Strait of Kerch.

The water depth in a typical narrow sea directly determines the size of ships and submarines to be employed, the ships' speed of advance, the use of underwater weapons, and the effectiveness of ASW sensors. Very shallow waters restrict or even preclude the employment of major surface ships and submarines, except for light combat craft, landing craft, and coastal and midget submarines. Shallow waters also influence the speed of large vessels, because of the proximity of the seabed, which refracts the waves, hindering a ship's movement.

The depth of water plays a great role in the employment of submarines in a typical narrow sea. A large diesel-electric submarine requires a minimal depth of water of about 72–82 feet. A medium-size submarine needs 55–65 feet, one of small size must have 40–50 feet, while a midget submarine requires 16–33 feet. A medium-size submarine with a hull height of about 43 feet requires about 33 feet of water under the keel for navigational safety.[14] Depending upon the water transparency of a respective area, several dozen feet have to be added to the submarine's operating diving depth to prevent its detection from the air. A medium-size diesel-electric submarine should have 82–98 feet of water above her hull's highest point or about 165 feet of water for safe sailing underwater.[15] This figure does not include the much greater depth required (if available) for the submarine to maneuver to avoid attack by surface ships and aircraft.

Most narrow seas are highly favorable areas for both offensive and defensive mine warfare. Moored mines generally may be laid down to a depth of 1,250 feet, while bottom mines could not be effective at depths greater than 660 feet. The former can be observed from the air to a depth of about 30 feet, and the latter are visible to a somewhat lesser depth. The shallowness of water can also restrict or even preclude the use of 533mm (21-inch) or larger caliber torpedoes. The 533-mm anti-shipping torpedoes normally require 60–75 feet depth of water at their launching point, and about 33 feet of water during their run to the target. Anti-shipping torpedoes fitted with an acoustic homing head normally require 60–75 feet depth of water during the search phase of their run.

In contrast to the open ocean, where the water's depth, salinity and temperature are the principal factors determining the propagation of underwater sound, in a typical narrow sea, the dominant roles are played by shallowness of water, the changing character of the seabed, and the proximity and configuration of the coast. Sound propagation is difficult to predict in shallow water because its losses are hard to estimate owing to multiple reflections of sound from the seabed to the surface and then back to the seabed. Shallow water acts like a 'duct' and hence it is possible that active sonars could achieve longer than normal detection ranges, even under negative gradient conditions. Also, target depth has little influence on sonar detection performance in shallow-water areas.[16] The main limitations in the performance of sonar sets especially those fitted on board surface ships come from the great variation and general unpredictability of the sea's temperature, salinity, surface conditions, tides, currents, sound reflection, and absorption caused by the diverse character and configuration of the bottom, and by background noise.

A large influx of fresh water into many enclosed seas often significantly lowers the saline content of the sea water. Sometimes, the structure of the sea's only exit can prevent the influx of saltier waters from the adjacent ocean or another sea. Variation in sea surface temperature and saline content, when combined with shallowness of the water causes great difficulties in the effective use of acoustic sensors. Sonar conditions in narrow seas at higher latitudes are generally better in the winter than in the summer. However, the sea state tends to be higher in the winter, and so can reverse this condition, especially if the sonar set operates at higher frequencies. While in the open ocean the vertical temperature structure of the water is more significant in determining the sonar's performance than the sea surface's roughness, the opposite is true in shallow waters.

In general, sonar conditions in shallow waters are better at night than by day, and better in the morning than in the afternoon, especially in the spring and summer. Again, there is little change in sonar conditions between these extremes, if 'whitecaps' are present.

Sound transmission in shallow water is principally affected by the characteristics and the flatness of the bottom, the depth of water, and thermal gradients in the surface layers.[17] The performance of sonar in shallow water is greatly affected by both reflection from and absorption of sound waves by the bottom. Sound transmission loss is least above a smooth sandy bottom, and greatest above soft mud (which is generally a good absorber of acoustic energy). A rocky bottom causes great scattering of sound waves, resulting in strong bottom reverberations. A seabed composed of both sand and mud has characteristics similar to a rocky bottom for transmission of underwater sound. Excessive roughness of a seabed also leads to a significant decrease in peak signal level, unless the sonar's projector is non-directional and a long pulse in sound emission is used. If the seabed is a very poor reflector of acoustic energy, as is mud, then the sound field should not differ significantly from that in the open ocean under similar refractory conditions. The reason for this is that the bottom-reflected sound makes only a slight contribution to the total sound field. When a positive thermal gradient exists, there is no difference in the propagation of sound between deep and shallow waters.

In general, the transmission of sound over all types of seabed is dependent upon the water thermal gradient. If a positive thermal gradient exists in shallow water, then little sound energy will reach the bottom and there will be little difference between deep-water and shallow-water transmission of sound. In the case of strong negative gradients and when the bottom is a poor reflector, shallow-water transmission will be similar to that in the shallow layers of deep water.[18] Therefore, poor sonar conditions prevail whenever there exists a negative thermal gradient causing downward refraction over a muddy bottom. Under these circumstances sound propagation is generally poorer at depths of water from 610 to 1,220 feet than at depths from 60 to 610 feet. In water less than 55 feet deep, and in the presence of downward refraction, the sound transmission anomalies increase with range. Good sonar conditions exist where there is a sandy seabed, and whenever the water is well mixed with a nearly isothermal overlayer. A negative thermal gradient, if it exists above a hard and smooth bottom, may form a sound channel (or 'duct') in the same manner as may occur in very shallow waters where a positive thermal gradient exists in the presence of a mixed surface layer. In both cases, sound is 'trapped' between the sea surface and the bottom. Such a sound channel may extend all the way to the sea surface. It can also occur even when there is no appreciable amount of sound refraction. Then the sound propagation losses are considerable, due to great scattering of sound. A so-called 'bottom effect' occurs in waters less than 90 feet deep and over a smooth seabed. In this case a sound beam bends down from the sea surface and bounces from the bottom, which acts as a mirror. Thereby spaces empty of sound or so-called 'skip-distances' occur, through which a submarine can pass undetected by shipborne sonar. In such a case the sonar contact will be lost even at short range. The distance in yards will be approximately equal to the depth of the submarine in feet at the moment when contact with her is lost.

Shallow waters present great difficulties in the effective use of shipborne sonars for sonar performance here is also influenced by the ambient noise, a disruption caused by disturbance of the surface of the water by wind, rain or snow, by marine life or by ship traffic. Ambient noise under some conditions is non-directional in character. However, under other conditions, the intensity of the ambient noise could be greater in the direction of the surface.[19]

Ambient noise level is considerably higher in narrow seas than in the open ocean. An abundance of marine life can greatly influence the sonar's performance. Man-made noise, caused by ship traffic in harbors and their approaches, by busy shipping lanes, and by coastal settlement activities, is, on average, 5–10 decibels (dB) higher in shallow waters close to shore than it is in deep waters, especially at frequencies greater than 500 Hertz (Hz).

Sonar detection varies greatly depending upon the proximity of the coast and the presence of river water. An indented coast fronted by numerous islands and islets usually makes classification of sonar contacts extremely difficult. The principal disadvantages in operating conventional sonar in shallow waters are the inability accurately to determine sound propagation and speed and automatically to obtain correct measurement data for sonar or ASW weapon fire control systems. Using a ship's variable depth sonar (VDS), results in a substantial increase in the detection range when it is lowered beneath the thermal sound velocity layer. However, VDS does not provide a reliable solution for the detection of submarines in shallow water owing to the sound transmission loss because of the contact with the sea bottom. Current sonar sets do not possess the ability to adjust their performance under diverse and rapidly changing oceanographic conditions.

A hard, smooth bottom is usually favorable for the laying of ground-type mines. By contrast, a seabed composed of soft mud may bury this type of mine after a period of time, and thus decrease the effectiveness of its firing mechanism. A smooth, sandy bottom allows a submarine to lie low, especially if there are wrecks nearby, to avoid attack by surface ships and aircraft.

Climate

The geographic position of a narrow sea in respect to its latitude generally determines the type of *climate* prevailing and the length of nights there. At high altitudes, the long hours of darkness prevailing in the winter combined with the short distance between the opposite shores offer very favorable conditions for the employment of one's surface ships, particularly for a side inferior in the air. The opposite is true, however, in the summer months. In June, the duration of the night at 54°N is about five hours, while at 60°N it is only one hour. Also the twilight lasts the entire night, while at 66°N the sun is continuously above the horizon. Hence, in the summer, the employment of one's naval forces would necessitate reliable air cover, because most of the missions would take place during daylight hours or at least hours of twilight.

Climatological factors, specifically, winds, waves, precipitation, ice, cloudiness, currents and air temperature greatly affect the employment of ships, aircraft and their sensors and weapons. From October through December, and January through March, westerly and southwesterly seas predominate in the Baltic. The frequent passages of severe storms add to the roughness of the sea. However, the latter is reduced in the winter due to extensive ice cover. Slight seas are frequent in all seasons, while rough seas prevail mostly in the fall, and are least frequent in the spring. High sea states in general influence the speed of all surface ships, affecting comfort and causing crew fatigue. This is especially the case for combat craft and amphibious craft. The employment in the Baltic of small surface combatants is seriously hampered on an average for 60 days of the year because of strong winds and high seas, rather than ice. The North Sea is characterized by violent northwestern storms in winter months that make navigation along the southeastern coast dangerous, and especially so along the Jutland Peninsula. Rain and fog occur frequently in all seasons in the Baltic and the North Sea.

By contrast, narrow seas at lower latitudes, such as the Arabian Gulf and the Red Sea, are more favorable to employment of small surface ships because of the generally less extreme weather conditions. In the Arabian Gulf, thunderstorms and fogs are rare, although dust storms and haze occur frequently in summer time. The most common and strongest wind (*shamal*) that blows from between northwest and west-northwest seldom reaches Force 6 and rarely Force 8. Combat employment of conventional-hulled combat craft becomes very difficult or even impossible at a sea state of five or higher. Then their speed must be drastically reduced, or waves may cause structural damage to the hull.

In narrow seas sonar performance is greatly affected by high or rough seas, while on the open ocean the vertical temperature structure of water is a more significant factor than the roughness of the sea surface. The opposite is true in shallow water where reverberations from the sea surface increase in intensity with an increase in sea state. High winds cause a large number of air bubbles, which absorb and scatter sound waves, emitted from the sonar's transducer. Under such circumstances a reduction in power of sound transmission would only proportionally raise the amplitude of echoes across the board. Reverberations decrease exponentially with range but more slowly than the target echo, because as the beam becomes wider, more scatterers fall within its scope.

Cloudiness, rain and fog can significantly decrease visibility over the sea surface, and thus substantially hinder the effectiveness of aircraft and surface ships. Ice-covered waters usually pose a significant obstacle to navigation even in ordinary winters in narrow seas at higher geographic latitudes, such as the Baltic and the Sea of Okhotsk. Ice conditions also greatly affect the performance of sonars and other acoustic sensors. The noise produced in ice-covered waters is notably different from that produced under ice-free conditions. In waters covered with ice, the intensity level and character of ambient noise is highly variable due to changes in wind speed, snow cover, and air temperature. Under rising temperatures, a continuous shore-fast ice cover has a very low noise level. When ice is not continuous, as in a brashy ice pack, the noise level is on the average 5–10 dB higher than that measured at the same sea state in ice-free conditions.[20] In ice-covered waters, the level and character of the noise are highly variable and depend upon ice conditions, wind speed, snow cover and air temperature changes.[21] Propagation under ice is significantly improved if there are no ridges from which the surface-reflected sound can scatter and diffuse.[22] Wind speed, through its own turbulence and the motion of drifting granular snow impinging upon the rough sea surface, causes a high level of noise. Wind noise is more prominent under non-continuous ice cover, when the noise level is independent of wind speed. Another source of noise is the bumping and scraping together of ice floes in non-continuous ice cover. Sonar performance in ice-covered waters is also greatly affected because ice makes the already-difficult problem of classifying contacts even harder to resolve.

The combined effect of air temperature and proximity of the landmass in a narrow sea frequently causes non-standard propagation of radio waves. This in turn greatly affects the range of radar and radio communications. Non-standard propagation exists when there is a considerable difference between the temperature of the air and that of the sea. Another phenomenon is super-refraction, which may occur when the relative humidity of the air steadily decreases instead of remaining constant, or when the air temperature decreases at a rate less than standard. Moreover, radio waves can then be bent down much more sharply, thus striking the sea surface, before being reflected upwards again, then curved down to the sea surface and so on continuously. This extreme case of super-refraction, known as 'ducting', occurs in conditions of temperature inversion, that is, a warmer layer of air is found above a cooler layer. Then the amount of refraction for the particles of air trapped within the 'duct' boundaries differs from the amount for those outside. Thus, 'ducts' or channels are created which conduct radio waves many miles beyond the assumed normal range.

The significance of the factor space in the planning and conducting of tactical naval actions and major naval operations in a typical narrow sea cannot be adequately emphasized. Many naval enterprises ultimately failed to accomplish their stated objective, because the factor space was either neglected or some of its key elements were improperly analyzed and hence led to a flawed operational scheme. The knowledge and understanding of the oceanography and climate is critical for the successful operations of ships, submarines, and aircraft in a given narrow sea.

NOTES

1 Anthony E. Sokol, 'Seapower in the Mediterranean 1940 to 1943', *Military Review*, 8 (August 1960), pp. 16–17.
2 Wolfgang Wegener, *The Naval Strategy of the World War*, trans. and introduction Holger Herwig (Annapolis, MD: Naval Institute Press, 1989), pp. 28–9.
3 See T. K. Derry, *The Campaign in Norway* (London: Her Majesty's Stationery Office [HMSO], 1952).
4 Mark W. Janis, 'The Law of Neutrality', in *The Law of Naval Operations*, ed. Horace B. Robertson, Jr (Newport, RI: Naval War College, 1991), p. 458.
5 Daniel Frei, 'Neutralisierte Zonen: Versuch einer strategisch-machtpolitischen Theorie', *Wehrwissenschaftliche Rundschau*, 12 (December 1969), pp. 669–70.
6 Janis, 'Law of Neutrality', p. 148.
7 I. F. E. Goldie, 'Maritime War Zones and Exclusion Zones', in *The Law of Naval Operations*, ed. Horace B. Robertson Jr, p. 169.
8 Ibid., p. 175.
9 Hartmut Behrendt, *Die Handlungsfreiheit der militaerischen Fuehrung-Moeglichkeiten und Grenzen aufgrund des heutigen Kriegsbildes* (Hamburg: Fuehrungsakademie der Bundeswehr, January 1968), pp. 6–7.
10 Alexander LaRoque, 'The Role of Geography in Military Planning', *Canadian Army Journal*, July 1955. Reprinted in *Military Review*, 2 (February 1956), p. 96.
11 Martin Blumenson, 'Beyond the Beaches', *Military Review*, 9 (September 1962), p. 72; the hedgerow is a fence, half-earth, half-hedge. The wall at the base is a dirt parapet varying in thickness from one to four or more feet and in height from three to 12 feet. Growing out of the wall is a hedge of hawthorn, brambles, vines, and trees, in thickness from one to three feet, in height from three to 15 feet.
12 LaRoque, 'The Role of Geography in Military Planning', p. 96.
13 John O'Brien, 'Coup d'Oeil: Military Geography and the Operational Level of War' (Fort Leavenworth, KS: School of Advanced Military Studies, US Army Command and General Staff College, 1991), pp. 38–9.
14 Miljenko Tesic, 'Fizička svojstva mora i podvodna detekcija', *Mornarički Glasnik* (Belgrade), 5 (September–October 1968), p. 665.

15 Milosav Kostić, 'Neki problemi protivpodmorničke odbrane', *Mornarički Glasnik* (Belgrade), 2 (March–April 1969), pp. 187–8.
16 Albert Cox, *Sonar and Underwater Sound* (Lexington, MA: Lexington Books, 1974), p. 32
17 Technical Report, Division 6, *Principles and Applications of Underwater Sound* (Washington, DC: NRDC, 1946, reprinted by Department of the Navy, Headquarters Naval Material Command, 1968), pp. 40–1.
18 Vernon M. Albers, *Underwater Acoustics Handbook – II* (University Park, PA: Pennsylvania State University Press, 1965), p. 83.
19 Ibid., p. 24.
20 Robert J. Urick, *Principles of Underwater Sound* (New York: McGraw-Hill, third edn, 1983), p. 224.
21 Ibid.
22 Tom Stefanick, *Strategic Antisubmarine Warfare and Naval Strategy* (Lexington, MA: Lexington Books, 1987), p. 341.

3

Positions

In the purely theoretical meaning of the term, a *position* consists of an indefinite number of geometrically connected and closely spaced points that allow one's forces to control and strike the enemy's forces. Admiral Mahan observed that among many positions only those which might decisively affect the outcome of war in a given theater of operations must be selected.[1] It is possession of power plus position that constitutes an advantage over power without a position, or, more instructively, equations of force are composed of power and position in varying degrees, surplus in one tending to compensate for deficiency in the other.[2]

Military value of a given place depends on many factors but principal among these are its position, its military strength, and its resources. A place may have a great military strength, but its position may be so unfavorable that it is not worth being held. Alternatively, a place may be excellently situated and have strong forces based there, yet lack resources in itself and in its surrounding territory. When a place enjoys an excellent position, has intrinsic strength and abundant resources, then it is likely to have a great strategic importance.[3]

A *strategic position* represents a series of points situated close to each other whose possession has the most direct effect on the outcome of war in a given theater. A favorable strategic position should allow rapid deployment and redeployment, and successful employment of one's fleet forces and aircraft. In general, there are fewer strategic positions in a given ocean or sea area than on land, and hence, they are so much more valuable. The strategic positions in the open ocean are few, and, in contrast to those in a typical narrow sea, can easily be avoided by taking a circuitous route. Admiral Mahan observed that, if a narrow sea provides not only the end or terminus of trade but is part of a continuous route – that is, if commerce not only comes to it but also passes through to other fields beyond – the number of passing ships is increased and thereby the strategic value of its controlling points.[4]

POSITIONS

A strategic position should possess a balanced defensive and offensive strength. It should be capable of resisting an attack from across the sea, from the air and from the land. A strategic position should allow one's forces to exercise control of the adjacent sea, and that means control of the trade routes. The northern trade route of Denmark, Sweden and Norway flows through the Kattegat and Skagerrak and runs from there into the Atlantic by way of the Shetlands. However, in World War I the Germans failed to take action to control this route.[5]

When a rough equality of strategic positions exists, each side can usually reach the vital objectives of its adversary at sea. However, if one side occupies a weaker strategic position initially, then disparity in positions exists and conflict will often end unfavorably for a side holding the weaker strategic position. In the past two major world conflicts, the German strategic position in the North Sea was inherently much weaker than that of Great Britain, because both exits to the open waters of the Atlantic were controlled by the Royal Navy. German Admiral Wolfgang Wegener aptly remarked that the British position in the northern part of Scotland dominated all the seaborne routes, while the German position in the Heligoland Bight commanded nothing.[6]

A strategic position can be a stretch of the mainland coast, a large island or archipelago, international straits, or an international waterway (artificial canal or estuary of a large river). Control of a certain part of the *mainland coast* often profoundly affects a strategic situation in a theater of operations or even in the entire theater of war. In World War I, the German Navy possessed a favorable strategic position in the Baltic. Its task was to exercise effective command of the sea by covering and protecting the trade route between Sweden and Germany, which was of vital importance after stalemate on the Western front. However, the German position in the Baltic would have been considerably enhanced by seizing the Aaland Islands and by subsequent capture of Libau. The Germans failed to do that because their strategic posture in the Baltic was too defensive.[7]

A country's strategic position can also be enhanced by obtaining additional territory through diplomacy. Failure to do so can cost dearly when the conflict breaks out, as the Germans learned in World War I when they were unable to pose a credible threat to the Royal Navy in the northern part of the North Sea because they failed to persuade Denmark to allow the free use of the Danish Straits for German naval vessels. Perhaps if the German fleet had held bases in the Kattegat, it would have posed a graver challenge to the English position in the North Sea.[8]

The Germans realized that a principal reason for defeat of their U-boats in World War I was the lack of bases allowing free access to the open waters of the Atlantic. Admiral Wegener wrote in the 1920s that if

Germany controlled Norway, Britain would be unable to maintain the blockade line from the Shetlands to Norway, but would have to withdraw to the less defensible Shetlands–Faeroes–Iceland line. In June 1940, at the end of their campaign in Norway, the Germans had greatly improved their initial position against Britain in the North Sea and northern Atlantic. The Norwegian coast was endowed with a large number of easily defended, deep natural harbors, and anchorages. The waters washing the Norwegian shores were extremely deep and mostly unsuitable for mining. The nearest British base in the Shetlands was 170 miles away from Bergen, while Scapa Flow was 240 miles from Stavanger. The route through the Norwegian Leads was protected by the *Luftwaffe* and large numbers of coastal gun batteries and minefields. At the same time, German opportunities toconduct raids against Allied shipping in the Northern Atlantic and the Barents Sea were greatly improved. Northern Norway's fiords provided the bases for deadly attacks by the *Luftwaffe*, submarines, and surface ships against the Russian convoys in 1942–43. The U-boats, which did not require the shelter of the Leads, also ranged freely from the pens built at Trondheim and Bergen.[9]

The value of a maritime position is greatly enhanced if one's naval forces and aircraft can strike the enemy's sea communications from short range. Its value is even higher if it lies near several crisscrossing shipping routes. After the fall of France in 1940, the Germans acquired naval and air bases flanking the British Isles. This complicated the problem of British home defense and most important of all weakened British control of the approaches to the Northern Atlantic. The occupation of France also meant that German naval forces and aircraft were within a few hundred miles or less of the main Allied trade routes, compared with the 1,500-mile passage each way they had previously had to undertake. Not only was British shipping in the English and Bristol Channels constantly subjected to attacks by the *Luftwaffe*, but the usefulness of the principal dockyards in Portsmouth and Plymouth was much diminished. By controlling the French coast, the Germans had also improved the chances of success for their heavy ships to strike Allied shipping in the open ocean. With Germany possessing bases in the Bay of Biscay, the Allies did not have much hope of success in preventing German heavy surface ships from operating in the Southwest Approaches. If British forces had been allowed to use a harbor in southwestern Ireland, which they could have equipped and defended as a fleet base, Britain's naval position would have been greatly enhanced, enabling them to intercept German surface raiders and to fight the U-boats more effectively. One British advantage was that the only major dockyard available to the Germans in the southwestern approaches, at Brest, was well within range of British bombers.[10]

Sometimes positions acquired in peacetime may turn out to be sources of great weakness rather than strength. Russia's strategic position at sea in the war with Japan in 1904–5 proved to be seriously flawed. The two territories in Russia's dispute with Japan, the Liaotung peninsula and Korea, were situated at the end of primitive lines of communication. Likewise, the two large Russian naval bases, Port Arthur and Vladivostok, were cut off from each other by the intervening landmass and the sea communications through the Korea Strait were controlled by the Japanese. Moreover, for the Russians to defeat the Japanese they would have been forced to invade the Japanese home islands. For the Japanese, in contrast, it was sufficient to seize Korea and the Liaotung peninsula and thereby accomplish their primary war objective. A prerequisite for this was control of the Sea of Japan and the Yellow Sea in the early stages of the war, when Japanese forces were to land in Korea and on the Liaotung peninsula. Control of the Sea of Japan was necessary for the Russians to send their armies across the sea to Japan's Home Islands. The Russians also made a blunder by dividing their fleet between Port Arthur and Vladivostok. As a result, the Japanese were left with a free hand to attack first one and then the other. The Russians apparently did not have any other choice, because their largest naval base (Port Arthur) lacked facilities to support more than a part of their Far Eastern Squadron.[11]

Prior to war with Italy in June 1940, France possessed a strong maritime position in the western part of the Mediterranean by having a large naval base at Toulon and numerous airfields on the coast of Provence and along the Rhône estuary. The most important route between Marseilles and Algiers was covered eastward partially by the island of Corsica and its naval base at Ajaccio. In the southern part of the Mediterranean, France controlled the coast of North Africa with its important and well-equipped naval bases and airfields from Bizerta (Bizerte) to Bône (Annaba), Phillippeville and Algiers to Oran. In the Eastern Mediterranean, French control was limited to its protectorate in Syria. Beirut was used as a fleet base. Britain's largest and best-equipped base in the Eastern Mediterranean was at Alexandria. Numerous airfields in the Nile Valley and elsewhere in Egypt secured airspace to the Libyan border and allowed Britain relatively easily to keep in check the Italian base at Tobruk. The British also controlled the Suez Canal and possessed naval bases and airfields in Palestine and Cyprus, thereby covering their maritime routes.

In 1940, Italy's position in the western Mediterranean was relatively strong because of its mainland's long coast and because Sardinia and Sicily almost made the Tyrrhenian Sea an enclosed sea. Although the French potentially threatened air and naval attacks from the island of Corsica, this threat was acute only against the Ligurian coast in the north and the naval base of Leghorn (Livorno) in the Gulf of Genoa. Italy's disadvantage was that its potential enemies, France and Great Britain, dominated its vital sea routes from the Strait of Gibraltar to the mainland coast.[12]

Map 4. THE KOREAN PENINSULA AND THE ADJACENT SEAS

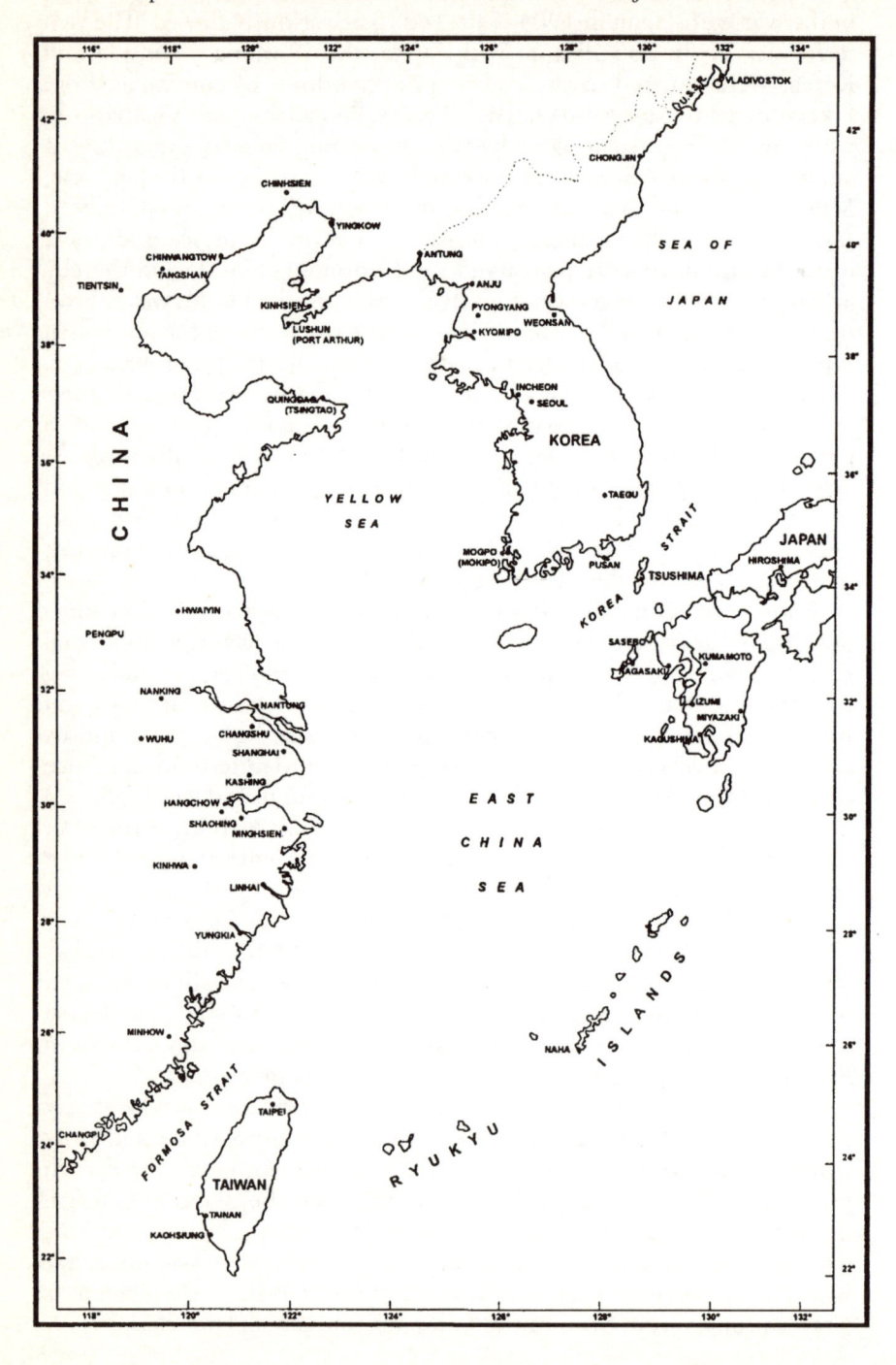

The maritime position of the Axis in the Mediterranean had been greatly improved after the fall of Greece and the island of Crete in May 1941. The Royal Navy, except for submarines, was no longer able to interfere with Italy's maritime traffic to the Dodecanese islands. Nor were the British able to prevent the Axis shipping from using routes in the Black Sea.

The Soviet position in the Baltic was dramatically weakened in the aftermath of the Bolshevik takeover of power compared with what it was prior to 1914. In the interwar years, the Soviets controlled only Leningrad and its immediate surroundings in the Gulf of Finland. This situation changed in June 1940 when the Soviet troops occupied the Baltic States and acquired the use of a number of bases for their Baltic Fleet. However, by September 1941, Germany's strategic position in the Baltic was greatly improved, because the Soviet Baltic Fleet had lost all its bases and was bottled up in the eastern part of the Gulf of Finland.

The Soviets considerably improved their naval position in the Baltic again after 1945 by virtue of their control of Poland and East Germany. Yet, in the aftermath of the East European Revolution of 1989, the Soviets lost control of East Germany and by the end of 1991 left their naval bases on the island of Ruegen and at Swinioujscie in Poland. By then all three former Soviet Baltic republics had obtained their independence and the Russian position in the Baltic had shrunk dramatically; only the Kaliningrad enclave and the eastern part of the Gulf of Finland remained in Russia's control.

A strategic position occupied by an *island* or *several islands* situated in close proximity to each other has always been one of the main elements on which rested the control of any narrow sea. The British Isles represent a 700-mile-long continuous physical obstacle extending from the Straits of Dover to the North Passage. Admiral Mahan noted that the British Isles exercise control of the North Sea similar to the control the island of Cuba has over the Gulf of Mexico. Their defensive value is the same.[13] The British position in the Channel facilitated the safety of the British trade in the Southwestern Approaches and made it difficult for their opponent on the continent to mount an invasion of the British Isles or of Ireland. Britain's geographical location allowed the Royal Navy to operate offensively in the narrow seas fronting western Europe, while the vast open spaces of the Atlantic Ocean protected the country's rear. The Royal Navy controlled the southern and northern approaches to the North Sea. In World War I, the main British trade routes in the Atlantic were beyond easy reach of the German fleet deployed in the North Sea. In contrast, the German routes could be easily intercepted in the English Channel and off Scotland. To defend this position, it was sufficient for the Royal Navy to stay on the strategic defensive, while the German High Seas Fleet would be forced to go on the offensive strategically in order drastically to change the situation in the North Sea and beyond.[14]

The strategic significance of the island of Malta lies mainly in its position commanding the midpoint between the western and eastern basins of the Mediterranean. This 15-mile-long island is situated at almost a midpoint of that sea and in the proximity of several crisscrossing routes. The distance from Malta to Gibraltar and Alexandria is 990 and 820 nautical miles, respectively. The island is situated only 50 nautical miles from Sicily, 190 nautical miles from the Libyan coast, and 175 nautical miles away from Cape Bon (Tunisia). Malta has played a vital role in the British strategy in the Mediterranean ever since it was captured by the British in 1800. Its importance was most dramatically shown in World War II, when it served as an air and naval base from which the Italian convoys to Libya were attacked. Field Marshal Erwin Rommel, upon taking command of the *Afrika Korps* in February 1941, prophetically remarked that without Malta the Axis could end by losing control of North Africa; that proved to be the case. In 1942 the German Naval Staff considered Malta as a prerequisite for the Axis offensive to take Suez.[15] Yet, before the war Malta had been written off as indefensible by the British Chiefs of Staff.[16] As long as the island remained in Allied hands, the advantage in supply to the Axis was largely neutralized. Allied aircraft and submarines based on Malta posed a most serious threat to the Italian sea routes to Libya as well as to the naval bases and airfields along the southern part of Italy's mainland coast.

In World War II, the 160-mile-long and eight to 36-mile-wide island of Crete played a critical role in the control of the approaches to the Suez Canal and the Turkish Straits. Crete forms a natural defensive line in the Aegean. It lies 350 miles away from the Turkish Straits, 200 miles from Athens, 150 miles from the islands of Rhodes and Cos (Kos) (in the Dodecanese) and 160 miles from the North African coast. Some of the principal maritime routes pass no more than 90 nautical miles from the island. In 1941 Crete offered the Royal Navy a refueling base 400 miles out from Alexandria. The island was only about 60 miles distant from the German airfields in Greece, while an Italian strip on Scarpanto was about 45 miles to the east. After the Germans seized the island in May 1941, the British problem became critical because the Germans controlled the 200-mile-wide passage between Crete and North Africa. The distance between Crete and Tobruk is only about 110 nautical miles, while the Nile Delta is some 300 nautical miles away. Whoever controlled Crete was also able to control the shipping routes in the Eastern and Central Mediterranean. The bombers based on Crete were able to reach targets in North Africa and oil fields in Romania.[17] Thus, if the Germans chose to build up their air strength on Crete and Cyrenaica, the Allied supply convoys from Egypt to Malta would have been exposed to grave, perhaps prohibitive risks. And if Malta could not have been supplied, it would have been no longer possible to attack the sea route between Italy and

Map 5. WESTERN MEDITERRANEAN

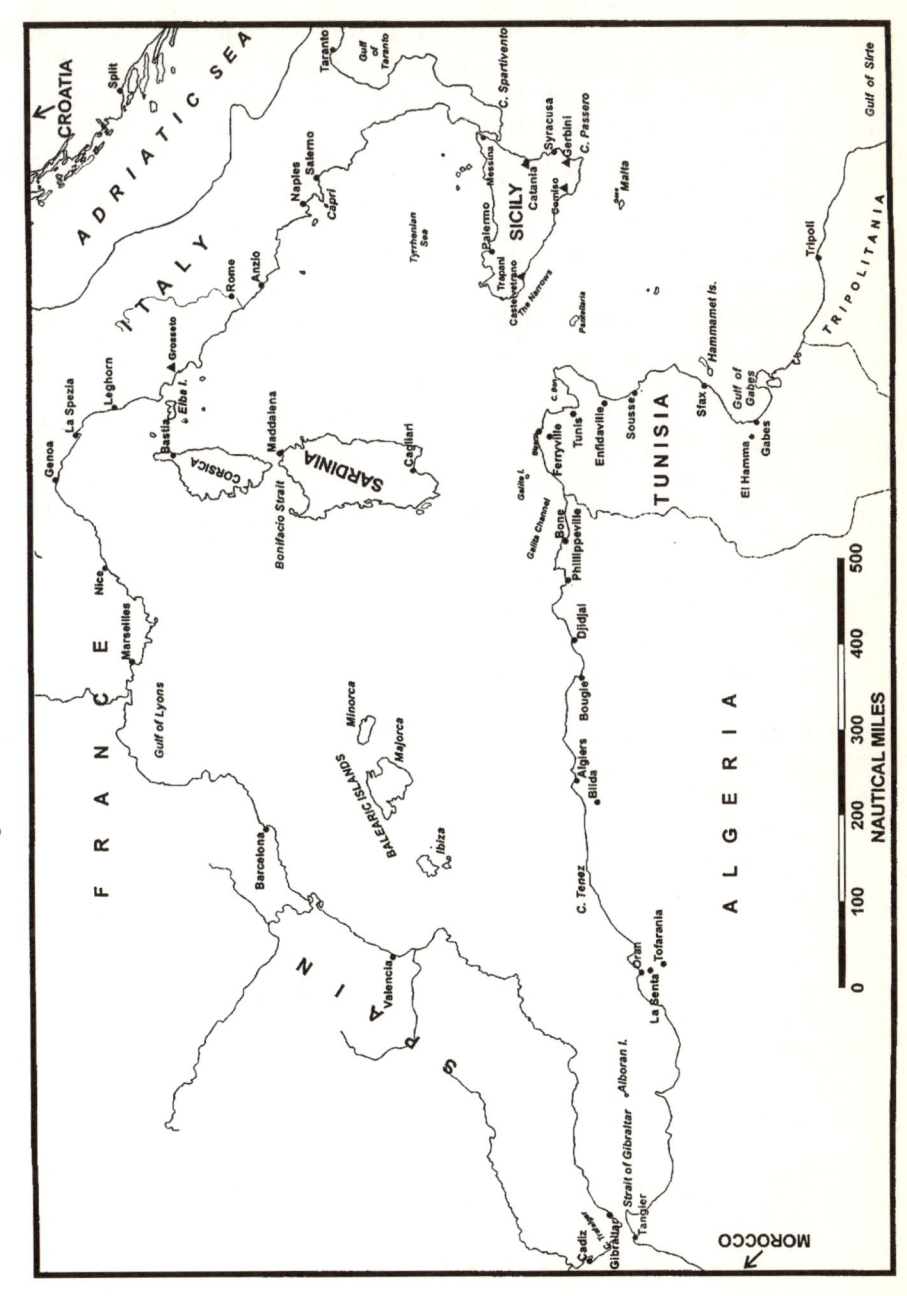

Africa with British surface ships and aircraft. Hence, the advance on Cyrenaica became most urgent for the Allies. The effort to reduce the hazards of running supplies to Malta was aimed to help to prevent the enemy from building up the forces that were threatening Egypt.[18]

In World War II the key to supremacy in the Aegean was the large Italian-controlled island of Rhodes. With a good harbor and two airfields, Rhodes proved to be a thorn in the British flesh from the moment Italy entered the war in June 1940. However, the British did not have the forces necessary to seize the island from the sea. The small island of Léros had one seaplane base and a harbor suitable for small warships, but no landing grounds for shore-based aircraft. Cos island had one airfield from which single-engine fighters could operate. Possession of the island of Casos (Kásos) would allow the British to shell Scarpanto airfield and the capture of that island would be of great value in an attack on Rhodes.[19]

The island of Corfu guards the southern approaches to the Strait of Otranto. In the medieval era Venice controlled Corfu and thereby the entrance to the Adriatic. Corfu played a critical role in the blockade of the Adriatic's exit by the Entente and their associated powers in 1917–18. Perhaps, had the Italians possessed Corfu in 1940, they might have been able to prevent the British air raid against their fleet at Tarent (Taranto), because Corfu could have been used for air surveillance of the surrounding waters.

Islands within an enclosed sea can also play a critical role in conduct of naval war there. For example, the military importance of the island of Lissa in the Adriatic Sea has long been recognized. The island is situated about 30 miles off Cape Ploče on Dalmatia's mainland and 100 miles off the tip of the Monte Gargano peninsula. By virtue of its geographic position, Lissa controlled the approaches to southern Dalmatia's archipelago and the mainland coast. It also flanked the routes to and from the Adriatic's exit. In the Prusso-Italian war against Austria in 1866 the Italian plan to land troops and occupy Lissa led to a decisive battle with the Austrian fleet under Admiral von Tegetthoff in July 1866. Lissa again played a significant role in the war in the Adriatic during 1943–44, when British light forces used it for their operations against the German controlled Dalmatian coast and offshore islands.

In another example, the Soviets in 1940–41 had control of the island of Hangö off the southeastern coast of Finland. Hangö flanked the German routes to Finland, and therefore the Germans put great importance on the need to eliminate this base, because its existence made the army supplies in the Baltic and the iron-ore traffic from Luleå much more difficult to protect. In addition, Hangö in Soviet hands tied up a sizeable number of German U-boats and S-boats.[20]

The 900-mile-long island of Cuba occupies the key strategic position in the Caribbean. It lies astride the sea communications to and from the

Gulf of Mexico and the Panama Canal. It also lies in close proximity to the route between the Central and South Atlantic and the US East Coast. Cuba controls the entrance to and from the Strait of Florida, the Yucatan Channel, the Windward Passage, and the Mona Passage. Jamaica, only one-fifth the length of Cuba, lies almost equidistant between Colón (Panama Canal) and Key West. Its military significance lies in its position flanking the sea routes from the Gulf of Mexico to Colón and those passing through the Windward Passage to Colón.

The channelization of movement on land is more varied than at sea. Terrain, unless highly mountainous, usually offers numerous passes and valleys as routes to certain objectives. By contrast, the avenues of approach at sea pass through one or more *straits*, which are usually the most vulnerable section of communications passing to and from the respective enclosed sea. The straits can be used effectively for blockading navies and merchant shipping. Because of their usually small width, they provide the most convenient place to invade the opposite shore.

The military importance of international straits has been demonstrated repeatedly throughout recorded history. This is especially true for straits or narrows which represent the sole exit from a narrow sea. Such straits are difficult if not impossible to avoid by using alternate routes. A power that controls one, and especially both shores of a strait used for international trade, enjoys an almost unassailable strategic position. The Strait of Gibraltar is the only passageway between the Atlantic and the Mediterranean. The strait is about 40 miles long and its width varies from eight to 24 miles. On the Atlantic side, the access to the Strait of Gibraltar is triangular, running from the Cape of Saint Vincent to Rota, to Tarifa and Ceuta. On the Mediterranean side, the approaches to the strait encompass the sea area between Cartagena, Melilla, Tarifa and Ceuta, while the island of Alborán and the Balearic Islands lie in the background.[21]

The Skagerrat and Kattegat (Danish Straits) are the only exit for all the shipping and naval forces transiting from the Baltic to the North Sea and other destinations overseas. The Skagerrak lies between the western coast of Sweden, the Jutland peninsula and the southeastern part of Norway. It extends for about 150 miles and is between 75 and 90 miles wide. The Kattegat separates the eastern coast of the Jutland peninsula and the southwestern part of Sweden. It is about 140 miles long and from 37 to 100 miles wide. The Danish Straits played an at times critical role in the policies and actions of the great European powers. In 1653 Denmark, then allied with the Dutch Provinces, closed the Sound to British ships, which cut Britain off from its main source of timber, hemp, and other vital naval supplies. Germany's control of the approaches to the Danish Straits in 1914–18 effectively prevented Britain from sending reinforcements to its Russian allies.

Map 6. THE CARIBBEAN SEA

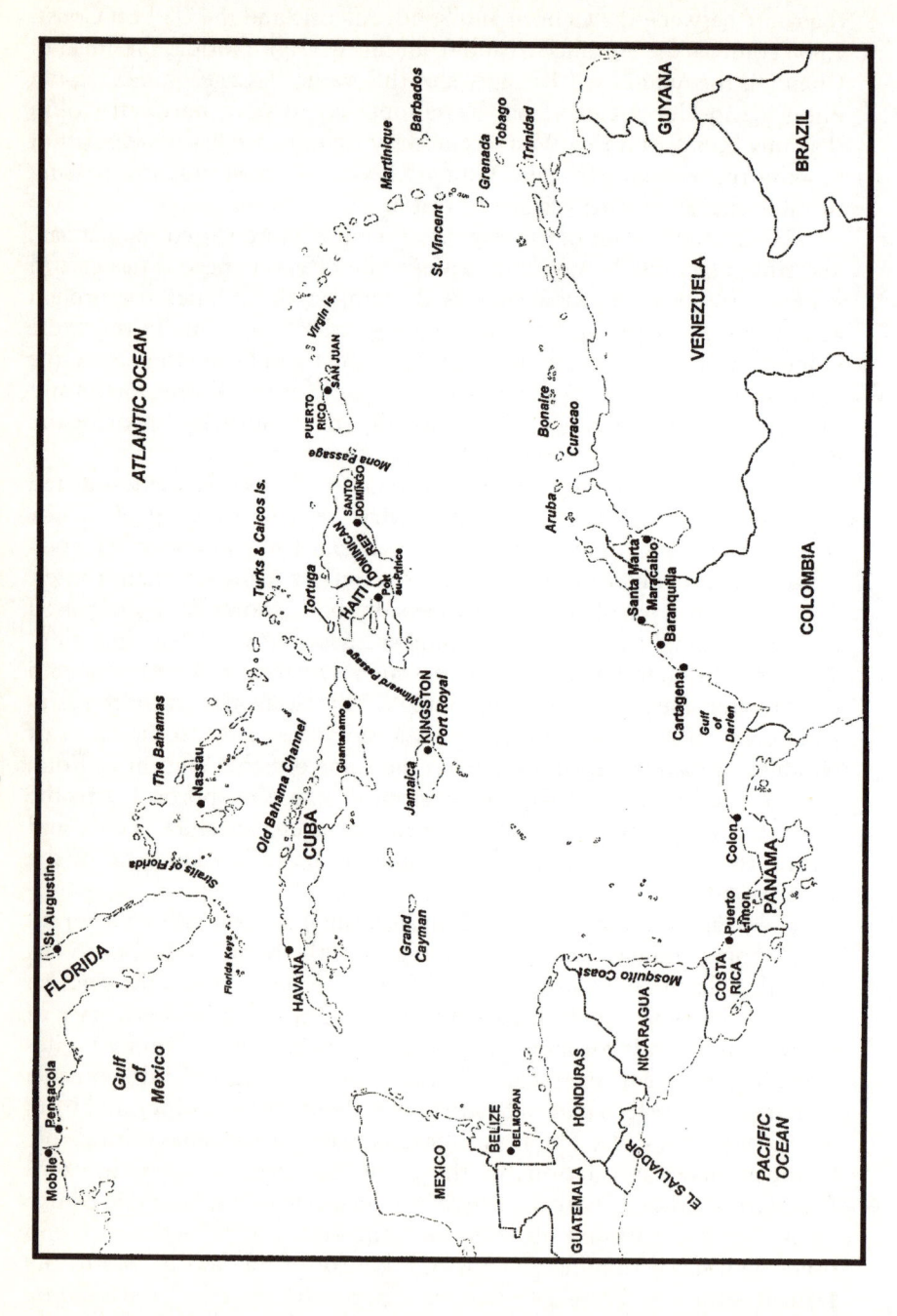

The strategic value of the Danish Straits was confirmed in World War II. Winston Churchill's pet project in late 1939 was to send a naval squadron into the Baltic to cut off the flow of Swedish ore to Germany (Operation *Catherine*). However, the difficulties of forcing the Danish Straits, the absence of bases, and the impossibility of screening the ships from the air led to the rejection of the plan.[22] The occupation of Denmark in April 1940 gave the Germans full control of the Baltic approaches. Thereby, the Royal Navy was cut off from the Baltic and the Germans were free to use the economic resources of the Scandinavian countries, especially Sweden's ores, to set up bases in Finland, and generally to improve conditions for the employment of their naval forces in the Atlantic.

The Strait of Hormuz connects the Arabian Gulf and the Gulf of Oman. It is about 170 miles long and from 30 to 50 miles wide. The strait is only 21 nautical miles wide at the northeastern end between Lārak Island and Quoins on the Omani side. The Red Sea's only exit to the south is Bab el-Mandeb. It is about 20 miles wide, but the island of Perim divides the waterway into two channels, 16 miles wide on the western side and two miles wide on the eastern.

Control of the sea's only exit is usually insufficient in itself to prevent the weaker fleet from having a freedom of action in some areas of a given enclosed or semi-enclosed sea. Hence, in addition to control of the sea's only exit, full or partial control of selected operational positions within such a sea should be obtained by a stronger fleet. In the Mediterranean, the Sicilian Narrows, the Strait of Messina, and the Strait of Otranto all have operational significance. The Sicilian Narrows are the natural passage between the eastern and western basin of the Mediterranean. The distance between Cape Granital (Sicily) and Cape Bon is only around 70 nautical miles. In addition, because of their shallowness, the Sicilian Narrows are easily mined.

In the Black Sea, the eight-mile-wide Kerch Strait is the only exit from the Sea of Azov and can be considered as having an operational significance. This strait and its shores were one of the most often and the longest fought over areas in the Black Sea in World War II. In the fall of 1941, defeated Soviet troops withdrew over the Taman Peninsula. By December 1941, however, the Soviets were back carrying out amphibious landings. In May 1942 those Soviet troops fled over the Taman Peninsula and German troops advanced to the same area in September of that year. One year later the German troops were forced to evacuate the Taman Peninsula, and by November 1943 Soviet troops had again crossed the Kerch Strait and taken firm control of the Crimea.[23]

The Indian and Pacific Oceans are connected by several straits of which the Strait of Malacca and the Sunda Strait are the most important. The Strait of Malacca is about 500 miles long and its width varies from

30 to 200 miles. All the shipping between the West and China in ancient times and in the medieval age had to pass through the Strait of Malacca. Hence, the powers which controlled one or both shores of the Strait of Malacca dominated the principal shipping route to and from the Indian Ocean. The rise of the Srivayayan Empire after the late seventh century AD was mainly due to its control of the Strait of Malacca. The empire was centered on Sumatra's port of Palembang and in Malaya. Srivayayan's rulers extended their control northwards along the Malayan peninsula to Kedah and several other northern isthmian ports along the Bay of Bengal. By their control of the Strait of Malacca they also had easy access to the Java Sea and the Moluccas archipelago. After the fall of the Srivayayan Empire, its successors and the foreign powers that ruled the Malayan peninsula virtually controlled the trade between China and the West. In contrast, the route via Sunda Strait was rarely used until the times of the Portuguese.

The strategic importance of the sea's exit is usually enhanced by the existence of a number of operationally significant island positions guarding its approaches from the open ocean. The strategic value of the Strait of Gibraltar could have been much greater if the British had also controlled the island of Madeira some 620 miles to the west and the Azores another 520 miles to the west. In the case of Gibraltar, it was critically important for the British to control the island of Alborán, some 130 miles to the east, and the Balearics, about 520 miles to the northeast, in the western part of the Mediterranean.

Large *rivers* offer similar features as military obstacles as mountains. They were usually regarded in the past as obstacles to advance not easily overcome, and became strategic frontiers when their principal passages were strongly held. However, defense of rivers has hardly ever been successful for any length of time. Rivers were great barriers to advance in some wars, such as, for example, the Rivers Niemen and Western Dvina (Daugava) in World War I. Some river estuaries, such as, for example, the Danube Estuary, also played a significant role in war in a narrow sea. If the river estuary and its adjacent land area represent the hub of regional trade, then it may even have a strategic significance. England went to war on several occasions in its modern history because of the prospect that its enemies on the continent might obtain a full control over the Scheldt (Schelde) River and the port of Antwerp.

Points

A strategic or operationally significant position contains a number of *points* or *localities* that, if held with adequate military strength, can force the opponent to modify or even abandon his selected course of action.

The Austrian Archduke Charles said that the possession of strategic points decides the success of operations in war. However, great care should be taken that not too many of such positions are held or controlled. Admiral Mahan has observed that it is a mistake to believe that holding every position adds to naval strength. Naval strength involves unquestionably the possession of strategic points, but its greatest constituent is a mobile navy.[24] Strategic points should be evaluated with reference to a given theater of operations, to their relationship to one another and to the fleet.[25] Other things being equal, the greater the distance the greater the difficulty of defense and of attack, and where there are many such points, the difficulty of defense increases in proportion to their distance, number, and dissemination.[26]

In general, the value of a strategic point in an enclosed or semi-enclosed sea is much greater than on the open ocean because of its more immediate effect upon the war on land. A strategic point may be located on the mainland coast, a peninsula, an island, or inside a strait. When a point is situated on the mainland coast it is more likely to have adequate resources for both offense and defense and good links with the country's interior. Because of its commanding position in the Sicilian Narrows, the naval and air base at Bizerte allowed the Germans to win the race for Tunisia in the spring of 1943. The Allies then blundered in deciding that it was too risky after capturing Bougie (Bejaïa) and Bône to continue their drive eastward along the coast and seize Bizerte.

A strategic point situated on a *peninsula*, such as Singapore, often greatly affects the war in a certain sea area. Its value depends not only on its location with respect to sea communications but also on whether it is well connected with the country's interior. A point on the peninsula is generally more vulnerable to a hostile attack from across the sea and from the land than a similar point on the mainland coast.

A strategic point on an *island* has inherently greater value than a similar point on the mainland, or on a peninsula, because it often occupies a more favorable position with respect to enemy sea communications. The nearer such a point lies to enemy sea communications, the greater is its value. A strategic point on an island is usually easier to defend from a hostile force than the one situated on the mainland coast or a peninsula. But if such a point is located too far from a friendly shore it is more vulnerable to enemy attack from across the sea and from the air.

A point situated *within a strait* or either end of an *international canal* has inherently a significant military value. The Iranians in 1971 occupied three islands in the Strait of Hormuz, Abū Mūsá and the Greater and Lesser Tunbs, thereby obtaining almost complete control of naval movements and merchant shipping traffic in and out of the Arabian Gulf. Likewise, the island of Perim in the Strait of Bab el-Mandeb controls the exit and the entrance to the Red Sea. The naval base of Valona represents a strategic point inside the Strait of Otranto, while Copenhagen occupies a similar position inside the Danish Straits.

A point on the *mainland coast*, but guarding the approaches to a strait, is normally more vulnerable than a point on an island, because it can also be attacked from the land side. During World War II the Germans and Finns blockaded the western end of the Gulf of Finland, thereby bottling up the Soviet Baltic Fleet, yet they were unable to seize the small island of Lavansaari guarding the approaches to Leningrad. This in turn allowed the Soviets to use the island as an air base to support minesweepers working to clear passages for the Soviet submarines through the Finnish–German minefields. The gun batteries on the island protected the Soviet small craft and together with the aircraft hampered the movements of German patrols. The *Luftwaffe* was too preoccupied with its support of the German troops on the land to assist the Finnish–German naval forces in the Gulf of Finland.[27]

Central vs Exterior Position

In general, a country or a territory or a force can occupy or operate from either a central or an exterior position. A force operating from a position interposed between hostile forces is said to occupy a *central* (or *interior*) position. Such a position offers some inherent advantages in both offense and defense. Normally, a force operating from a central position can concentrate more quickly at a selected point within its effective striking range than a hostile force moving along the periphery.

A central position allows one's forces to concentrate quickly – simultaneously or sequentially, and at short distances with local superiority – against forces operating on the periphery. If a force operates from a central position, the enemy cannot easily concentrate, except by bypassing the mass of that force or by moving in a circle around it.[28] The British Isles occupy what amounts to a central position with respect to potential opponents at sea or in the air. West European and Scandinavian overseas trade had either to go through the 21-mile-wide Dover Strait and the more than 110-mile-wide English Channel or to undertake the longer and more difficult route north of Scotland. These routes were open to attack by the Royal Navy operating from a central position.

During World War I, the Royal Navy, by concentrating superior forces in home waters, controlled the oceanic supply lines to and from the north and west coast of Europe. At the operational level, the British Grand Fleet, by its disposition in the Firth of Forth, occupied a central position in respect to the German High Seas Fleet and its bases, had the latter moved either to the northern exit or south past the Texel–Yarmouth line. Likewise, the High Seas Fleet, by occupying a central position in regard to two narrow seas, prevented the Grand Fleet from operating in the Baltic Sea.

Japan occupies a central position with respect to the adjacent Asian landmass and any hostile force approaching from across the Pacific. In World War II, the Japanese-controlled Philippines and the Netherlands East Indies also occupied a central position in the Western Pacific. The island of Luzon dominated the sea and air routes from the South China Sea northwards toward Formosa (Taiwan) and Japan. Also, the Japanese control of the islands of Java, Timor and Sumatra provided them with a strong central position in using their air and naval forces against any Allied force approaching the archipelago from the south or the west.

A country occupying a central position usually has the difficult task of ensuring the security of overseas lines of supply once hostilities start. The British, without a shot being fired, essentially cut off Nazi Germany's sea lines of communication.[29] Seemingly, this problem can be resolved only if additional adjacent land or maritime area is gained by conquest to obtain sufficient resources for the wartime economy. If a country occupying a central position is attacked simultaneously from several directions, its forces will be dispersed.[30] A side that controls a small space but occupies a central geostrategic position usually cannot effectively employ large forces and is also in danger of having its lines of operations and lines of supplies cut off by the enemy.

A central position offers potentially a great defensive strength, because there are no gaps or weak links in lines of communications used for logistical support and sustainment of one's forces. Yet a central position is of little use if the enemy is much stronger than one's own forces operating from it.

A country (or group of countries) occupies an *exterior position* if it lies on the periphery of a strategic center. Such a position *de facto* corresponds to a strategic flanking position. An exterior position allows one's forces to strike a wide range of physical objectives along the enemy's periphery. To be successful, however, forces operating from the exterior position must be numerically larger and more mobile than the forces opposed to them. Forces operating from an exterior position can also threaten or attack from multiple directions or axes. They can conduct a tactical offensive, even if the country's overall posture is strategically defensive. Another advantage of the exterior position is the potential for one's forces to draw the opposing force away from its assigned physical objective.

Because of the short distances involved, exterior positions are relatively more valuable in a semi-enclosed or enclosed sea than on the open ocean. The effect of the German occupation of France in 1940 on the British shipping routes in home waters was considerable and almost immediate. Because German naval and air bases flanked the British routes in the

Map 7. SOUTH EAST ASIA AND ADJACENT SEA AREAS

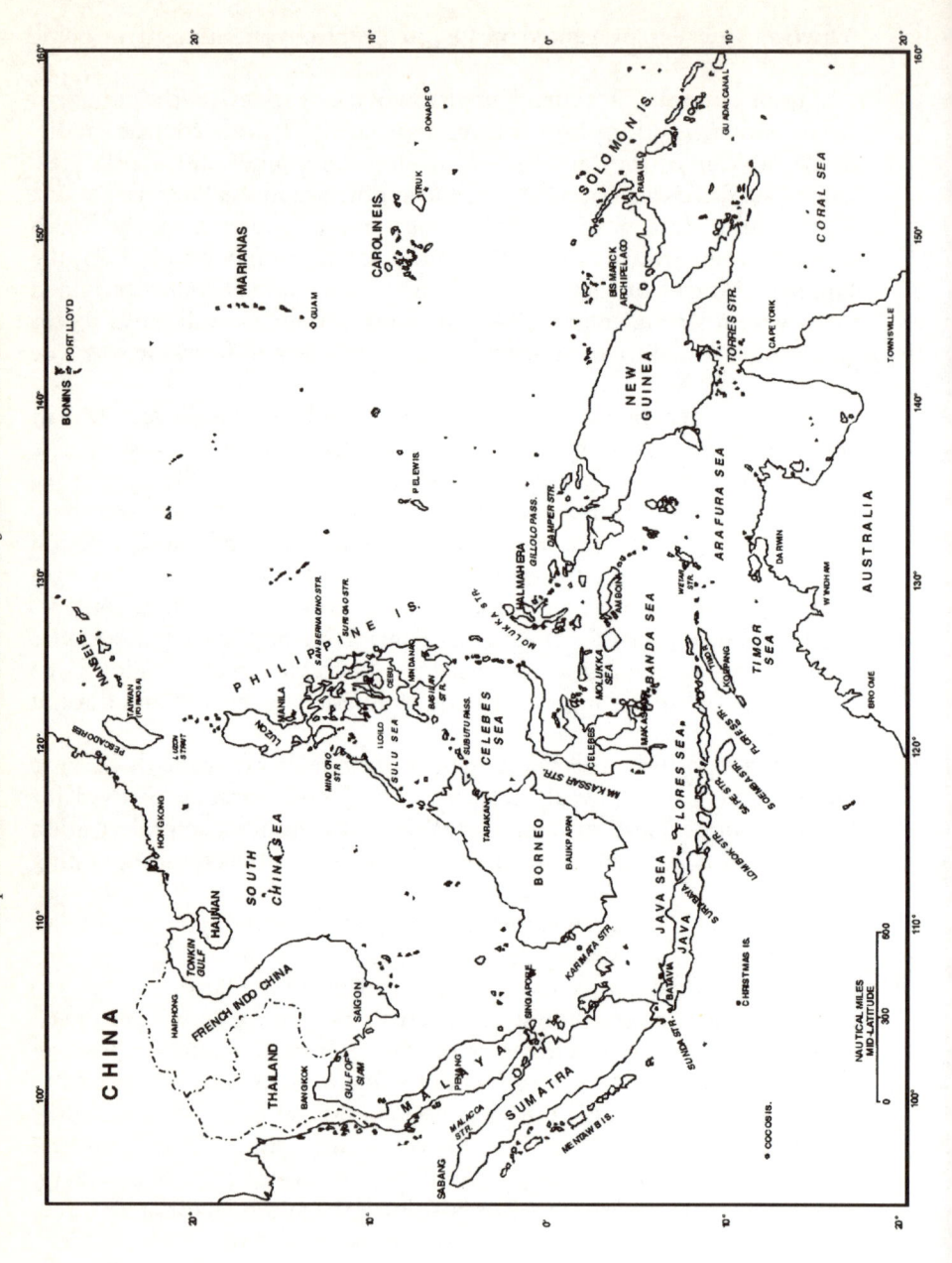

English Channel and along the country's eastern coast, the Channel could no longer be used, except for a trickle of coastal trade.

The importance of strategic positions and basing areas whether acquired peacefully or by force of arms should never be overstated. It is invariably the force that counts the most not the geography. However, it is equally wrong to underestimate the role that geography has played in the past, plays now, and will play in the future in the employment of naval forces and aircraft. Therefore, it is critically important to know and to understand the advantages as well as the disadvantages of given strategic or operationally significant positions and points.

NOTES

1 Alfred T. Mahan, *Naval Strategy Compared and Contrasted with the Principles and Practice of Military Operations on Land* (Boston, MA: Little, Brown, 1919), p. 131.
2 Alfred T. Mahan, *Naval Strategy* (Westport, CT: Greenwood Press, 1972), pp. 53, 241.
3 Mahan, *Naval Strategy Compared and Contrasted with the Principles and Practice of Military Operations on Land*, p. 134.
4 Ibid., p. 135.
5 Wolfgang Wegener, *The Naval Strategy of the World War*, trans., introduction and notes Holger H. Herwig (Annapolis, MD: Naval Institute Press, 1989), p. 36 (originally published by E. S. Mittler & Son, Berlin, 1929).
6 Ibid., p. 26.
7 Ibid., p. 24.
8 Ibid., pp. 28–9.
9 T. K. Derry, *The Campaign in Norway* (London: Her Majesty's Stationery Office, 1952) p. 229.
10 John Creswell, *Sea Warfare 1941–1945* (Berkeley, CA: University of California Press, revised edn, 1967), p. 50.
11 Donald Macintyre, *Sea Power in the Pacific: A History from the 16th Century to the Present Day* (London: Military Book Society, 1972), pp. 130–1.
12 Karl Gundelach, *Die deutsche Luftwaffe im Mittelmeer 1940–1945*, Vol. I (Frankfurt a.M: Peter D. Lang, 1981), pp. 18–19.
13 Mahan, *Naval Strategy Compared and Contrasted with the Principles and Practice of Military Operations on Land*, p. 129.
14 Gundelach, *Die deutsche Luftwaffe im Mittelmeer*, Vol. I, pp. 16–18.
15 Michael Salewski, *Die deutsche Seekriegsleitung 1935–1945*, Vol. II: *1942–1945* (Frankfurt, a.M: Bernard & Graefe, 1975), p. 67.
16 N. H. Gibbs, Vol. I: *Rearmament Policy*, in John P. W. Ehrman, ed., *Grand Strategy* (London: Her Majesty's Stationery Office, 6 Vols, 1956–76), p. 484.
17 Thomas Kempf, 'Der britische Rueckzug aus Kreta im Mai 1941', *Truppenpraxis*, 7 (July 1981), p. 584.
18 Creswell, *Sea Warfare 1941–1945*, p. 101.
19 Salewski, *Die deutsche Seekriegsleitung 1935–1945*, Vol. II, p. 67.
20 Ibid., *Die deutsche Seekriegsleitung 1935–1945*, Vol. I (Frankfurt a.M: Bernard & Graefe, 1970), p. 422.

21 Vicente Blay Biosca, 'Defending the Strait of Gibraltar: Spain's Role is Vital', *International Defense Review* (September 1985), p. 1401.
22 J. R. M. Butler, Vol. II: *Sept. 1939–June 1941*, in John P. W. Ehrman, ed., *Grand Strategy* (London: Her Majesty's Stationery Office, 6 Vols, 1956–76), p. 77.
23 Juerg Meister, *Der Seekrieg in den osteuropaeischen Gewaessern 1941–45* (Munich: J. F. Lehmans Verlag, 1958), p. 265.
24 Mahan, *Naval Strategy Compared and Contrasted with the Principles and Practice of Military Operations on Land*, p. 127.
25 Ibid., p. 175.
26 Ibid., p. 178.
27 Friedrich Ruge, *The Soviets as Naval Opponents 1941–1945* (Annapolis, MD: Naval Institute Press, 1979), pp. 25–6.
28 Rudolf Heinstein, *Zur Strategie des Mehrfrontenkrieges: Das Problem der 'inneren und ausseren Linie', dargestellt am Beispiel des Ersten Weltkrieges* (Hamburg: Fuehrungsakademie der Bundeswehr, November 1975), p. 6; Antoine Henri de Jomini, *The Art of War* (London: Greenhill Books, repr. 1992), p. 331.
29 Rudolf Boehmer, *Die Massnahmen des Deutschen Reiches vor und waehrend des zweiten Weltkrieges fuer Schuetz und Kontrolle der Deutschen Handelsschiffahrt* (Hamburg: Führungsakademie der Bundeswehr, September 1973), pp. 14–15, 21.
30 Heinstein, *Zur Strategie des Mehrfrontenkrieges*, p. 5.

4

Bases

One of the principal tasks of naval strategy in peacetime is to build or acquire a sufficient number of naval and air bases to allow one's forces to accomplish national and military strategic objectives in wartime. A fleet cannot sustain its strength unless it possesses an adequate basing area from which to deploy or redeploy, and to which when necessary retreat. Bases are the principal centers of all naval and air activity in a given sea area. A large number of geographically favorable bases capable of accommodating the major part of a fleet significantly facilitates its employment in combat. Even an inferior fleet retains its value as long as it possesses a sufficient number of well-protected bases, because they offer freedom of movement.[1] Seizing, controlling, and protecting one's naval bases was an important element of naval strategy in all eras. One of the prerequisites for conducting a successful blockade is the possession of a sufficient number of naval bases favorably located to keep the enemy fleet in check.

Any part of the mainland coast, or an island enjoying a favorable geographic location, which possesses an anchorage and enough flat terrain for an airfield, can be developed into a usable operating base. In general, the possession of large numbers of bases offers significant advantages for deployment, maneuver, and redeployment of one's fleet forces. It is invariably a bad thing to rely on the use of a single major base in wartime, because then one's own forces use the same line of operation for both attack and retreat. The enemy can with relative ease operate against a fleet that uses a single major naval base in a given area of operations as the German experience in the North Sea in 1914–18 convincingly shows.[2]

The lack of good bases invariably provides a source of grave weakness for the employment of one's fleet forces. If the Russians had possessed a base in the Sea of Japan during their conflict with Japan in 1904–5, their fleet might not have suffered the disastrous defeat it did, and they might have prevented full command of the sea by the Japanese fleet. The Royal Navy did not possess a sufficient number of large naval bases in the North

Sea in 1914. The only first-class base there was at Rosyth, Cromarty being only a second-class base.³ The only naval anchorage of any significance was at Harwich, while others, such as Grimsby, Hartlepool and Aberdeen, were defended commercial ports.

The main reason for the poor performance of the Soviet Baltic Fleet in 1941 was the loss of its bases and ports in the Baltic, all of which, except Leningrad and its immediate vicinity, fell into German hands within ten weeks of the start of the war. Likewise, the Soviet Black Sea Fleet lost all its large bases by July 1942. Shortly after the fall of Sevastopol', the Germans seized Novorossiysk, so that the Soviet Black Sea Fleet was left with only the small naval bases of Tuapse, Poti and Batumi on the Caucasian coast, which were not capable of providing support for major surface combatants and submarines.⁴

By the summer of 1944, the loss of the Atlantic bases forced the German Naval Staff to continue the U-boat war from bases in the North Sea, which were within effective range of Allied bombers. This in turn made it critically important for the Germans to retain control of their bases in the Baltic, which served as a testing and training area for the new U-boats and their crews. The German Naval Staff insisted on holding the left flank of Army Group North, where in the area of the Hel Peninsula and Libau were concentrated 50–60 per cent of all German U-boat training facilities and some 25,000 sailors. The loss of the Gulf of Danzig resulted in a reduction in the training capability of 12 to 15 U-boat crews per month. Moreover, all the other training procedures had to be reduced by as much as 50 per cent.⁵

Purpose

The main purpose of a naval base in general is to provide shelter, repair, and supply for the ships and rest and recreation for their crews, to act as a focus of power as close to the potential adversary as possible, to serve as part of the screen to warn against an enemy attack, and to ensure control of one's own shipping.

Naval bases can serve for offense, defense and offense-defense. They can be permanent or temporary. Permanent bases are built on one's own territory. Their main purpose is to extend the effective radius of one's fleet or to neutralize the enemy's major base in the area. Major naval bases usually serve for both offense and defense. The German main naval base at Wilhelmshaven was used primarily for offense in 1914–18. The Russian base at Port Arthur served for both defensive and offensive purposes in 1904–5, as did the Austro-Hungarian naval base at Pola in the Adriatic in World War I and Britain's base at Malta in 1940–43.

Minor naval bases built or acquired in the proximity of the potential scene of action are normally used for offensive purposes. A minor naval base at the Bay of Cattaro served for forays by the Austro-Hungarian light forces in the southern part of the Adriatic in World War I. It was also used for offensive actions by German and Austro-Hungarian submarines in the Mediterranean.

Sometimes naval bases are built to provide timely warning of any potential threat by enemy forces in a certain part of a narrow sea. The German naval base at Kiel was used to protect approaches to the Kiel Canal and to observe hostile movements in the western part of the Baltic. The British base at Scapa Flow at the northern exit of the North Sea had a similar role, as did the Italian bases at La Maddalena (Sardinia) in the Bonifacio Strait and at Otranto in the Strait of Otranto.

Many naval bases were built for the explicit purpose of controlling one's maritime traffic. The British naval bases at Gibraltar, Malta, Alexandria, Suez and Aden were seized or acquired to control the key choke points of shipping. Bases flanking friendly or enemy shipping routes, as the examples of Scapa Flow and Dover in World War I and Malta in World War II clearly show, provide great advantages in both offense and defense in wartime.

Characteristics

A naval base should provide a roomy and deep (but not too deep) anchorage to accommodate a large number of ships, including those with the greatest draft. The opening from the anchorage to the sea should be narrow and preferably lie nearer a river's estuary. A good naval base should have small variations between low and high tide. It should provide protection from high waves and bad weather. It should be ice-free to allow all-year round use, and it should be surrounded by hilly or mountainous terrain rising at some distance from the anchorage.

A naval base should be situated at not too great a distance from the essential sources of supply and it should have good communications with the interior of the country or territory. A naval base should be protected from attacks on the land side, from the sea, and especially from the air. Otherwise, its value is drastically diminished. The threat from the air for the ships in bases within a typical narrow sea have become progressively more serious since the end of World War I. This is an especially acute problem today and it cannot be resolved easily, especially in the case of a small country that lacks adequate air strength. Optimally, the most important bases should be located beyond the effective range of enemy aircraft. Main naval bases should be located as far as possible from those enemy airfields that can pose a constant threat from the air.[6] Also, one's naval bases should be located within effective range of one's aircraft.

Generally, narrow seas with a mountainous and cliffed coast allow construction of bases underground for small surface combatants such as missile and torpedo craft, minesweepers, and submarines. Shelters are usually built of concrete and fitted with heavy steel doors that provide total protection against any air attack, even including one with the use of nuclear weapons. They also can include a range of repair facilities and allow ship's crews accommodation for several weeks. Sweden has built along its coast probably the most extensive and sophisticated underground shelters able to accommodate even destroyer-size ships. Its largest underground facility is on the island of Muskö in the southern part of the Stockholm archipelago (some six miles northeast of the city of Nynäshamn). It is capable of providing the full range of services for ships and submarines of all sizes. The former Yugoslav navy also built a number of underground shelters along Dalmatia's coast. Specifically, two shelters each for use by missile and torpedo craft were dug out in the Ploče-Ston area and on the island of Lastovo.[7] Also, a number of underground shelters capable of accommodating missile craft and minesweepers were built along the southern shore of the Bay of Cattaro.

A good naval base should provide more than one line of operations. A base situated on an island or peninsula fronting the open sea usually offers multiple lines of operations, while the opposite is true in the case of a base located at the inner part of an enclosed sea or large gulf.

A mistakenly selected naval base is often a source of grave weakness. The Russians' selection of Port Arthur in 1899 as a large base for their fleet was in retrospect a big mistake. Although the port was ice-free, its communications with China's and Russia's Far Eastern ports were dangerously close to the islands of the Japanese Empire. Also, Port Arthur lay too far from Russia's center of power in the Far East, and the Russians were also unable to sustain communication with that base by either land or sea.[8] When the Russians tried to obtain a more favorably located base in the Korea Strait in 1899 the Japanese threatened war. Russia was not ready to fight and was forced to accept an unfavorable position brought upon itself by the acquisition of Port Arthur.[9]

Type

In general, major, minor and advanced naval bases may be differentiated. As the term implies, a *major naval base* is developed to accommodate a major part of the fleet. It should possess facilities and installations of all kinds, including ship repair facilities, depots for fuel, ammunition and supplies, and a hospital. It should also be fortified to provide protection from all kinds of attack.

Minor naval bases are built to provide shelter for only a few ships and have limited capacity for ship repairs and other support functions. A naval base situated near the potential scene of action is called a *forward* or *advanced base*. It is established either permanently or temporarily near the prospective theater of operations, but in either case should be within supplying distance from a major naval base. The use of the Elliot Islands (Changshan, Qundao, Korea Bay) by the Japanese in their war with Russia in 1904–5 impressed naval opinion in England and persuaded the Royal Navy to use less frequented but spacious bases such as those at Scapa Flow and at Cromarty Firth as advanced bases. Scapa Flow was selected because a fleet based there controlled the northern approaches to the North Sea. By 1912 the British had decided that both Scapa Flow and Cromarty were so important that they needed fixed defenses.[10]

An advanced base marks the outer extent of the hitherto effective striking power of the fleet and its lines of supplies. Any island or point along a coast accessible by sea, and which possesses an anchorage, could be turned into an advanced base rather quickly.[11] Advanced bases are usually temporary and are normally established in wartime. If a fleet is on the strategic offensive, new advanced bases are usually acquired. As a consequence, the previously acquired or built advanced bases will necessarily lose their importance. If a main naval base is overcrowded or vulnerable to enemy attack from the sea, then an advanced naval base can be built in peacetime and in the relative vicinity of the main naval base. The Austrians had built an advanced base on the island of Lussin Piccolo, only about 30 nautical miles southeast of their major naval base in Pola. The Italians also built a large number of advanced naval bases in the Adriatic between their major naval bases at Venice and Brindisi in the years preceding the outbreak of war in 1914. Likewise, the British by mid-October 1943 had established two advanced naval bases for their coastal forces in the Adriatic, one on the island of Vis and the other at Termoli on the Italian coast.

The military value of a base depends greatly on the adequacy of its resources and of surrounding territory. Obviously, the more abundant resources that a naval base has the higher is its military value. The availability of resources affects the size of a base, the number and capacity of its ship's repair or docking facilities, and the size of its supply establishment. A naval or air base built on a relatively small island with very few or no resources is generally inferior to one on the mainland because it has to be supplied from a distance. It also requires relatively large forces for its defense. Those built near a large commercial port have greater significance, since the resources necessary for the ships' repair and supply are ready at hand.

A naval or air base has much greater value if it commands good and secure communications by land and by sea. A base connected with its sources of supply by several rail tracks and roads is generally superior to one with weak or non-existent links with the country's interior. The Italian base at Venice in 1914 was superior to Pola because it had excellent communication links with its hinterland. Likewise, the Austro-Hungarian naval bases in Sebenico and the Bay of Cattaro were inferior to Pola, not only because of their small size, but also because of weak communications links with the interior of the country. The Bay of Cattaro was almost entirely dependent on its supply by sea and thereby was vulnerable. Valona was the best located naval base in Albania in 1914, but the coastal road between Durazzo and Valona was little more than a mule track.

A fleet operating in a typical narrow sea may establish bases not only in its home waters but also on the territory of another riparian state in the respective sea to improve its ability to operate offensively. The Soviets used the Polish base Świnioujście and East German bases on the Island of Ruegen to deploy permanently some of their fleet forces in the Baltic. Thereby the Soviets in effect moved their Baltic fleet to about 80 miles from their probable objectives in wartime.

A blue-water navy may decide to establish some of its overseas bases at the exit or inside a narrow sea. Both the US Navy and the Royal Navy used Otranto and Corfu as advanced bases to blockade the Strait of Otranto in World War I. Since World War II, the Royal Navy and then the US Navy have used Bahrain as a base for their forces deployed in the Arabian Gulf. The US Navy and other Western navies have used the naval bases and airfields of several friendly Arab riparian countries during, and in the aftermath of, the Gulf War of 1990–91.

Location

A naval base should be selected as far as practicable within a probable theater of operations in time of war.[12] It should be located within a striking distance of the potential scene of action. The British in 1914 selected bases along the country's east coast that allowed the Royal Navy to control both the northern and southern exits from the North Sea. Thus, the Grand Fleet based in Scapa Flow (the Orkneys) held the control of the Northern Passage, while forces based at Harwich secured control of the southern part of the North Sea.[13]

In general, a balance must be found between selecting a base that is close to the prospective theater of operations and one that is further away but provides more security from enemy attack. Optimally, a fleet should operate along short lines of operation. The radius of action is also directly related to the distance between the fleet's base of operations and the area of combat. The larger that distance, the longer are lines of operations along which a fleet must move to reach its prospective combat area and, hence, the shorter its effective radius of action will be.

Sometimes a major naval base is selected to meet changed circumstances in a given theater. In the first decade of this century, the growing strength of the German Navy rendered the naval bases in the southeastern part of England unfavorably located for use by the Royal Navy for its actions in the North Sea. This was the reason that the Royal Navy began to build naval bases along the central and northern part of the North Sea coast. Rosyth (across from Edinburgh) in the Firth of Forth was selected as the main naval base for the close blockade of Germany. A large number of weapons and victualling depots, dockyards, and hospitals were built there. However, after the plan for a distant blockade was adopted it was necessary to choose new bases that lay closer to the Northern Passage between Scotland and Iceland. Hence, two anchorages were selected, Invergordon in the Moray Firth and Scapa Flow.[14] All three British services agreed in 1938 that Rosyth in the Firth of Forth should be selected as the main naval base in the North Sea. Rosyth was well placed to intercept German ships returning to their bases from forays in the northern part of the North Sea. Also, its air defense could be combined with the defenses of the cities of Edinburgh and Glasgow. Eventually, however, it was agreed that Rosyth could not meet the changed circumstances of the war with Germany, being badly placed to intercept German ships attempting to break out into the open waters of the North Atlantic. Moreover, its long approaches were vulnerable to mining. Thus, Scapa Flow, which lay about 150 miles nearer to a position between the Shetlands and Norway, came to be selected as the main naval base. From there the Home Fleet could more easily carry out the tasks of protecting lightly armed ships of the Northern Patrol and of enforcing British contraband control in the northern waters. However, a few weeks after the war started, the Home Fleet left Scapa Flow and moved to Loch Ewe.[15] This base was one of the best natural bases on either coast of Britain. One of its great advantages was that it offered the shortest route to the deep waters of the ocean.

The value of a naval base depends both on its distance from the potential military physical objective and on the facilities it offers. However, the geographic location of a base always had a dominating influence on naval strategy. The military importance of a naval base within an enclosed or semi-enclosed sea depends on its position within its respective theater of operations, the distance to the country's land frontiers, its position relative to the enemy's sea communications, the adequacy (or inadequacy) of one's resources as well as those of the surrounding territory, its facility of use, and the security of its land and sea communications.

The closer a base lies to the prospective theater of operations, the shorter the lines of operations it offers. Obviously, a fleet in a narrow sea has significant advantages over a blue-water navy, because its bases are situated so much closer to the area of prospective combat. The naval base of Sevastopol' is only about 305 nautical miles from the entrance to the Turkish Straits. The Russian (formerly Soviet) naval base at Liepāya lay about 460 nautical miles from Skaens Odde (Denmark) via the Sound, or 600 nautical miles via the Store Bælt, while the former East German naval bases at Sassnitz (Island of Ruegen) and Rostock were about 95 and 230 nautical miles respectively away from the sea's exit.

In the Adriatic, the Bay of Cattaro was an invaluable base for the Austro-Hungarian and German U-boats operating in the Mediterranean during World War I because of its proximity to the Strait of Otranto. The centrally located major naval base at Taranto and its relative nearness to the scene of action played a great role for the Italian Navy in both world wars. Taranto is only about 320 nautical miles away from Malta, 145 nautical miles from Valona and 520 nautical miles from Tunis. In 1914–18, the large British base on Malta and the French bases at Toulon and Bizerta proved to be of little operational significance because they were too far from the scene of actions in the eastern Mediterranean. Hence, the British and French navies seized control of several islands off the entrance to the Dardanelles and turned them into advanced bases. The Greek harbor on Mudros was used for supplies, while Soúdha Bay (Crete) and Famagusta (Cyprus) opened up for the Entente after the Greek King fled the country and Greece changed its foreign policy posture. Afterwards, the ports on Corfu and Cephalonia (Keffalinīa) in the Ionian Sea were used as advanced naval bases.[16]

Control and protection of one's maritime trade can be accomplished more easily if naval bases lie close to focal points of merchant shipping. The German bases in the North Sea flanked the British–Norwegian convoy route in World War I. Likewise, the Axis' ability to interrupt Allied trade in the central and eastern part of the Mediterranean in 1940–43 was largely due to the string of naval and air bases from Crete to Sardinia and along Libya's coast that flanked the main route from Gibraltar and Alexandria to Malta.

The operational value of a naval basing area or cluster of airfields is greatly diminished if it is located close to the land frontier of a potential opponent at sea. The Austro-Hungarian base at Pola was only about 75 miles away from the country's border with Italy. The value of Trieste as a fleet base was even less than that of Pola because it was situated only a few miles from the Austro-Italian border. In another example, the Iraqi naval base at Umm Qasr lies about 65 miles up the Khor Abdullah lagoon west of the Shatt el-Arab, and is highly vulnerable to enemy attack due to its proximity to both the Kuwaiti and Iranian borders. Kuwait dominates the approaches to it from two islands located at the entrance to the lagoon, Būbiyān and Warba. Umm Qasr is also situated too far from the Gulf's only exit, because it lies about 500 miles away from the Strait of Hormuz.

BASES

The military importance of a base depends upon its position within a theater of operations. A naval base within an enclosed or semi-enclosed sea may be located in a strait guarding the exit from the respective sea, at the opposite end of a narrow sea, or between these extremities. A naval base that is centrally located usually has much greater value than a base situated at the extreme end of a given narrow sea. Sebenico, situated in the central part of the Dalmatian coast, was an almost ideal fleet base for operations by Austro-Hungarian surface forces in any part of the Adriatic theater during World War I.

A base situated far from the sea's exit may be selected as a main naval base. Normally, such a base is large in size and serves as the point of assembly for a major part of one's fleet forces. The value of rear bases was illustrated in the summer of 1942 when the advance by the *Afrika Korps* towards the Egyptian border had alarmed the British. By the last day of June 1942, the Eighth Army had withdrawn to the defense lines at El Alamein, only 60 miles away from Alexandria. The British moved six destroyers and two auxiliary vessels south of Suez. The remainder of the British forces then based at Alexandria was divided between Haifa and Port Said, except for the First Submarine Flotilla, which moved to Beirut. The British also made preparations to demolish stores and facilities at Alexandria and to block the harbor as well. If the British had not possessed their rear bases in Egypt, Palestine, and Syria, there would have been no alternative but to withdraw the entire fleet through the Suez Canal. The existence of these bases gave the British more room to maintain their tenuous hold on the eastern part of the Mediterranean.[17]

A disadvantage of naval bases located in the rear is that they often complicate the employment of one's naval forces because they are too far removed from the main scene of action. The Austro-Hungarian main naval base at Pola lies about 210 nautical miles away from the Strait of Otranto, and hence was suitable only as a base for operations in the northern and central part of the Adriatic. This situation would have greatly changed if Austro-Hungarian troops had seized the Albanian base of Valona. Then, an Austro-Hungarian fleet based there would have been a serious threat to the Italian naval squadrons in Brindisi, about 60 miles away.

If a naval base flanks the enemy line of operations or sea routes, then a force based there would usually have a greater chance to inflict damage on its adversaries. The naval base at Zeebrugge on the coast of Flanders, seized by the Germans in World War I, with its flotilla of submarines and destroyers flanked the British routes between the Thames River and Holland. In World War II, Malta not only served as a way station for ships passing between Gibraltar and Alexandria, but also sat squarely athwart the sea route between Italy and Libya. Fighter aircraft and bombers based on Malta forced the Axis shipping to make wide detours, thereby greatly reducing the efficiency of transport to North Africa. Used as a fleet base,

Malta threatened the safety of Italian convoys, forcing the Italian fleet to provide heavy escorts and to accept actions under unfavorable circumstances.[18]

The military value of a naval/air base is greatly enhanced whenever it is situated in the proximity of a crisscrossing of several maritime routes or near the terminal points of shipping. Likewise, naval forces and aircraft based on a large island flanking high-volume shipping traffic lanes, such as Crete or Cyprus in the eastern Mediterranean or Cuba in the Caribbean, are in an excellent position to inflict heavy damages on hostile maritime trade.

The value of a base depends upon its location relative to the adjacent land area and the sea. Thus, a base situated on the mainland coast is generally more vulnerable to enemy invasion by land. A base on an island but located at some distance from the mainland coast is relatively safe from attack across the land but is, at the same time, highly vulnerable to an invasion from across the sea. A naval base established on the tip of a long and narrow peninsula is usually vulnerable to enemy attack from land and from across the sea. In the modern era, any base, regardless of its position with respect to the adjacent land area and the sea, has been vulnerable to enemy attack from the air. Yet, if a naval base fronts a large number of offshore islands, its depth of defense against air or surface attacks is larger than that for a base located on an outlying island.

The number of airfields in the coastal area, especially in the vicinity of naval bases, considerably enhances one's ability to attack ships or to defend one's naval forces and troops on the ground. The physical seizure of airfields, or of positions where airfields can be built significantly speeds up and reinforces the hold over a specific sea or ocean area. The importance of Malta as an airbase was apparent soon after the beginning of hostilities with Italy in June 1940. However, at the time it was more important for the British to secure the airfields in Cyrenaica, from where it would have been possible to extend air activities over the Central Mediterranean.

The rather speedy Japanese occupation of the Netherlands East Indies would not have been possible without control of a large number of airfields. The Japanese offensive in South East Asia was directed principally at seizing airfields at selected points in the archipelago from which another leap was made, usually within the 400-mile effective range of its land-based fighter aircraft. Also, Japan's ability to control the Bismarck Sea and the adjacent landmass of New Guinea was due to the large number of airfields it established in the vicinity of its naval base at Rabaul (New Britain). In the fall of 1942 between 50 and 100 fully loaded freighters were found at times in Rabaul. This base was also the originating point for the *Tokyo Express* running down the 'Middle Course' (or the 'Slot' as it was called by the Americans) bringing men and materials to the Japanese troops in the Solomons.[19] By August 1943, the Japanese were using five airfields in Rabaul and its vicinity.[20] The Allied New Guinea (Papuan) Campaign in 1942–44 revolved in essence upon who controlled a number of tropical harbors (little known to the outside world) and their adjacent airstrips.

After the loss of Guadalcanal in January 1943, the Japanese had no intention of leaving the Solomons because they still had relatively strong positions in the archipelago, four airfields near Rabaul and five more built on other islands reaching down the Solomons as far south as New Georgia. The Japanese held a central strategic position in the area and their reinforcement routes from the Marshall Islands and the Japanese mainland, although long, were more than adequate. By December 1943, however, Allied troops had landed at four points on New Georgia, and on adjacent islands. By then, the Americans had in use six airfields on Guadalcanal and the Russell Islands, and hence landing forces could be effectively covered. These Allied forces neutralized the Japanese bases at Munda and Kolombangara before the assault, and the enemy could therefore only reply by sending strike forces from Rabaul 400 miles away.

Naval bases and airfields in general are a critical element for the effective employment of one's own naval forces and aircraft in a given maritime theater. A coastal navy may have only a few major and many minor naval bases, but these collectively may allow for full support for their surface ships and submarines. In contrast to major navies, small navies operating in a narrow sea can use even small bays or gulfs as anchorages for basing their small surface combatants. Because of the short distances involved the land-based aircraft operating over a typical narrow sea can be deployed far inland and still be within effective range of their prospective targets. They also do not need to be established in clusters but can be widely dispersed. The use of rotary-wing aircraft is considerably more effective in narrow seas than on the open ocean. Almost any flat surface on the coast or on an island can be used for the operations of helicopters.

NOTES

1 Oscar di Giamberardino, *Seekriegskunst*, trans. E. Mohr (Berlin: Verlag 'Offene Worte', 1938), p. 99.
2 Ibid., p. 100.
3 M. G. Cook, 'Naval Strategy', 2 March 1931, Air Corps Tactical School, Langley Field, VA, 1930–31, Strategic Plans Division Records, Series, Box 003 (Washington, DC: Naval Operational Archives), p. 18.
4 Friedrich Ruge, *The Soviets As Naval Opponents 1941–1945* (Annapolis, MD: Naval Institute Press, 1979), p. 77.
5 Michael Salewski, *Die deutsche Seekriegsleitung 1935–1945*, Vol. II: *1942–1945* (Frankfurt, a.M: Bernard & Graefe Verlag fuer Wehrwesen, 1975), pp. 448, 462.

6 Giamberardino, *Seekriegskunst*, p. 100.
7 Milan Vego, 'Die jugoslawische Marine 1945–1977', *Marine Rundschau*, 9 (September 1977), p. 232.
8 Alfred T. Mahan, *Naval Strategy Compared and Contrasted with the Principles and Practice of Military Operations on Land* (Boston, MA: Little, Brown, 1919), p. 188.
9 David Woodward, *The Russians at Sea: A History of the Russian Navy* (New York: Frederick A. Praeger, 1966), pp. 120–1; Macintyre, *Sea Power in the Pacific*, pp. 126–7.
10 Julian S. Corbett, Vol. I: *To the Battle of the Falklands, December 1914, Naval Operations*, 5 Vols (London: Longmans, Green, 1920), p. 4.
11 M. R. Browning, 'Advanced Naval Bases', *Military Review*, 6 (June 1947), p. 45.
12 Cook, 'Naval Strategy', p. 17.
13 James Stewart, 'The Evolution of Naval Bases in the British Isles', *Proceedings*, 7 (July 1957), p. 758.
14 Ibid., p. 757.
15 Stephen W. Roskill, *The War At Sea 1939–1945*, Vol. I: *The Defensive* (London: Her Majesty's Stationery Office, 1954), p. 77.
16 Walter Gadow, 'Flottenstuetzpunkte', *Militaerwissenschaftliche Rundschau*, 4 (April 1936), p. 519.
17 Roskill, *The War At Sea 1939–1945*, Vol. II: *The Period of Balance* (London: Her Majesty's Stationery Office, 1956), p. 74.
18 Elmer B. Potter, *The Great Sea War: The Story of Naval Action in World War II* (Englewood Cliffs, NJ: Prentice-Hall, 1960), p. 59.
19 James C. Shaw, 'The Rise and Ruin of Rabaul', *Proceedings*, 6 (June 1951), p. 626.
20 Samuel E. Morison, *Breaking the Bismarck Barrier, 22 July 1942–1 May 1944*, Vol. VI of *History of the United States Naval Operations in World War II* (Boston, MA: Little, Brown, 1950, reprinted 1975), p. 393.

5

Theater Geometry

Any space gained must be militarily organized to facilitate the planning, preparation, and conduct of major operations or campaigns. With the advent of massive armies and large navies in the late nineteenth century it became necessary to delineate and militarily organize a given space that might be used by one's forces in terms of the 'theater of war' or 'theater of operations'. Today, a theater of war in the generic definition of the term is understood as an area of land or sea and the airspace above it that might become directly a scene of military action. Land and maritime theaters of war may be differentiated. A *land* theater is usually established to control a major part of the landmass, and its associated airspace, vital to the sustenance of a nation or nations, or to destroy the enemy's means of exercising such control. In generic terms, a *maritime theater of war* encompasses the entire surface of a given ocean or sea, including the sub-surface, adjacent coastal areas, islands, archipelagoes, and the airspace above it. Military actions in a maritime theater may vary from the limited use of forces for peacekeeping and nation-assistance to major operations intended to destroy or neutralize enemy naval forces or major joint operations to control the principal land areas. The terms 'land' and 'maritime' theaters do not pertain to any particular service but are used to describe the predominant feature of the physical environment in which one's forces operate or are to operate. In each type of theater, all services will be employed. Hence, the theater commands are inherently joint (multi-service) in character, and often combined (multi-national).

In a general conflict, or in the case of two or more local conflicts, one theater of war will be selected as the *primary* theater of war, while all others will be *secondary*. Normally, the primary theater of war is also the *theater of main effort*. Such a theater will receive the highest priority in regard to guidance and the assignment of one's forces. However, if both belligerents are deterred from fighting in a primary theater, the secondary theaters may acquire a greater importance.

Either a part of or the entire theater of war can be subdivided into a number of *theaters of operations* – an area of land or sea and the airspace above it capable of accommodating combat forces and their support to accomplish a single strategic objective. For a major country, such as the United States, involved in a major regional conflict, a part of the theater of war may be formally or implicitly declared to be a land or maritime theater of operations. For a riparian country bordering an enclosed or semi-enclosed sea, that sea area will normally comprise a maritime theater of operations. If a country borders two narrow seas separated by a strait or intervening landmass, as Turkey and Germany do, then one of them will be declared the primary, and the other a secondary theater of operations.

In general, boundaries of theaters of war and theater of operations are arbitrarily determined and are usually expressed in terms of geographic longitude and latitude or some distinctive geographic feature of the area. The length and width of a given theater are influenced primarily by the need to provide a sufficiently large base of operations and supporting infrastructure to accomplish the assigned strategic objectives. In the event of hostilities in a typical narrow sea, the entire sea area and the adjacent coastal area and the airspace above it would usually be declared a theater of operations.

Based on the physical objectives to be accomplished, a theater of operations may be further formally subdivided into separate land, maritime, and airspace areas of operations. A joint area of operations for multi-service employment might be formally declared as a part of a given theater of operations. The size of each of these areas is arbitrarily determined. However, any area of operations should be large enough for component commanders to accomplish their missions and protect their forces. Specifically, a maritime area of operations encompasses a sea or ocean area with adjacent coast, islands, and the airspace above it, large enough to allow space for the conduct of major operations. Such an area can also be established in one's own coastal waters where a major part of the fleet routinely conducts tactical actions to obtain and maintain a favorable operational situation. For example, in the Gulf War of 1990–91, the Arabian Gulf and the Red Sea each represented a separate maritime area of operations. A maritime area of operations is usually further subdivided into several *naval* (or *maritime*) *combat zones* (or *sectors*) where tactical actions are conducted. In the Iran–Iraq War, 1980–88, or the Gulf War of 1990–91, the northern, central, and southern parts of the Gulf each represented a maritime combat zone.

Any maritime theater contains a number of randomly distributed physical (tangible) or 'abstract' (intangible) features that can be either 'natural' or 'man-made'. These features, arbitrarily called 'elements' considerably affect the employment and sustainment of naval forces and aviation in general. The principal elements of any maritime theater are: base of operations, physical objectives, decisive points, lines of operations, and lines of communications. In a typical narrow sea, these elements are spaced rather densely because of the much shorter distances involved. The importance of each element stems from its relative position or geometry with respect to the others, and especially in relation to friendly and the enemy's forces. The commander and his staff should properly appreciate these distinctions in employing subordinate force elements. However, the importance of physical features of the theater should never be considered absolute. While geography considerably affects the conduct and outcome of major operations or campaigns, it is rarely the most important factor in a success.[1] Warfare was in the past, and will continue to be in the future, a complex interplay of human decisions and actions, where each natural or man-made physical element is only one of the factors to be considered in the combat employment of forces.

Base of Operations

The ability of naval or aviation forces to strike targets at sea and on the coast depends on the length and the directional orientation of their *base of operations*, an imaginary line connecting naval bases, anchorages, and air bases used by one's own and friendly forces. A base of operations should not be confused with a naval shore establishment or a forward-deployed logistical support; rather it is a sea or ocean area from where movement of a naval force toward the selected objective originates.

The importance of one's base of operations should be evaluated not only with respect to its purely military value, but also in regard to distances, directional orientation, and routes to adjacent bases. The main elements in evaluating the importance of a base of operations are its length, shape or configuration, maneuvering space, size and mix of forces, number of lines of operations and lines of supplies, and its distance from the enemy base of operations. Generally, a *longer base* of operations is preferred to a shorter base. The longer base offers greater maneuvering space for the movement of one's forces, because a fleet can use, simultaneously or sequentially, a larger number of lines of operations. It also allows a larger force to be deployed. The longer base of operations usually offers a larger number of the enemy's physical objectives that can be selected. In addition, one's forces cannot be easily enveloped during their movement towards the assigned physical objective. A force operating from a longer base of operations can shift its lines of operations and turn the flanks of the position without sacrificing the security of its lines of supply. Then one's force may seek safety from either damage or destruction by retreating along one of the adjacent lines of operations, assuming it has air superiority or freedom from attack by enemy aircraft.

A long base of operations is also more favorable than a short one because a larger number of ports can be used for disembarkation of troops and *matériel*. The long French coast with numerous ports in the English Channel and the Bay of Biscay allowed within a few months in 1918 the disembarkation of more than one million US soldiers despite the presence of German U-boats.[2] The German Navy operated initially from a relatively short base of operations in the Baltic in 1914 by controlling a string of bases from the eastern coast of the Jutland to Danzig and Pillau. However, in the course of war, the Germans extended their base of operations to include Riga and the Aaland Islands.[3]

A long base of operations requires relatively large forces for defense. Also, coordination of movements of one's forces is more complicated. An example of a long base of operations was the string of naval and air bases along the eastern coast of England and Scotland from the Strait of Dover to Scapa Flow in both world conflicts. In World War II, the *Kriegsmarine* enjoyed the advantages of a long base of operations stretching from the North Cape to the southern part of the French Atlantic coast. The German U-boats, heavy surface ships, and land-based aircraft obtained almost unhindered access to the open waters of the Atlantic. Moreover, the British naval position came in danger of being enveloped from both north and south. Likewise, the Germans enjoyed a longer base of operations in the Baltic in 1939 where they controlled about 745 miles of the coast, while their potential opponent, the Soviet Union, controlled only about 93 miles along the coastline in the eastern part of the Gulf of Finland.[4]

A *short base* of operations is generally easier to defend and protect than a long one. At the same time, such a base of operations offers a smaller area for maneuver of one's forces, and few lines of operations. Lines of operations are spaced more closely to each other. Then one's forces are in danger of being enveloped or cut off from their supply by an enemy force. Sometimes, a short base of operations severely limits or even precludes the operational employment of friendly forces.

A fleet operating from a longer base of operations has significant strategic advantages against an opponent possessing a short base of operations. This was one of the key disadvantages of the German High Seas Fleet in the North Sea in World War I. Any action by German surface forces originated from the very short base in the Heligoland Bight and the British always knew that the fleet must return there. This disadvantage was especially acute in the employment of the U-boats, because the Allies blocked both exits from the North Sea. The German situation would have been much improved if the U-boats had been able to use the long French

Atlantic coast that flanked the main routes of German enemies.[5] In other words, the High Seas Fleet was able to operate tactically, but not operationally, against the Entente's maritime trade in the Atlantic.

The shape or configuration of a base of operations may also affect the combat employment of forces on both sides. In general, a base of operations for naval forces or aircraft may extend in a straight line, a 'broken' line, an angular line, a curved line, or in the form of a triangle. A base of operations extending in a roughly *straight* line generally offers no military advantage from its configuration. Such a base can run parallel, athwart, or at some angle to the enemy's base of operations. For a force operating from a base of operations astride that of its opponent, the front of operations is parallel with the line of operations. Thereby, one's force can strike relatively easily at the enemy's lines of advance and retreat. A straight base of operations extending at some angle or oblique to the enemy's base of operations allows one's forces to threaten the enemy's flanks and sea lines of communications.

If the imaginary lines that connect the adjacent naval bases successively run at some angle in respect to the enemy's base of operations, that base of operations is said to extend in a *broken* line. A good example of such a base were the former Soviet/Warsaw Pact naval bases and airfields along the eastern coast of the Baltic. A base of operations extending in a broken line offers similar advantages and disadvantages as an oblique base of operations. However, if a route from one base to another runs between offshore islands or an archipelago, that route provides greater security (although not necessarily greater speed of advance) than a base of operations extending along a uniform and exposed coast.

The *angular* or *salient* base of operations exists when it projects toward the opposing coast, as for example the string of naval bases on the Crimean Peninsula or the Jutland Peninsula. Such a base of operations consists of two *sides* or *flanks* and an *apex*. The smaller the angle at the apex, the longer the lines of operations are. A salient base allows a concentrated force to operate from interior and short lines. Also, the adversary would generally have difficulty deciding whether the main thrust will come from the salient's apex or the flanks. A salient base of operations confers the offensive power. However, it usually offers a few lines of operations and, therefore, is ill suited for defense.

If a base of operations runs in a *curved* line, but away from the enemy's coast, for example the southern coast of the Baltic, it is said to have *re-entrant* shape. Such a base of operations affords the choice of several lines of operations. If the two sides are almost perpendicular to each other, then a *double base* exists. Then control of both sides is secured and two widely separated lines of retreat are formed. In general, a re-entrant base of operations makes it easier for one's forces to change their lines of

Map 8. THE BALTIC SEA

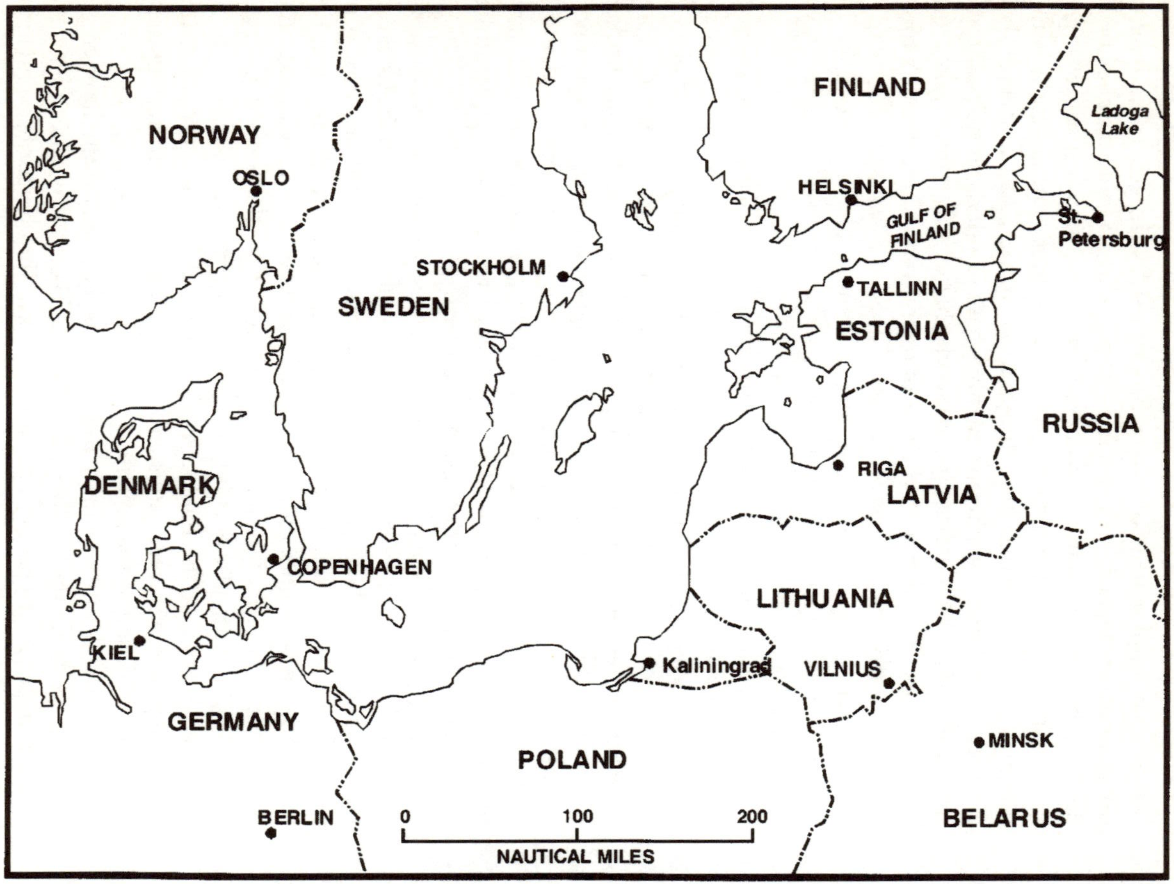

Map 9. THE BLACK SEA

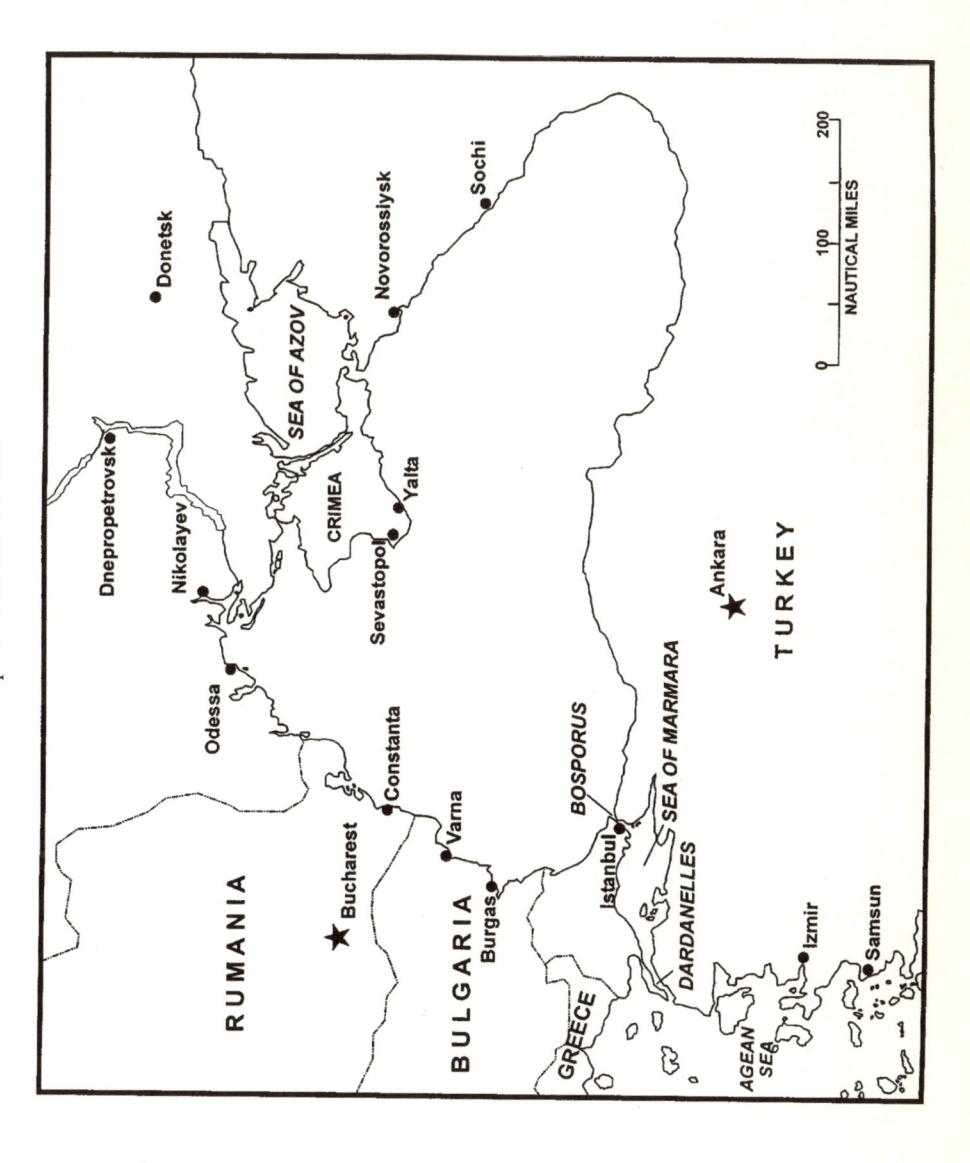

operations from one area of operations to another. Also, a force operating from such a base can attack enemy maritime trade, while protecting its own maritime trade. However, the inherent advantage of the re-entrant-like base of operations accrues only if used with initiative and determination.

The *triangle*-like base of operations exists when one's naval bases are situated on a relatively small peninsula or within a large river estuary fronted with one or several large islands. Such a base of operations usually offers a single line of operations and line of retreat. It also usually favors defense from attack from across the sea and from the air. In the North Sea in 1914, the German Navy's base of operations took the form of a triangle with short sides. The principal bases were Wilhelmshaven, Heligoland, Cuxhaven, Brunsbuettel, and Emden. During the war the German base of operations was extended to include the Belgian bases at Ostend and Zeebrugge (but not the Dutch coast). Yet, the Germans failed to seize control of the French ports on the Channel coast to further significantly extend their base of operations.[6]

Physical Objectives

Any maritime theater contains a large number of objectives to be seized, held, controlled, neutralized or destroyed/annihilated. The physical objective is the point upon which one's efforts are aimed or directed.[7] It can also be defined as that 'concrete thing in possession of the enemy against which our effort is to be directed.'[8] Tactical, operational, and strategic objectives are differentiated with respect to the military importance they might have in a given theater. Tactical and operational objectives are normally physical in their character, while strategic objectives encompass both physical and intangible elements.

In generic terms, the accomplishment of a *major tactical objective* should result in a drastic change of the situation in a given area of operations. Examples of major tactical objectives are the destruction of a carrier battle group, an amphibious task force, a convoy, seizing or holding a large naval base or port, or an airfield complex. The accomplishment of a *minor tactical objective* should lead to a drastic change in the situation in a given *combat zone/sector*. The destruction of a single or group of aircraft, small surface combatants, transports, or C3I nodes provide examples of minor tactical objectives.

The accomplishment of an *operational objective* should lead to a drastic change of the situation in a given *theater of operations*. Examples of operational objectives in a maritime theater are the seizing or holding of a large island or archipelago, straits/narrows, a large naval base/commercial port, a large base/airfield complex, or destroying or neutralizing an enemy task force or large convoy. Control of Guadalcanal was the first operational objective to be accomplished in the Allied Solomons Campaign, 1942–43, and its subsequent control led to a drastic change of the situation in the southern Pacific for both the Allies and the Japanese. A similar effect on the situation in the area followed the Allied capture of other large islands in the Solomons archipelago: Bougainville and New Georgia.

The accomplishment of a *strategic* objective should result in a drastic change of the situation in a given *theater of war*. In any maritime theater of operations there is usually a single strategic objective to be accomplished. Depending on the degree of 'maturity' of the theater a strategic objective contains both tangible and intangible elements (political, economic, social, psychological, ethnic, etc.). Examples of strategic objectives are destruction or neutralization of the entire enemy fleet or the seizing/holding or defense of a large part of a coast or large archipelago. Generally, the less 'developed' (in terms of infrastructure) a maritime theater is, the more likely it is that strategic objectives would contain almost exclusively military aspects. One of the strategic objectives in the Pacific Ocean Area (POA) in the summer of 1942 was the seizure of control of the Solomons Archipelago and part of the northeastern coast of New Guinea. This eventually required the accomplishment of several operational objectives, specifically, the initial seizure of Guadalcanal and adjacent islands, subsequent occupation of New Georgia and Bougainville, and finally the seizure of New Britain and the Japanese stronghold at Rabaul. The strategic objective in the Southwest Pacific Area (SOWESPAC) in 1942–44 was the destruction of the Japanese forces defending New Guinea and its adjacent islands. In the Philippines Campaign of 1944–45 the strategic objective for Allied forces was the liberation of the Philippines.

Decisive Points

Any theater contains a large number of geographic or man-made features called 'decisive points' that may exert a decisive influence on the course and outcome of a military action. Decisive points could also include other physical elements such as enemy formations, command posts, and communications nodes. There is a close relationship between a given 'decisive point' and the physical objective to be accomplished. Note that what is for higher echelon planners a 'decisive point' becomes for the next lower command echelon a physical objective to be seized, controlled, held or defended.

In general, terrain or geographically oriented and maneuver- or force-oriented decisive points may be differentiated. They can be tactical, operational, and sometimes strategic in their military importance. *Geographically oriented* points are fixed in place and, hence, have more enduring importance. However, their value is largely situational; that is, when a force moves to another area, the importance of a given geographic decisive point changes as well. Typical geographically oriented decisive points in a maritime theater are straits, narrows, islands, river estuaries, or sea/ocean areas where several major shipping routes converge. Straits and narrows are usually the most important geographically oriented decisive points in both offense and defense. In the Leyte Operation in October 1944, the Bungo Suidō (the only southern exit from the Japanese Inland Sea), the San Bernardino Strait, and the Surigao Strait were geographically decisive points for both the Japanese and Allied naval forces.

Force-oriented decisive points are usually mobile and, hence, change their location in time and space. Their importance in a given situation is usually transitory. Also, their value depends primarily on their relative position and combat power with respect to the opposing force. In general, operational planners have great difficulty in predicting accurately the movements and actions of force-oriented decisive points.[9] Examples of force-oriented decisive points in a maritime theater are naval bases/anchorages, naval air bases, logistical support elements ashore, air defenses of a naval/air base, C3I nodes, a convoy screen, screen of a carrier battle group (CVBG), amphibious task force (ATF), underway replenishment group (URG), etc. In the information age, a specific part of the computer system or internet system that can be penetrated and may lead to disruption or significant degradation of the C4I system can be called *cybernetic* decisive points.

Line of Operations

The imaginary line along which a force moves from its base of operations towards a given physical objective is called the line of operations. In land warfare, these lines often pass through selected decisive points, but this is usually not the case in the open ocean. In a typical narrow sea, lines of operations can run across open waters, or pass through geographic decisive points, specifically straits, channels and narrows. They can also run along the coast, especially if one's force operates under the constant threat of a much stronger force. Napoleon I's invasion fleet in order to reach the point of assembly at Boulogne had to move along the coastline to avoid encounter with British ships.[10] The Germans in their planned invasion of England in 1940 were faced with essentially the same problem.

The choice of a line of operations is one of the fundamental tasks of the commander in planning a military action at any level. In war at sea, lines of operations may coincide in part with lines of communications. However, they are not lines of supply as is often mistakenly thought. Also, in contrast to these, lines of operations may also be lines of retreat. A change in the line of operations in the midst of a combat action is a delicate step that should be avoided whenever possible. Lines of operations should be flexible so that they can be closed together for mutual support as soon as the enemy force renders it necessary, or so that they can be opened out and made to meet at a point beyond the originally intended physical objective.[11]

Long or short, single, double or multiple lines of operations are differentiated. In general, a force operating along shorter and multiple lines of operations has a distinct advantage over the opposing force moving along the longer and fewer lines of operations. A centrally located base of operations offers shorter and more secure lines of operations than the one situated on the flanks of the enemy's strategic center. However, long lines of operations are more vulnerable to enemy attack and make the covert movement of forces more difficult.

Lines of operations on the open ocean are usually long and multiple, and offer greater flexibility in selecting or changing them in the course of combat. In contrast, lines of operations in a typical narrow sea are generally much shorter, less numerous and more rigidly laid than those on the open ocean. They can also be quite circuitous if the coast is fronted by a large number of offshore islands.

A *single* line of operations allows simpler command and control (C2) and logistical support and sustainment. A force moving along a single line of operations is vulnerable to enemy attack because it moves in a limited space. Its objective is also more apparent to the opponent. A hostile force could with relative ease cut off the line of retreat and defeat one's forces in detail. The vulnerability of a single line of operations is especially high if it passes through a restricted area, such as a strait or narrows.

Double lines of operations are those used by major parts of a single force or two independent forces, but are widely separated in terms of space and time. They are more advantageous than a single line because they allow more space for the movement of one's forces. They also offer some advantages in countering a numerically stronger force, because the opponent would have difficulties in discerning the real objective assigned to friendly forces. However, a force operating from double lines of operations requires more precise synchronization of their elements than if they move along a single line. Also, command and control and logistical support and sustainment are usually more complex than when a single line of operations is used.

Multiple lines of operations have similar advantages and disadvantages to those of double lines of operations. Movement of one's forces is generally more difficult to coordinate. However, force protection is greatly enhanced if one's forces move within the mutual supporting distance of each other. Establishment of one's lines of movement beyond mutually supporting distances is advisable only if each prong is stronger in combat power and mobility than that of the opponent or the space–time factor is extremely favorable for one's forces. An example of the successful use of multiple lines of operations was the Imperial Japanese Navy's support and covering actions during the invasion of the Netherlands East Indies in 1941–42. Yet, the Japanese Navy violated this principle in the Leyte Operation, when several fleet elements moved toward the central Philippines beyond mutual supporting distance. Normally, it is a serious error simultaneously to use several lines of operations, between which there is an intervening landmass, choke point or some navigational obstacle. This exposes a force to the danger of being defeated in detail, should the enemy advance more rapidly than expected.[12]

Directional orientation of lines of operations is influenced primarily by the length and configuration of a friendly base of operations, the number and location of decisive points, the distance to the assigned physical objective, intervening physical obstacles between one's base of operations and the objective, and the size and mix of one's forces. The task of the commander is to select lines of operations in reference to the base of operations and the movement of forces so as to seize the enemy's communications without imperiling one's own. This is one of the most important and most difficult problems in strategy.[13]

As the term implies, parallel lines of operations run at some fixed lateral distance from each other. When multiple and parallel lines of operations are spatially dense, shifting the direction of movement in the course of combat from one line to another is relatively easy and can be executed quickly. Then the opponent would have greater difficulties in cutting off the line of retreat of one's forces. Also, multiple but spatially widely separated parallel lines of operations greatly complicate the synchronization of one's force movements.

Converging (or *concentric*) lines of operations start from widely separated points and their spatial separation is steadily diminished as they approach a given physical objective. Converging lines offer great advantages to one's forces moving from widely dispersed start-up positions to a point of concentration before the arrival of a hostile force. They are usually advantageous in controlling the movements of two parts of a force in danger of being defeated separately by a stronger hostile force. These lines of operations also offer great advantages by covering one's sea lines of communications. The attacker can accomplish decisive results by either converging on the enemy force or cutting off their line of retreat. By operating on divergent lines, the defender can use the principle of movement to concentrate the mass of his force sequentially on parts of the converging force and defeat that force in detail.[14] The chief disadvantage of operating on converging lines of operations lies in the synchronization of the movement of one's forces. The enemy can relatively easily discern the ultimate objective of opposing forces as they approach their objective. Also, the converging forces must concentrate before the hostile force can strike each part of one's force separately. The Japanese naval forces used converging line of operations in their approach to the Philippines prior to what is called the Battle for Leyte in October 1944.

Divergent (or *eccentric*) lines of operations originate from a central position and relatively short base of operations. The distance between adjacent lines widens steadily as one's forces approach their assigned physical objective. A force moving on divergent lines does not make clear its final destination to the enemy. Divergent lines offer some advantages if the enemy's forces are widely separated due either to loss of time or engagement with one's own or friendly forces. They can be used by a force starting from a given base of operations and moving eccentrically to divide and separately to destroy a hostile force moving along exterior lines. However, the mutual support among one's forces moving simultaneously along divergent lines is complicated to plan and execute. There are also considerable problems in organizing command and control, forces' synchronization, and logistical support and sustainment.

Interior vs Exterior Lines

With regard to the relative position of a base of operations and the physical objectives, a force can move along interior or exterior lines. A force moves along *interior lines* when it runs between those of the enemy's lines of operations. Interior lines always originate from a central position. They are formed from a central position prolonged in one or more directions or they can also be understood as a series of central positions linked with one another. Interior lines in general allow concentration of one's forces against one part of the enemy force, while holding the other in check with a force distinctly inferior in strength.[15]

In any given situation, interior lines are much shorter than the lines used by a force moving on the periphery. They can be used effectively in both offense and defense. Interior lines allow one's force to strike a part of the enemy's force sooner than the adversary can reinforce it. They enhance the ability to concentrate one's forces against either of the two parts of the hostile force while a much inferior force holds the other part of the hostile force in check. Interior lines allow simultaneous concentric actions from many directions against the enemy's center. The enemy is forced to fight simultaneously from several directions, and his forces are dispersed.[16] The effect of interior lines is more pronounced if the commander forces a division of the hostile force. Then, the enemy force operating on exterior lines will usually have great difficulties in threatening one's own center from multiple directions.[17] Also, a force operating from interior lines should not allow itself to be attacked by several enemies simultaneously. Interior lines facilitate shifting forces to meet an external threat, maintaining communications and covering distances to approach the enemy's force.

The inherent advantages of operating on interior lines have been shown repeatedly in war at sea in all ages. In the War of 1812, for example, a naval power sustained upon the Great Lakes controlled the lines of communications between the East and West. Control of the Great Lakes also conferred the advantage of interior lines, that is, shorter distance both in length and in time to move from point to point of the lake's shore close to which lay the scene of action. The advantage of the interior lines was also shown in England's war with Napoleonic France. The Royal Navy maintained adequate forces off minor French bases, and in the event of any of the French detachments breaking out of port, orders were given for the outlying British squadrons to fall back on their strategic center (operational center in today's terms) off Brest, so as to follow the French either to Ireland or up the Channel. In such an event, the French also had to reckon with a British force in the Downs and with the North Sea squadron. In short, no matter what the French might do to avoid the vigilance of the blockading squadrons, a powerful British force was always ready to bring them to action as they approached the vital point.[18]

The great value of operating from a central position was shown in the Russo-Japanese War of 1904–5 at sea. Prior to the battle of Tsushima in May 1905, Russian Admiral Zinovi P. Rozhestvenskiy chose the course which was the most direct and most obvious. Theoretically, he could steam either eastward of Japan and enter the Sea of Japan through the narrow Tsugaru Strait used by the Vladivostok cruisers for their sortie into the Pacific in July 1904, or through La Perouse Strait further north. But either course would have added more than 1,000 miles to the distance steamed and have involved coaling several times near the enemy coast. Moreover, the Japanese Fleet commander, Admiral Heichichiro Togo, had the interior lines and would almost certainly have been able to interpose his fleet between Rozhestvenskiy and Vladivostok no matter which way the Russians had entered the Sea of Japan.[19]

A force operating on interior lines can accomplish decisive results only if the enemy's line of retreat is cut off by a maneuver or some obstacle or unless exceptional speed is used in the pursuit. Field Marshal Helmuth von Moltke, Sr wrote that the advantages of the interior lines of operations are valid only as long as one retains enough space to advance against one hostile force, gaining time to beat and pursue him, and then to turn against the other enemy force. However, if the operating space of one's forces is narrowed down to the extent that one cannot attack one enemy without running the risk of meeting the other who attacks from the flank or rear, then the strategic advantage of interior lines turns into the tactical disadvantage of encirclement.[20]

The main disadvantage of operating on interior lines in defense is that the reduction of space endangers movement of one's forces. Also, the attacker can force the defender to maintain a defense along the entire periphery. Also, the entire theater is a combat zone, and lines of supply from the outside are endangered. Another disadvantage of operating on interior lines is that they can be enveloped on the flanks. The enemy force can threaten one's lines of retreat. In the end, a force operating on interior lines can be confined to a fairly narrow space and then overwhelmed by a stronger hostile force.[21]

In general, the effective use of interior lines requires that one's force move faster than the hostile force. Interior lines should not be abused by extending them too far. Their inherent advantages can be considerably degraded, neutralized or even nullified by the extremely unfavorable space–force factor. To maximize the advantages of interior lines, the commander must have a thorough grasp of the geography and how it affects the movement of one's large forces. To understand fully the theoretical and practical constructs of operating on interior lines, one requires excellent intelligence regarding the opposing forces. Also, higher mobility of one's forces is a definite advantage when operating on interior lines.

The choice of operating on interior lines should not be considered as the solution for all the problems associated with one's numerically inferior force. If the enemy has unity of command and employs superior forces in a coordinated manner against one's own force occupying a central position, then possibly he can overwhelm the inferior force. Therefore, interior lines in themselves do not confer a marked advantage. There have been many cases where a force operating on interior lines was defeated by a force operating along exterior lines.

A force is said to operate along *exterior lines* when its lines of operations are longer than the shortest line the enemy force can use. These lines originate from the exterior position. The chief requirement for operating on exterior lines is superiority in one's numbers and/or combat power, fast, secure, and above all reliable communications between the individual parts of the force, and high speed in executing the movements. A force moving along exterior lines can select the point of attack along the enemy's periphery. The inherent advantage of a force moving along exterior lines is that it can threaten the enemy with an envelopment. Normally, a force operating from the exterior position moves along longer lines of operations than a force using a central position. The longer the distance between the base of operations and the attack objective, the longer the lines of supplies. Thus, logistical support and sustainment of one's force operating along exterior lines is complicated. There is also a constant danger for one's forces of being defeated in detail, unless each force is stronger than the enemy force opposed to it.

At the operational level, the Japanese fleet moved along multiple and exterior lines in its invasion of the Philippines, Malaya, and the Netherlands East Indies in 1941–42. In the Java Sea, the Japanese advanced toward the island of Java simultaneously along two lines, while the ABDA's force under Admiral Karel Doorman was in the middle. Because each Japanese prong was stronger than the entire force under Admiral Doorman's command, the Allied fleet was doomed. Japanese naval invasion forces poured southward in two main streams divided by the landmass of Borneo. With brief pauses at the key ports to secure the surrounding land area and establish an air base, the advance continued unchecked. The opposition by Dutch and American submarines took a certain toll but could not stop the Japanese flood.[22] Also, Japanese naval forces moved along exterior lines in the execution of the plan for defense of the Philippines that led to the battle for Leyte.

Sea Lines of Communications

Communications in general are understood as lines of movement aimed to support and sustain a military force, be it land, naval, or air. Communications dominate any employment of a combat force. Any maritime theater contains a number of sea lines of communications (SLOCs) – designated lines of movement at sea between two points over which maritime trade or troops and military supplies flow. Yet, this term is not only inaccurate in itself, but also represents a mathematical abstraction – the lines drawn upon a chart. It also apparently ignores the obvious, in that it is the ships, ports, storage depots, and other elements of maritime trade which must be protected and defended, not the imaginary lines along which the ships move. Probably, the better term to use when referring to shipping at sea is *maritime route*.

Although pretty obvious, the differences between sea and land lines of communications should be highlighted. Land lines of communications (LLOC) start from widely separated bases and approach each other from more or less opposite directions until they meet. One's army covers its links with the rear as it advances. In contrast, sea routes are usually common to both belligerents and neutrals. In land warfare, to strike at the enemy's communications it is necessary in some way or other to move round the enemy's flank and away from one's true line of advance. In doing so the tendency is to uncover one's own communications. Consequently, it is usually open to question whether or not to attack the enemy's communications. In a war at sea that situation rarely arises because the important sea routes are identical or lie in close proximity to each other. Therefore, one's communications normally cannot be successfully made secure without attacking those of the opponent.[23]

In general, any maritime trade route consists of two *ends* (a port of departure and a port of destination), the *middle*, and the *flanks*. A port of departure also serves sometimes as the destination point for ships on their return voyage. Ships usually have greater latitude in the choice of their routes than vehicles on land. As the ships approach their destination port their routes tend to converge in a *terminal area* – an area in the proximity of each large port of destination where ships need to close on each other.

A *focal area* is an area through which shipping must pass without actually going into a port. They are usually created off a cape, strait, narrows, or some large island. It is there where major maritime routes converge. Examples of focal areas of maritime trade are the western mouth of the English Channel, the sea area off Cape Finisterre, the Strait of Gibraltar, the Strait of Hormuz, and the approaches to international canals, notably the Kiel Canal, the Panama Canal, and the Suez Canal, to name just some of the better known. Some international straits, such as the Strait of Hormuz, Bab el-Mandeb, and the Strait of Malacca, play a crucial role in the uninterrupted flow of oil traffic to major industrial countries of the world. Any prolonged closure of any of these straits in peacetime or in time of war would have dire consequences for many Western industrial countries and the oil exporting countries in the Gulf.[24] Closure of these straits would require use of longer land routes over Saudi territory.[25] Likewise, the closure of the Bab el-Mandeb Strait for shipping traffic would have some significant repercussions not only on oil but also on non-oil traffic through the Suez Canal.[26]

The Strait of Malacca and the Indonesian straits (Selat Sunda and Selat Lombok) also play an extraordinary role in the flow of the world's shipping. Any disruption of shipping traffic through these straits would have a major impact worldwide. If the Strait of Malacca is closed, an alternative route to the South China Sea through Selat Sunda, Selat Lombok, and the Makassar Strait may be used, but in terms of distance, these detours are very significant. However, in practice, the closure of the Malacca Strait would force nearly half of the world's merchant fleet to sail further north. All the excess capacity of the world's fleet might be absorbed, depending on the number of straits closed and the duration of their closure – the effect being strongest on crude oil shipments and dry bulk such as iron ore and coal. Moreover, as the experience of the closure of the Suez Canal in the aftermath of the Yom Kippur (Ramadan) War of 1973 showed, any closure of the Straits of Malacca would immediately raise freight rates to as much as five times the level they are today.[27]

A new focal area of trade can emerge as the result of the enemy's action or even a threat of it. At the beginning of World War II, because of the threat posed by the *Luftwaffe*, Great Britain moved all trade to its northern and western ports, where, although the shipping was more protected, the new circumstances created an excess load on rail and road transport. This situation also created a new focal point of trade at the northwest approaches to the British Isles and thereby more targets for the U-boats.

Protection of one's maritime trade routes is a difficult and complicated task. The most vulnerable parts of lines of communication are terminals and focal areas of merchant shipping. *Long routes* require more escort ships, possibly leading one's fleet to become overextended. They require the use of more commercial ships, because of the long transit time and therefore longer turnaround time. The opponent at sea has more opportunity to attack, hence, more escorts are needed. This is true especially if the routes pass through restricted waters. The lengthening of sea lines of communications also requires additional bases, both main and advanced, which must be equipped for defense, repair and supply. The lack of screening forces may be compensated for in some degree by using diverse routes. The only assurance against overextension would be to secure strong intermediate positions from which local control can be maintained and in which shipping can seek safety.

In general, *short routes* are preferred to long ones, multiple to single, and centrally located routes to those in one's rear or on the flank. Short routes offer faster turnaround time, thereby increasing the effectiveness of available shipping. They also generally require fewer forces for their protection.

Multiple routes are more advantageous than single ones because they are inherently more flexible and secure. At the same time, these routes require larger forces for their protection. Centrally located routes are more favorable than flanking or rear routes because the first are usually shorter and confer greater security.

With respect to their directional orientation, longitudinal and lateral sea lines of communications may be differentiated. *Longitudinal routes* extend along the length of a given sea or ocean area. They are usually long and potentially vulnerable to enemy attack along their entire length.

Longitudinal routes require longer time for transit so that more merchant vessels on a given route are required. They also require larger forces for their protection. A typical example of longitudinal sea routes were the British routes from Gibraltar to Alexandria via Malta or the German routes from the southern part of the Baltic to Sweden's ports in the Gulf of Bothnia and to Finland's ports during World War II.

Lateral routes run across the width of a given semi-enclosed or enclosed sea. They have advantages and disadvantages that are the reverse of those for longitudinal lines. One example of a lateral sea route across a narrow sea was that used by the British in World War I in the English Channel. Shortly after the outbreak of the war in August 1914, a long stream of ships to transport British troops to France under Royal Navy escort started. Another example of lateral sea lines of communications were the Axis' routes from Italy to North Africa. In the early spring 1941 the normal route for Axis convoys between Italy and North Africa ran round the west of Sicily and close inshore down the eastern Tunisian coast to Tripoli and occasionally to Benghazi (Banghāzī). To relieve the pressure on road transportation, some supplies unloaded at Tripoli were moved to subsidiary harbors farther east in small coastal craft, a number of which were sunk or damaged by British submarines. A fighter escort could not be easily provided for the middle part of this voyage, and it was usual for the Axis convoys to transit this part at night. The convoys averaged four vessels each, and followed one another at intervals of two or three days. They consisted of German and Italian ships. The German cargoes were carried mainly in German ships. Surface escorts were entirely Italian, and the whole movement was under Italian naval direction.[28]

Shipping routes should not be *shifted* in wartime. Such a shift is invariably fraught with dangers even for the side having complete local command of the sea. It is also a time-consuming task, new off-loading facilities must be established, the organization of transport becomes disrupted, and, in most cases, a new system to defend one's sea routes must be created. This is even more dangerous after an exhausting retreat on land, such as that of the British Army in France in the late summer of 1914. With heavy losses to replace, the change of sea routes in the English Channel meant shifting the base of operations from narrow seas to the more open waters of the Atlantic. There was not a port in the Channel, not even Cherbourg, that was able to make a new line reasonably secure. British General Sir John French decided on 29 August 1914 that a new base needed to be established in St Nazaire at the mouth of the Loire River. Hence, the British Admiralty was faced with a completely new task because the transport of the Expeditionary Forces was not yet complete. The task was not simple. The main sea routes were to run not in narrow seas, but outside and round Ushant into the Bay of Biscay. It was difficult to patrol these routes and there was nothing to draw upon except the British Channel Fleet.[29]

Many enclosed and semi-enclosed seas have played disproportionately large roles in transporting troops and/or raw material in wartime. German maritime traffic in the Baltic during most of World War II was substantial because of the need to transport troops and *matériel* to the Eastern front and iron ore from Sweden to Germany. To avoid major industrial breakdown, Germany had to import during the first year of the war at least 750,000 tons of iron ore monthly from Sweden. The chief Swedish iron-ore field was the Kiruna-Gallivare district in the north, from where the ore was shipped partly via Narvik and partly via Luleå. This port was normally closed by ice from mid-December to mid-April, while Narvik was ice-free. There was a small iron-ore field northwest of Stockholm, as well as southerly ports of which the most important were Oxelösund and Gävle, but the maximum monthly rate of delivery during the winter from these ports was 500,000 tons, the limiting factor being the capacity of the railways.[30]

The most important route ran from the German ports in the southern part of the Baltic to Sweden's port of Luleå. It ran for most of its 900-nautical-mile length within Swedish waters, and hence was beyond the reach of Soviet submarines. The second route, also about 900-nautical-miles long, ran along the southern and eastern shore of the Baltic and was used almost exclusively to transport German troops and *matériel* to the front. During the war an estimated 20–30 million tons of merchant shipping crossed the area in which Soviet submarines operated.[31]

The main differences between maritime routes on the open ocean and those in narrow seas concern their number and length. Shipping routes on the open ocean are usually numerous, long, broad, and crisscross each other. They are also more vulnerable and require relatively large forces both for their protection and in making attacks on them. The routes in an enclosed sea in contrast are usually much shorter, less numerous and more *channelized*, especially along a coast fronted with numerous islands and islets. They in fact resemble more those on land, because they are more or less rigidly laid. There are also few if any alternative routes available. Trade routes in narrow seas can be open to hostile attack if the routes are along a coast with only a few or no offshore islands along the route. However, protection is much enhanced if one's routes run among island chains fronting the shore. Another advantage of sea communications in narrow seas is that the movement of troops and *matériel* can be conducted by using even the smallest and the least seaworthy ships and craft. Mines and gun or missile batteries can protect coastal routes running between islands. A good example of a protected coastal shipping route is the Inner Leads along Norway's coast. In World War II, German ships entered

Norwegian territorial waters at remote points well inside the Arctic Circle and moved under the protection of coastal gun batteries almost as far as the entrance to the Skagerrak, where the proximity of German air and submarine bases made the rest of the voyage comparatively safe from British interference or attack.

Land Communications

One important difference between a war in narrow seas and one on the open ocean is the influence of land communications on the planning and conduct of major operations and campaigns. A well-developed road and railroad network in the littoral area reduces, although it does not eliminate, the need for sea transport in support of the war effort. A coast with a great number of lateral and longitudinal roads and railways makes it easy to transport troops and war *matériel*. It also makes it difficult for the adversary to interrupt traffic for any extended period of time. However, such a coast is more vulnerable to invasion from across the sea because it renders a rapid advance of an opposing army into the country's interior easier. In contrast, a coast with poor land communications requires relatively greater reliance on coastwise traffic for transport of troops and *matériel*; also traffic can easily be interrupted for a long period of time especially if the main roads run close and parallel to the coast and are backed by steep high mountains, as is the case along both coasts of the Adriatic.

A coast without any or with only a few lateral communications favors defense because it offers few access routes into the country's interior for an army successfully landed on the coast. In the Korean War, the North Korean rail system was divided by the mountainous backbone in the two principal zones. The eastern and western network consisted of six rail lines crossing the Yalu and Timen rivers southward from Manchuria, three on each side of the peninsula. The western rail network was able to deliver from 500–1,500 tons per day to the battle area for the Communists. The eastern network encompassed about 1,140 miles of track, 960 bridges and causeways and 230 tunnels. The very large number of bridges and tunnels was due to the mountainous character of the terrain. The peacetime capacity of the eastern rail network was estimated at 5,000 tons per day. However, as a result of the US Navy's interdiction effort, its capacity was reduced to less than 500 tons per day and in certain periods to almost nothing.[32] The second North Korean supply network was the highway system. None of the roads was good by Western standards, but proved to be even more difficult targets for US aircraft than the railways. The very primitiveness of the roads was an advantage to the Communists because it made them unprofitable targets for air assault. The road network roughly paralleled the rail system and the entire area was crisscrossed with roads wherever the mountains permitted. About 2,000 miles of roads were estimated for each half of North Korea, and the logistical capacity of this network was estimated at more than 1,500 tons nightly (and probably a great deal more). The third supply system consisted of horses, mules, and manpower. The fourth element, the sea, was closed by naval blockade by the UN forces.

The key to the successful employment of one's forces at the operational level is to know and fully understand the time–space–force factors. This means, among other things, comprehending the concept of the theater of operations, its structure, and its geometry. The operational commanders and their staffs must know and understand the key elements of a maritime theater. The geometry of the situation, while important, should never be evaluated outside of the context of other elements of combat power. However, the elements of a theater must be analyzed in terms of the advantages and disadvantages that each element offers. The significance of positions and basing areas whether acquired peacefully or by force of arms should never be overstated. It is always a force that counts the most, not the geography. However, it is perhaps equally wrong to underestimate the role geography has had in the past, has now and will have in the future in the employment of one's forces in a typical narrow sea.

NOTES

1 Carl Von Clausewitz warned not to give too much emphasis to the geometry of the battlefield. He observed that in strategy ('operational art' in today's terms) the number and scale of the engagements won are more meaningful than the pattern of the major lines connecting them; Carl von Clausewitz, *On War*, ed. and trans. Michael Howard and Peter Paret (Princeton, NJ: Princeton University Press, 1984), p. 215.
2 Rudolf Stange, 'Kuestenlaenge und Flottenstaerke', *Marine Rundschau*, 4 (April 1934), p. 148.
3 Walter Gadow, 'Flottenstuetzpunkte', *Militaerwissenschaftliche Rundschau*, 4 (April 1936), p. 517.
4 Dieter Seebens, *Grundlagen: Auffassungen und Plaene fuer eine Kriegfuehrung in der Ostsee 1935–1939* (Hamburg: Fuehrungsakademie der Bundeswehr, August 1971), pp. 6, 13.
5 Stange, 'Kuestenlaenge und Flottenstaerke', p. 147.
6 Gadow, 'Flottenstuetzpunkte', p. 516.
7 Richard L. Connolly, 'The Principles of War', *Proceedings*, 1 (January 1953), p. 3.
8 'Principles of War and Their Application to Naval Warfare', July 1925, p. 5; Record Group 4, Box 23, Folder 1018A, Naval War College Archives (Newport, RI: US Naval War College).
9 Walter A. Vanderbeek, *The Decisive Point: The Key to Victory* (Fort Leavenworth,

KS: School of Advanced Military Studies, US Army Command and General Staff College, April 1988), pp. 5–6, 37; Thomas M. McGinnis, *Jomini and the Ardennes: An Analysis of Lines of Communications and Decisive Points* (Fort Leavenworth, KS: School of Advanced Military Studies, US Army Command and General Staff College, May 1988), p. 4.
10 Alfred T. Mahan, *Naval Strategy Compared and Contrasted with the Principles and Practice of Military Operations on Land* (Boston: Little, Brown, 1919), p. 165.
11 Colmar von der Goltz, *The Conduct of War: A Short Treatise on its Most Important Branches and Guiding Rules*, trans. G. F. Leverson (London: Kegan Paul, Trench, Trubner; New York: E. P. Dutton, 1917), p. 110.
12 Ibid., pp. 110–11.
13 Joseph Rodriguez, Jr, *How to Maximize the Advantages of Interior Lines at the Operational Level* (Fort Leavenworth, KS: School of Advanced Military Studies, US Army Command and General Staff College, 1987), p. 8.
14 Rudolf Heinstein, *Zur Strategie des Mehrfrontenkrieges: Das Problem der 'inneren und ausseren Linien' dargestellt am Beispiel des Ersten Weltkrieges* (Hamburg: Fuehrungsakademie der Bundeswehr, 10 November 1975), p. 5.
15 Mahan, *Naval Strategy Compared and Contrasted with the Principles and Practice of Military Operations on Land*, p. 31.
16 Heinstein, *Zur Strategie des Mehrfrontenkrieges*, p. 5.
17 Ibid., p. 6.
18 G. J. Marcus, *A Naval History of England*, Vol. II: *The Age of Nelson: The Royal Navy 1793–1815* (New York: Viking Press, 1971), p. 243.
19 David Woodward, *The Russians at Sea: A History of the Russian Navy* (New York/Washington, DC: Frederick A. Praeger, 1966), pp. 144–5.
20 Ibid., p. 23.
21 Ibid., p. 6; Friedrich von Bernhardi, *On War of To-Day*, Vol. II: *Combat and Conduct of War*, trans. Karl Donat (London: Hugh Rees, 1913), p. 91.
22 Donald Macintyre, *Sea Power in the Pacific: A History from the 16th Century to the Present Day* (London: Military Book Society, 1972), p. 204.
23 Julian S. Corbett, *England in the Seven Year War: A Study in Combined Strategy* (London: Longmans, Green, 1918), Vol. I, p. 309.
24 In 1995 about 14 million barrels per day (b/d) of oil passed though the Strait of Hormuz mainly destined for Japan, United States and Western Europe.
25 These include the 4.8 million b/d Petroline and 2.2 million b/d IPSA pipelines 1 and 2 and the Abqaiq-Yanbu natural gas liquid line.
26 The oil traffic through Bab el-Mandeb in 1995 amounted to 3.4 million b/d. In case of its closure oil could be transported by using the east–west pipeline (capacity 5.0 million b/d) across Saudi Arabian territory. Nevertheless, the southbound traffic (some 600,000 b/d in 1995) still has to pass through the strait. More important, closing the Bab el-Mandeb would effectively block non-oil shipping from using the Suez Canal, except for a limited trade within the Red Sea.
27 John H. Noer, 'Southeast Asian Chokepoints. Keeping Lines of Communications Open' (Washington, DC: Institute for National Strategic Studies, National Defense University, Strategic Studies, No. 98, December 1996), pp. 1, 2–3.
28 I. S. O. Playfair et al., *The Mediterranean and the Middle East*, Vol. II: *The Germans Come to the Help of their Ally (1941)* (London: Her Majesty's Stationery Office, 1956), p. 525.
29 Julian S. Corbett, Vol. 1: *To the Battle of the Falklands, December 1914, Naval Operations History of the Great War Based on Official Documents* (London: Longmans, Green, 5 vols, 1920), pp. 125–6.

30 J. R. M. Butler, Vol. II: *Sept. 1939–June 1941*, in John P. W. Ehrman, ed., *Grand Strategy* (London: Her Majesty's Stationery Office, 6 vols, 1956–76), p. 91.
31 Office of Naval Intelligence (ONI), Report: 'Soviet Submarine Operations – Soviet Submarine Operations in the Baltic during World War II' (Washington, DC, Report No. 404-S-49, 21 September 1949), p. 5.
32 Malcolm Cagle and Frank A. Manson, *The Sea War in Korea* (Annapolis, MD: United States Naval Institute, 1957), pp. 231–2.

6

Strategic Objectives and Fleet Distribution

The selection of an *objective* – a mission, aim, or purpose of one's efforts – is the first and the most important step in planning for war or for undertaking any military enterprise. It is the most important principle both in a major war and in a low-intensity conflict. The primary objectives of any war are not at sea or in airspace but on land. Yet, naval warfare can contribute to the accomplishment of the overall aim of the war and may render enemy forces on the coast impotent.[1] Nowhere is this truer than in a war in a typical narrow sea. In general, the ultimate object of naval warfare is to keep the sea open for the unrestricted use as a means of transportation for one's forces, while denying its use for such purpose to the enemy. Thus, the proper objective of a navy operating in a narrow sea is to defeat the enemy fleet. However, a success over the enemy fleet is meaningless unless it contributes to the war's overall objective. After victory has been won and peace has been established it is still necessary to enforce the conditions of peace and both the politicians and the military leaders must keep this in mind at all times. All efforts must be directed to accomplish the selected or assigned objective. Without a clearly stated and attainable objective, the entire military effort becomes essentially pointless.[2] However, determination of the objective is usually the most difficult part of any military operation.

A strategic objective encompasses a wide range of elements, from political and military, to economic, social and even environmental. A clear distinction should be made between a political and a military objective. Nations do not go to war because of military reasons alone. Hence, the military objective should be governed by the political objective. At the same time, however, the policy should not demand what is militarily impossible.[3]

The single most important decision of a war, is the selection of a national or alliance/coalition *political objective*. The *political purpose* of a war must be clearly defined and must be attainable with the means available. There is probably no greater mistake than the pursuit of open-ended commitments and unfocused political aims. Once political objectives of a war are determined, the *strategic military objectives* must be identified as well. They are derived from national policy and strategy. It is the task of strategy to ensure enforcement of the national policy through the application of physical force. A military objective must be consonant with the political aims of a war. Too much emphasis on either the political or the military aspect of a given strategic objective would usually create serious imbalance and that, in turn, will usually have an adverse effect on the accomplishment of the strategic objective as a whole. The war should not be conducted with the sole aim of defeating enemy military forces, but to accomplish the overall political objective of the war. Moreover, it is essential never to lose sight of what the political, diplomatic, economic and other aspects of a strategic situation should be once the fighting stops. History is replete with examples of a military victory being achieved that fell short of gaining the object of policy.[4] Sometimes an exclusive emphasis on political goals may point to a strategic objective that, while highly desirable, is clearly unattainable with the military means on hand or becoming available. The result can be a strategic blunder of the first magnitude that will result in a waste of sorely needed resources, as the ill-fated Allied expeditions to Norway in April 1940 and to Greece in the spring of 1941 shows. The British were also lucky that Prime Minister Winston S. Churchill's plan for the Baltic expedition (Operation *Catherine*) in 1940 never took place.

A military objective in general should be something specific or concrete. A condition, or state of affairs, such as control of the sea, freedom of action, or control of sea lines of communications are not in themselves practical objectives to be accomplished, unless they can be obtained or maintained by the accomplishment of certain physical tasks, that is by achieving the necessary physical objectives.[5]

Any mismatch or serious imbalance between the objective to be accomplished and the means on hand will invariably lead to a failure of the entire military enterprise. Also, any vagueness in the assigned military objective would usually result in a poor application of the principle of mass and economy of effort. The US Navy was given an unclear objective in the Arabian Gulf in 1987–88. Instead of completely neutralizing Iran's ability to lay mines in the Gulf, the US chose to sweep continuously the mines laid by the Iranians in international waters. The US reacted to other hostile acts by Iran by conducting practice firing against the Iranian oil platforms. It in effect followed an unfocused and open-ended commitment in the Gulf, and repeatedly failed to use force available in the most efficient way.

Optimally, one should have a single strategic or operational objective at any one time to increase the opportunities of success and lessen the chances of failure and defeat. However, the accomplishment of an ultimate strategic or operational objective would require selection of several intermediate operational or tactical objectives that are properly sequenced in terms of time and space.

To secure the accomplishment of the main objective it is necessary to envisage one or more secondary or alternative objectives and thereby create the preconditions for the application of another important principle of war – flexibility. A plan must be flexible in case something goes wrong during its execution, although the focus should invariably be on the accomplishment of the principal objective. No secondary objective should be pursued unless it directly contributes to the main purpose, regardless of its attractiveness in terms of time gained or possible losses inflicted on the opponent.[6] Admiral Carlo di Persano at Lissa in 1866 became so involved in the attack on the fortified positions on the coast that he lost sight of his primary objective, the defeat of the Austrian fleet. As a result he was tactically surprised at the appearance of Admiral Tegetthoff's squadron and ultimately was decisively beaten.

The initially selected main objective should not be adhered to, if the operational or strategic situation has been so changed that it cannot be accomplished. One of the secondary objectives might be selected as the principal objective in the case that something goes wrong during the execution of a military action. Secondary objectives may also be used to deceive or distract the opponent from one's main objective.

There is probably no greater mistake than to formulate and pursue the accomplishment of several strategic or operational objectives simultaneously. Such a course of action may normally be adopted only when one's forces possess an overwhelming strength against any conceivable combination of the opposing forces.

The objective should not be too obvious, because a clever and agile opponent could frustrate one's own plans and complicate the entire military action or even possibly prevent the completion of the objectives.

Strategic Offensive versus Defensive

In general, a military objective can be offensive, defensive, or a combination of the two. Whether a war in a narrow sea is fought primarily offensively or defensively depends, among other things, upon the geostrategic position on land and at sea, the initial balance of strength at sea and in the air, and the overall war aims. A country on the offensive strategically on land would usually act offensively at sea, as Japan did in its war with Russia in 1904–5 and Nazi Germany in its invasion of Poland in September 1939.

In most cases, the progress of the war on land will determine whether one's fleet is employed predominantly offensively or defensively. The

Soviet Black Sea Fleet operated offensively once the Soviet ground forces started their long awaited offensive against the *Wehrmacht* in the aftermath of the battle of Stalingrad in January 1943. Likewise, the Soviet Baltic Fleet took the offensive after the Soviet ground troops broke through the siege of Leningrad and began their advance southward along the Baltic coast in September 1944.

The German Naval Staff's plans in 1938–39 for a war in the Baltic were based on the assumption that the Soviet Baltic Fleet would assume a passive posture, because of its poor state of training despite its greater numerical strength over German naval forces deployed in the area. Hence, the *Kriegsmarine* would be on a tactical offensive, a plan similar to the one prepared by Germany's Imperial Navy before 1914 and one drawn up by the *Reichsmarine* in the 1920s. In both cases the Germans after a long internal discussion decided to conduct offensive war against Russia even with inferior forces. The course of the war in the Baltic in 1941–44 proved that this decision was soundly based. The Germans and the Finns established a close blockade of the much larger Soviet fleet in the Baltic and held control there until the summer of 1944. This success was achieved despite the small forces the Germans had available in the Baltic for most of the war.[7]

An inferior fleet will normally operate defensively in a war with a much stronger force. In the war between Austria and France against Sardinia in 1859 the small Austrian Navy went on the defensive against the much stronger French fleet operating in the Adriatic. After the Austrian Army had invaded Piedmont on 3 May and France entered the war the Austrian Navy's Commander, Archduke Ferdinand Max, assembled his best ships at Pola, while the remainder of his small fleet was deployed among the islands of Dalmatia. Several Austrian ships were intentionally sunk at Venice and in other harbors to block their entrances to the French squadron. On 2 June the French declared a blockade of Venice and shortly thereafter also occupied the island of Lussin to use it as a base for an assault against the mainland coast. However, the war had already been decided on land after the Austrian defeats at Magenta and Solferino.[8]

The German High Seas Fleet was on the strategic defensive in the North Sea after the outbreak of hostilities in August 1914. One reason for this was that it was the German General Staff which decided that the Navy's principal role was protection of the Army's flank against possible British or Russian landings on the coasts of the North Sea or the Baltic. The influence of the German Army was also decisive in tying the Navy's hands during the transit of the British Expeditionary Force (BEF) to France. Germany's generals were confident that they could destroy it on land, and therefore did not need to risk the fleet to accomplish the same objective. The German Navy's final operation order for war in 1914 envisaged that the main objective should be to inflict damage with mines on the British forces patrolling and blockading the Heligoland Bight. Meanwhile, the U-boat offensive would be extended to the British coast and, after the balance of forces had been changed in Germany's favor, the aim then would be to engage the Grand Fleet, under favorable circumstances, in a decisive battle.[9] However, as Admiral Wolfgang Wegener observed, no tactical offensive could transform a poorly defined defensive plan into an offensive one. Simply put, it was not possible for the Germans to obtain sea control by seeking a decisive battle in the North Sea. The Heligoland Bight was 'a dead angle in a dead sea' and regardless of how many sorties the High Seas Fleet made from the Heligoland Bight, it remained on the defensive strategically.[10] Britain's Grand Fleet, although much superior, was also strategically on the defensive. The fact that the opposing fleets in the North Sea remained strategically essentially on the defensive throughout the war was probably the main reason for the lack of decisive results in that theater.[11]

In contrast to the North Sea, the Imperial German Navy accomplished its defensive tasks in the Baltic even with small forces because the Imperial Russian Baltic Fleet remained essentially on the defensive in the Baltic. The main objective of the Russians was to prevent German naval forces penetrating the Gulf of Finland.[12] The naval situation in the Baltic hinged on the fate of the German High Seas Fleet in the North Sea. For the Germans, the greatest naval threat was not the Imperial Russian Navy, but the British Grand Fleet.[13]

Prior to war with Nazi Germany in June 1941, the Soviets, despite their huge superiority in naval strength over the German–Finnish forces in the area, chose to be on the strategic defensive in the Baltic. Their plan focused on defense of the Bay of Riga, a complete closure of the Gulf of Finland from Hangö to Odenspholm-Reval, and a submarine and air war in the central part of the Baltic. Cruisers and destroyers would be used for mining, while heavy and older ships would defend the straits along the newly acquired coast of the Baltic States.[14] The Soviet Baltic Fleet, except for occasional forays by its submarines into the open waters of the Baltic, was on the defensive until the summer of 1944 when the advance of the Red Army along the Baltic coast allowed the fleet to play a more prominent role.

Sometimes a relatively strong fleet may decide to be on the defensive from the outset because its leadership lacks offensive spirit, even though it has superiority in numbers over its main adversary. A country or a force that is inferior to its opponent in the key elements of strength is usually forced to stay on a defensive until the balance of forces shifts in its favor. A superior force that lacks offensive spirit would usually take a passive posture and pursue defensive strategic military objectives.

In both world conflicts, the Italian Navy took a distinctively defensive posture despite its considerable numerical strength compared with its main opponents at sea. The Italian Navy's mission after the country entered the war on the side of the Entente powers in May 1915 was to keep the fleet ready to engage the Austro-Hungarian fleet, to protect the Italian coast against enemy raids by using mines, torpedo craft, small submarines, armored trains, and light cruiser-destroyer squadrons, and to establish an effective watch in the Strait of Otranto to prevent the transit of Austro-Hungarian and German submarines.[15] Likewise, in 1940, the Italian Navy limited its role to defense of the country's coast and the protection of sea routes to Libya. The Italian High Naval Staff (*Supermarina*) envisaged offensive actions only in the eastern Mediterranean, while a defensive posture would be taken in the Western and Central Mediterranean.[16] The *Supermarina* did not see any possibility of carrying out intensive offensive actions against the British fleet. It hoped, unrealistically, that the war would be short, and that several important positions such as Malta and Crete might fall into its hands without a fight.

The Iraqi Navy failed to operate offensively against the Iranian Navy in the Iran–Iraq War of 1980–88, choosing to retreat to its bases at Umm Qasr and Basra, and the Khor Abdullah waterway during the first days of the war. Moreover, it never deployed significant forces in the Gulf until the 1988 cease-fire.[17]

In contrast to a blue-water navy, the tasks of a coastal navy in a narrow sea are usually a combination of offensive and defensive tasks. The Imperial German Navy's tasks, as they evolved over the course of World War I, were to secure the western and central part of the Baltic and especially the vital iron ore traffic from Sweden, to protect the North Sea coastline and, under favorable conditions, whittle down the British fleet through aggressive mining and the use of U-boats.[18] Likewise in 1941, the *Kriegsmarine*'s tasks in the Baltic were to protect the German-controlled coast, to prevent a breakthrough by Soviet naval forces into the Baltic, and to eliminate Soviet bases and the Soviet Baltic Fleet. The Germans also envisaged transporting two or three divisions of troops to their northern war theater. The navy's main task in the course of the land campaign was to protect German commerce and iron ore traffic in the Baltic.[19]

A fleet facing significant naval threats in two adjacent maritime theaters would usually maintain a defensive-offensive posture so long as the hostile fleet's threat is not sufficiently neutralized or eliminated. Then, one maritime theater would usually be the theater of main effort, while the other would be the theater of secondary effort. Depending on the balance of forces, such a fleet might take an offensive posture in the main theater and be on the defensive in the secondary theater while, in different circumstances, strategic defensive objectives may be pursued in both maritime theaters. The German Navy in 1914 did just that. A major part of its fleet was concentrated in the North Sea, while the remainder was assigned to the Baltic, where its task was essentially defensive – to prevent offensive actions by the Russian Baltic Fleet. The Germans did not try to penetrate into the Gulf of Finland realizing that this might lead to serious losses of ships. By contrast, the main theater for the German *Kriegsmarine* in World War II was the Atlantic, while the Baltic was a sideshow.

The Imperial Russian Navy planned to take a defensive-offensive posture in the Black Sea in its plans for war drafted in 1907. Its initial strategic task was to prevent the German and Austro-Hungarian fleets from operating in the Black Sea. Afterwards, the main task was to seize the Turkish Straits and thereby gain access to the Mediterranean.[20] Because of the changes in the naval situation in the Black Sea, the Russians planned in 1909–13 to obtain control of the Black Sea and prevent hostile fleets from operating there. This objective would be accomplished by sending a strong force of Russian battleships, cruisers, and destroyers to the Bosporus, laying minefields off the eastern entrance of the Straits, deploying a force of surface ships off the Danube Estuary, and protecting the eastern coast of the Black Sea with torpedo boats and patrol ships. Because of the strengthening of the Turkish Navy, the Russians in their last plan prior to August 1914 envisaged taking a strategically defensive position in the Black Sea, the major part of their fleet being concentrated at Sevastopol' (Crimea) to fight a decisive battle with an opponent superior in strength. The Turkish fleet would be weakened in advance of the decisive battle by night attacks by torpedo boats and offensive mining.[21]

Normally, a stronger fleet should avoid being on the defensive both in home waters and overseas. The Royal Navy violated the principle of offensive during the American War of Independence (1776–83) by being on the defensive everywhere. It remained on the defensive in home waters, which was the prudent thing to do, but at the same time tried to cover all the vulnerable points in the Americas. The result was predictable. Not only were the American colonies lost but much of the West Indies and Minorca in the Western Mediterranean were lost as well. Even the security of Gibraltar and England itself was endangered. In home waters, too, it was indecision and defensive-mindedness by the French rather than British strategy that saved England from an invasion.[22]

In the Russo-Japanese War of 1904–5, the Japanese fleet assumed the offensive strategically with the objective of gaining control of the sea, which was disputed by the Russian Far Eastern Squadron (ex-Baltic Fleet). After the defeat of that squadron, the Japanese assumed the defensive strategically and awaited the arrival of the Second Russian Pacific Squadron, which eventually was decisively beaten in the Battle of Tsushima in May 1905.[23]

Another example of an offensive-defensive objective was the Italian Navy's plan for war at sea prior to June 1940. The Italians planned to close off the Adriatic and Tyrrhenian Seas to British forces and to maintain safe communications with Libya and the Dodecanese. The offensive objectives were the interruption of French sea communications to North Africa and the keeping open of certain sea routes for the eventual dispatch of troops to the enemy territory. However, the French thought that an attack on Tunis, considered to be a key to the control of the Mediterranean, was unlikely because of the Italian inferiority in the air.

Fleet Distribution

Among other things, the principal task of naval strategy is to determine the distribution and composition of the fleet in various maritime theaters. First, based on the strategic objectives to be accomplished, principal and secondary maritime theaters of war or theaters of operations should be selected. Afterwards, the distribution of the fleet should be determined in consonance with the objectives of strategy. This means, in practice, the faithful application of economy of effort.

The peacetime deployment of one's naval forces is dictated by the current strategic estimate of the situation and decisions are based upon the national policy and an estimate of the most probable theater of operations in case of war.[24] Distribution of available naval strength based on the peacetime and probable wartime objectives is an especially critical problem for the navies that operate in narrow seas and for a blue water navy whose home bases are located within one or several narrow seas. It is normally unwise to distribute a fleet equally among the various theaters as the British and Dutch did in 1652–54, when they divided their fleets evenly between the North Sea and the Mediterranean. Two centuries later, the Imperial Russian Navy made a similar mistake when its fleet was roughly divided between the Baltic and Port Arthur.[25]

Britain has always been faced with the problem of how to distribute its naval forces so that its source of strength at home was fully protected while, at the same time, defending its overseas possessions and the free flow of trade to the mother country.[26] The British traditionally maintained the 'two-power standard' of naval strength from 1770 (when the policy was first enunciated) until the end of nineteenth century. Around the turn of the twentieth century, this policy rapidly became obsolete because of the steady increase of the strength of the German Navy concentrated in the North Sea. A new policy, the 'two-power standard with a margin' was declared on 1 March 1904. The British Admiralty then considered Germany and Russia or France and Russia as the most likely opponents at sea. By 1907 the British planned to obtain a margin of at least 10 per cent in battleships over any of these two combinations. As result of the new policy, the Royal Navy was drastically reorganized in the last decade prior to 1914 and its main strength shifted from the Mediterranean to home waters. The Mediterranean was then a critical link between Britain and its possessions in the Far East. A fleet based there was usually composed of heavy and fast ships capable of quickly reinforcing the fleet in the English Channel and in the North Sea or the fleets in the Indian Ocean and the Far East.[27] In 1904 the Royal Navy reduced its force of battleships in the Mediterranean from 12 to eight. Two years later an additional two battleships were moved from the Mediterranean and redeployed in the North Sea. Between 1907 and 1912 there were only four British battleships deployed in the Mediterranean.[28]

By the spring of 1912 the British Admiralty decided to reduce further the fleet in the Mediterranean because of the need to concentrate as much as possible of the Royal Navy's strength in home waters to meet the German challenge. A major organizational change effective 1 May 1912 was the establishment of the Home Fleet responsible for naval defense of home waters. It consisted of three fleets totaling 33 battleships in full commission and eight more with skeleton crews. These 41 battleships were thought to be superior to the 25 German battleships ready for war.[29]

Under the new scheme, only two or three battle cruisers, a squadron of armored cruisers, one destroyer and two submarine flotillas (one based at Malta and the other based at Alexandria) were deployed in the Mediterranean. The rest of the Royal Navy was serving on colonial duties in the East Indies, at the Cape, in West Asia and Australia, and along the east coast of the Americas. The British fleet in the Mediterranean was believed to be large enough in combination with the French Mediterranean fleet (six dreadnoughts and 14 pre-dreadnoughts in commission plus seven dreadnoughts to be completed in 1913–15) to counter the Austro-Italian combination (17 pre-dreadnoughts and ten dreadnoughts by 1915).[30] The result was that no British battleships, except for a squadron of pre-dreadnoughts temporarily deployed during the Balkan wars, were deployed in the Mediterranean.[31] In essence, the French fleet was to defend British interests in the Mediterranean. The British Admiralty was not confident that in the event of war with Germany and at least one other hostile Mediterranean power it would be able to ensure the safety of British sea communications in the Mediterranean until the situation in the North Sea had been clarified, and it estimated that this would take several months from the outbreak of hostilities.[32]

Prior to 1914 France faced the problem of maintaining a strong fleet in the Mediterranean to protect the country's coast and its possessions in North Africa. It had to deal with the threat of a combined Italian and Austro-Hungarian fleet. Reassured as to the safety of its Channel and Atlantic coasts by the Entente agreement in 1912 France began to concentrate its fleet in the Mediterranean where it had to protect the vital lines of communications to its possessions in Africa, from which a considerable troop contingent (about 100,000 men) for its army would come on mobilization.[33]

In the early 1930s the French Navy's tasks included protection of supplies in the Mediterranean. Thus, the First Squadron in Toulon was tasked to protect sea routes in the Mediterranean. Control of the Marseille–Algiers–Bizerta triangle was considered critical for control of the Western Mediterranean. By the mid-1930s the French began to change the focus from the Mediterranean to the Atlantic. Hence, the main route for troop transports shifted from Bizerta–Toulon to Casablanca–the French Atlantic coast. This was the reason for the steady strengthening after 1933 of the Second Squadron based in Brest.[34]

The Imperial German Navy was also faced with the task of distributing its strength between two narrow seas, the Baltic and the North Sea. After 1890, the Germans considered France to be their main potential enemy while Britain would stay neutral in a conflict. By 1900 a new Naval Act envisaged a large construction program aimed primarily to challenge the British supremacy at sea. Therefore, all the plans of the German Navy prior to 1914 were aimed against the British Navy as the main opponent and hence the North Sea was selected as the principal maritime theater. This change of policy was also reflected in the German fleet distribution. By August 1914, the Germans had deployed a major part of their navy in the North Sea, 26 battleships (including 14 dreadnoughts) organized in six battle squadrons, four battle cruisers, 21 cruisers, about 90 destroyers and torpedo boats, and 17 U-boats.[35] In the Baltic, the Germans planned to be on the defensive strategically and, hence, only nine cruisers, ten destroyers and torpedo boats, four U-boats, and about 30 miscellaneous ships were deployed there. In the Mediterranean, the Germans had deployed one battle cruiser and one cruiser and a patrol ship (*Missionschiff*) at Constantinople.[36]

The former Imperial Russian and the Soviet Russian Navy were also confronted with a perennial problem as to how to divide their total strength among several maritime theaters. The strength and composition of each fleet varied over time due to the different strategic objectives to be accomplished in a given theater. In 1881, Russian policy aimed to obtain naval superiority in the Baltic over Germany and in the Black Sea over Turkey. Yet in both theaters the Russian strategic objectives were essentially defensive. In the Baltic, the Russians intended to conduct an active defense of the area and to prevent a blockade of the country's coast. The Russian naval forces in the Far East were also tasked to defend naval bases and the coast.[37]

The Baltic Fleet was the largest of the Russian fleets prior to the outbreak of World War I. It was, however, much weakened in the aftermath of the Russo-Japanese War of 1904–5. The major part of that fleet, renamed the Second Pacific Squadron, was destroyed in the battle of Tsushima. In 1906–7, the Baltic fleet consisted of only one battleship, two armored cruisers, four cruisers, 20 destroyers (former mine-laying cruisers), and about 60 torpedo boats. In addition, on the eve of World War I, a large number of battleships, cruisers, and destroyers were under construction.[38]

Between the two world wars, the Soviets deployed four fleets: the Baltic, Black Sea, Pacific and Northern Fleets (formally established in 1937). The Baltic Fleet, followed by the Black Sea Fleet, was the largest and contained the most modern surface combatants and submarines. This situation changed in the aftermath of World War II and especially after the late 1950s when the Soviets embarked on a policy to challenge US/Western dominance in the world's oceans. By the late 1980s the Soviets deployed most of their nuclear-powered submarines, heavy surface combatants, and naval aircraft with the Northern and Pacific fleets – the two fleets that potentially had the easiest access to the open waters of the ocean. Thus, it was not surprising, that the Northern and the Pacific fleets provided most of the ships for the Soviet forward presence in the peacetime. Most of the Soviet large surface combatants and attack submarines were assigned the task of protecting ballistic-missile nuclear-powered submarines in their sanctuaries in the Barents Sea/White Sea and the Sea of Okhotsk. The other two fleets, the Baltic Fleet and the Black Sea Fleet, were composed predominantly of smaller surface combatants, diesel-electric attack submarines, amphibious ships, mine warfare ships/craft and ASW fixed-wing aircraft and helicopters.[39]

To sum up, the first step before operational planning begins is to determine the objectives that can be accomplished with the means available. Policy always dominates the use of military force at the strategic level. Both the political purpose of a war and the military strategic objectives must be in consonance. Any imbalance between these two will cause severe problems in the accomplishment of the stated strategic objectives. However, the policy should not be allowed to influence the attainment of operational or tactical objectives, unless some extraordinary circumstances require it. The objective to be achieved should be clear-cut and any ambiguity should be avoided. The objective laid down should not be rigidly pursued. The application of the principle of flexibility requires that beside the main objective, several secondary or alternative objectives should be selected. No greater mistake can be made than that of trying to accomplish several objectives simultaneously or in rapid sequence.

The highest politico-military leadership of a country or of an alliance/coalition bears the responsibility of determining whether naval forces should be on the offensive, the defensive or a combination of these two in the case of hostilities at sea. This in turn will determine which maritime theater should be the main and which secondary in importance. Accordingly, the overall size of the fleet and its distribution among maritime theaters must be in consonance with the strategic objectives to be accomplished in wartime. Therefore, the size of a fleet and its force-mix in peacetime are a good indicator as to what probable strategic objectives in a given narrow sea will be pursued in wartime.

NOTES

1 'Principles of War as Applied to Naval Warfare', 30 July 1926, Naval War College Archives, Record Group 4, Box 28, Folder 1144 (Newport, RI: Naval War College), p. 5.
2 C. R. Brown, 'The Principles of War', *Proceedings*, 6 (June 1949), p. 624.
3 B. H. Liddell Hart, 'The Objective in War: National Object and Military Aim', lecture delivered at the Naval War College, 24 September 1952, p. 1.
4 Ibid., p. 2.
5 'Principles of War as Applied to Naval Warfare', p. 1.
6 Ibid., p. 4.
7 Michael Salewski, *Die deutsche Seekriegsleitung 1935–1945*, Vol. I, *1935–41* (Frankfurt a.M: Bernard & Graefe, 1970), p. 365; idem., Vol. II: *1942–1945* (Frankfurt a.M: Bernard & Graefe, 1975), p. 574.
8 Lawrence Sondhaus, *The Habsburg Empire and the Sea: Austrian Naval Policy 1797–1866* (West Lafayette, IN: Purdue University Press, 1989), pp. 191–2.
9 Ivo Nikolai Lambi, *The Navy and German Power Politics, 1862–1914* (Boston, MA: Allen & Unwin, 1984), p. 422.
10 Arthur Marder, *From the Dreadnought to Scapa Flow: The Royal Navy in the Fisher Era, 1914–1919*, Vol. II: *The War Years: To the Eve of Jutland* (London: Oxford University Press, 1965), p. 43; Wolfgang Wegener, *The Naval Strategy of the World War*, trans. Holger Herwig (Annapolis, MD: Naval Institute Press, 1989), pp. 21–2.
11 M. G. Cook, 'Naval Strategy', 2 March 1931, Air Corps Tactical School, Langley Field, VA, 1930–1931, Strategic Plans Division Records, Series, Box 003 (Washington, DC: Naval Operational Archives), p. 9.
12 N. B. Pavlovich, ed., *The Fleet in the First World War*, Vol. I: *Operations of the Russian Fleet* (Moscow: Voyenizdat, 1964; trans. and published for the Smithsonian Institution and the National Science Foundation, Washington, DC, by Amerind Publishing, New Delhi, 1979), pp. 22–3.
13 Wegener, *The Naval Strategy of the World War*, p. 24.
14 Juerg Meister, *Der Seekrieg in den osteuropaeischen Gewaessern 1941–45* (Munich: J. F. Lehmans Verlag, 1958), p. 11.
15 Marder, *From the Dreadnought to Scapa Flow: The Royal Navy in the Fisher Era, 1914–1919*, Vol. II, p. 330.
16 S. W. C. Pack, *Sea Power in the Mediterranean: A History from the Seventeenth Century to the Present Day* (London: Arthur Barker, 1971) p. 181; Michael Salewski, *Die deutsche Seekriegsleitung 1935–1945*, Vol. I: *1935–1941*, pp. 228, 341. The German Naval Staff, by contrast, believed that it was in Italy's interest to concentrate actions by all three of its Services to neutralize British dominance in the Eastern Mediterranean. Specifically, this meant providing mutual support in blockading the British Mediterranean Fleet, seizing Egypt, and eliminating British influence in Palestine and Turkey. The Germans thought that the reduction of the British position in the Mediterranean was a key prerequisite for the successful outcome of the war at sea. Both the German Naval Staff and the Supreme Command of the *Wehrmacht* considered Suez/Egypt and Gibraltar to be the strategic pillars in the Mediterranean. The Germans also tried very hard to

persuade the Italians to take a more offensive posture, and thus change the situation in the Mediterranean in favor of the Axis. However, all these attempts remained unsuccessful; ibid., pp. 294–5, 297, 324.
17. Anthony H. Cordesman, *The Iran–Iraq War and Western Security 1984–87: Strategic Implications and Policy Options* (London: Jane's, 1987) p. 102.
18. Marder, *From the Dreadnought to Scapa Flow*, Vol. II, p. 45.
19. Salewski, *Die deutsche Seekriegsleitung 1935–1945*, Vol. I, pp. 366–7.
20. Pavlovich, ed., *The Fleet in the First World War*, Vol. I: *Operations of the Russian Fleet*, p. 23.
21. Ibid., pp. 33–4.
22. Geoffrey Till, ed., *Maritime Strategy and the Nuclear Age* (New York: St Martin's Press, 2nd edn, 1984), p. 116.
23. Cook, 'Naval Strategy', p. 8.
24. Ibid., p. 16.
25. Alfred T. Mahan, *Naval Strategy Compared and Contrasted with the Principles and Practice of Military Operations on Land* (Boston, MA: Little, Brown, 1919), p. 115–16.
26. Kupfer, 'Die strategische Verteilung der Hauptflotten im Hinblick auf ihre Friedens- und Kriegsaufgaben', *Marine Rundschau*, 6 (June 1936), p. 291.
27. Ibid., pp. 291–2.
28. Paul G. Halpern, *The Naval War in the Mediterranean 1914–1918* (Annapolis, MD: Naval Institute Press, 1987), pp. 1–2; Arthur J. Marder, *From the Dreadnought to Scapa Flow: The Royal Navy in the Fisher Era, 1904–1919*, Vol. I: *The Road to War 1904–1914* (London: Oxford University Press, 1961), pp. 123–4.
29. Ibid., pp. 287–8.
30. Ibid., p. 288.
31. Halpern, *The Naval War in the Mediterranean 1914–1918*, p. 2.
32. Marder, *From the Dreadnought to Scapa Flow: The Royal Navy in the Fisher Era, 1904–1919*, Vol. I, p. 288.
33. Cook, 'Naval Strategy', p. 16.
34. Kupfer, 'Die strategische Verteilung der Hauptflotten im Hinblick auf ihre Friedens- und Kriegsaufgaben', p. 295.
35. See Table 12, 'Kriegsgliederung der britischen Flotte im Juli 1914', in D. Groos, *Der Krieg in der Nordsee*, Vol. I: *Von Kriegsbeginn bis Anfang September 1914*, ed. E. von Mantey, *Der Krieg zur See 1914–1918* (Berlin: E. S. Mittler & Son, 1922).
36. Ibid; Pavlovich, ed., *The Fleet in the First World War*, Vol. I: *Operations of the Russian Fleet*, p. 74.
37. Ibid., p. 11.
38. Ibid., pp. 45–6.
39. Bruce W. Watson and Susan M. Watson, eds, *The Soviet Naval Threat to Europe: Military and Political Dimension* (Boulder, CO: Westview Press, 1989), pp. 335–8.

7

Sea Control and Sea Denial

The main objective of any fleet in the past was to attain and maintain what was called *command of the sea* (or *mastery of the sea*). In the modern era this term has been substituted in the West with the term *sea control*; in contrast, the Soviets/Russians use the term *sea mastery* (*morskoye gospodstvo*). There are, however, some differences between these terms and they must be explained. Command of the sea essentially pertained to control of sea communications for a specific purpose, that is, one's ability to move across the sea without considerable opposition, while denying the same to the adversary at sea. Sir Julian Corbett observed that the object of naval warfare should always be aimed directly or indirectly either to secure command of the sea or to prevent the enemy from securing it.[1]

Control of the Sea

Control of the sea involves control of sea communications, which essentially means that one side possesses superiority over its opponent in a given theater or area of operations. By gaining control of the sea we can move ships, continue to conduct overseas maritime trade, interrupt or cut off the enemy's seaborne trade and if necessary project power onto the enemy shore. On the open ocean, command of the sea has been rarely, if ever, complete. It has usually meant only control of maritime communications over a certain sea or ocean area by one side in a conflict.

In contrast to the open ocean, in a typical narrow sea with its small size, and hence more favorable ratio of space–force, a blue-water navy can obtain almost absolute and permanent command of the sea surface and airspace and, possibly, even control the subsurface. In 1991 the Coalition's naval forces obtained almost absolute control of the northern part of the Arabian Gulf on the surface and in the air in the first two weeks of the conflict with Iraq. Although the Iraqis did not have submarines, they contested control of the subsurface by laying mines off the Kuwaiti coast.[2] Command of the sea in an enclosed sea theater is normally obtained and maintained by the closest cooperation among all the services of a country's armed forces. It is most directly affected by the situation on the land front and in the air. Yet, that does not mean that one's fleet forces play a negligible role. Air power can also be decisive in winning the war at sea, but the war cannot be won without offensive employment of one's fleet forces.

The term *sea control* is perhaps more accurate than command of the sea because it reflects more realistically the fact that it is more difficult to keep control of the sea in an age of aircraft, missiles, torpedoes, and mines. Today, even the smallest navy can make sea control complicated for a major navy. The term sea control conveys the meaning that it is not possible except in the most limited sense totally to control the seas for one's use or to deny them completely to an opponent. Sea control essentially means the ability of one's fleet to operate with a high degree of freedom in a sea or ocean area, but for a limited period of time. On the open ocean, sea control may encompass only a relatively narrow part of the ocean surface and subsurface around a moving carrier battle group, amphibious task force, or convoy, or the ocean area where sea control is exercised may be more or less geographically fixed, as for example in operating areas of one's ballistic missile submarines. This is markedly more true in a narrow sea, where the concept of sea control means full control of one's coastal waters.

Sea control is a very dynamic term. The attainment, maintenance, and exercise of sea control on the open ocean may involve sequential operations of considerable size over a longer period. These actions are aimed at obtaining control not only of the open ocean, but also of adjacent marginal seas and enclosed seas. Control of a typical narrow sea has in contrast a broader meaning than that on the open ocean, because it involves control of the entire theater and adjacent landmass important to sea control. It should allow one's naval forces and aircraft to strike enemy forces not only at sea, but also on the adjacent coast as well. Consequently, the term command of the sea is more appropriate when referring to a narrow sea than in regard to the open ocean. Also, sea mastery can be obtained and maintained in a typical narrow sea, but it is almost impossible to do the same on the open ocean. When one side possesses sea mastery, this means that its main opponent cannot use his submarines, aircraft, nor sometimes even mines.

In terms of scope and extent, sea control by one side can range from general to local to non-existent. With respect to degree it can be absolute (or uncontested), disputed (or contested), while its duration can be permanent or temporary. In practice, sea control on the open ocean has never meant control that was either complete in degree or unlimited in space. It was always relative, that is, incomplete and imperfect. Even if one side obtained control of an ocean, that did not mean control of all the marginal seas, and even less of enclosed seas bordering a particular ocean.

General command or *full command* on the open ocean was usually a decisive factor in a war against an opponent heavily dependent on commerce or vulnerable to naval blockade. A blue-water navy may possess general command of all the oceans, but not the contiguous narrow seas. The examples of the Royal Navy during the Seven Years War (1756–63), the wars with Revolutionary France, and the Napoleonic Wars clearly showed that this objective is difficult or nearly impossible to accomplish. In World War I, the Entente Powers controlled the surface of every ocean and of all the strategically important narrow seas except for the Black Sea, the Baltic, and southeastern rim of the North Sea.[3] In World War II, the British Royal Navy controlled the oceans but not the Baltic, the Adriatic, and the Black Sea, while the control in the Mediterranean in 1940–43 was in dispute. At the same time, Germany's inability in its war with the Soviet Union to obtain full control of the Black Sea and the Sea of Azov had a significantly negative effect on the operations on land.

For a regional naval power, complete command of the adjacent narrow seas can be accomplished by using ground troops only, as the Germans did in the Baltic in 1941. By September 1941, German Army Group North seized the entire Soviet-controlled coast with the exception of the eastern part of the Gulf of Finland including the large Soviet naval bases at Leningrad and Kronstadt. However, the Germans and Finns did not plan to – nor could they – operate in the Bay of Kronstadt to prevent Soviet submarines from breaking through into the open waters of the Baltic and destroy the remainder of Soviet surface ships.[4] It can be argued that if the *Kriegsmarine* had obtained complete sea control in the Baltic in 1941 that would most likely have led to faster advance of German troops on the coast. At the same time, the speedier withdrawal into the depths of the country's territory would probably result in much lower Soviet losses and leave them with a larger force for the winter offensive.[5]

In 1941, the Germans relied on the army's advance along the Soviet Black Sea coast to obtain control of the sea. By the fall of 1942, the Germans had seized from the landside most Soviet naval bases and major commercial ports forcing the Soviets to operate their remaining surface combatants and submarines from small and inadequately equipped naval bases in the eastern part of the Black Sea. Nevertheless, the Soviets were able to challenge German control of the Black Sea.[6]

Complete Soviet control of the Baltic and the Black Sea would most likely have had severe consequences for the German conduct of war in both theaters. The Swedish iron-ore transport to Germany would not have been possible and Sweden would most likely have been strictly neutral in the conflict. Most serious of all, the Germans would have been forced to deploy large number of troops to defend the coast because of the threat of Soviet amphibious landings.[7]

The Germans also hoped that a successful land campaign in Africa and the Middle East would lead to control of the sea in the Mediterranean and the adjacent smaller narrow seas. Their ultimate aim was to eliminate the British position in the Mediterranean and the Middle East by concentric attacks on Egypt from Libya, via Bulgaria and Turkey, and, if the situation allowed, also from Transcaucasus through Iran, then to close the western entry to the Mediterranean by seizing Gibraltar, and finally to clean up North Africa.[8]

Not only the Army but also coastal defense forces can be used to attain absolute local, though often temporary, control of straits or river estuaries. This was, for example, the case in the Strait of Messina in August 1943 where the Axis' gun batteries on both sides of the strait obtained local and temporary command of the surface and the accompanying airspace. Likewise, the Germans controlled the Kerch Strait (Sea of Azov) in the fall of 1943 by mounting strong coastal defenses on both shores.

In a conflict with a small coastal navy a blue-water navy can gain general control of a narrow sea, or its only exit (to another narrow sea, or a marginal sea of an ocean) relatively easily. However, a weaker navy can also obtain full control of the sea in part of a theater of operations. The Germans in World War I indirectly exercised full control in the western part of the Baltic, while their control in the eastern and central part of that sea was in dispute because of the continued existence of a relatively strong Russian fleet. By contrast, the Germans had almost complete control of the Baltic in 1941–44. Only Soviet submarines contested that control by their attempts, though infrequently, to break through the Gulf of Finland to reach the open waters of the Baltic. In 1943 the German Navy Commander, Grand Admiral Karl Doenitz, considered control of the Baltic vitally important for testing the U-boats and the training of their crews. Hence, the Soviet Baltic Fleet would be blockaded, while Sweden would be prevented in all circumstances from siding with Germany's enemies.[9]

Local sea control exists when one side possesses superiority in one part of the sea for the purpose of executing a specific mission. Sometimes, local control must be obtained to ensure the safety of a large convoy, the conduct of an amphibious landing, the bombardment of the enemy's coastal installations/facilities, or to carry out hasty raids. Consequently, drastic changes of a situation are common and the duration of local control is often only temporary. Before the advent of aircraft and missiles, local sea control did not mean that a stronger fleet was able to carry out a protracted effort giving it complete influence on the course of the war.[10] In the modern era, however, local sea control makes it possible to mount a large-scale expedition across short distances. Temporary control of the Skagerrak and the adjacent sea areas allowed the Germans to move their warships and transports to invade the southern part of Norway in April 1940 and similarly Crete in May 1941.

The importance of local sea control was shown in several naval conflicts in the past. After the fight off the Kentish Knock in September 1652, the English held local command in the Channel and the southern part of the North Sea. However, shortly thereafter they committed a serious blunder by detaching a considerable part of their fleet to the Mediterranean on trade protection duties, thereby allowing the Dutch to regain command of England's home waters. In another example, the Russian squadron under Admiral Dmitry Senyavin obtained local control of the southern part of the Adriatic shortly after the renewal of the hostilities with France in September 1806. This made it possible for the Russian squadron to blockade Ragusa from the sea side. The French, however, mounted an overland drive and laid siege to the city of Cattaro. Though stalemate on land continued, Senyavin started another series of attacks on the southern Dalmatian islands. The Russians seized Curzola for the third time in the war while the island of Lesina was attacked.[11]

In the Sino-Japanese War of 1894, both sides depended on supplies transported across the sea. Although Korea was connected to China at its landward frontier, road communications were so primitive as to be of little use for the dispatch of reinforcements or supplies. Therefore, local control of the Yellow Sea for either side was the key to success in the campaign on the land front.[12] Likewise, the Spaniards in 1898 depended greatly on the security of their communications between Cuba and Spain. To obtain local command of the area of operations, the Spanish fleet would need to destroy, drive off, or at least threaten US naval forces in the area. And the same was true for the US Navy.[13]

The weakness in numbers and *matériel* of the Romanian and German naval forces prevented any activity at sea during the Axis' advance toward the Crimea Peninsula in 1941. The much stronger Soviet naval forces based in Sevastopol' controlled the central part of the Black Sea and prevented the movement of small craft to the ports in the eastern part of the Crimea seized by German troops in the fall of 1941.[14]

When a fleet is only barely stronger than its opponent, its main objective should be to obtain and maintain command of the sea in the most critical area, even if that means leaving freedom of action for the opponent to operate elsewhere. In the Russo-Japanese War of 1904–5, the Japanese fleet held local command in the sea area between Vladivostok and Port Arthur, but left to the Russian fleet undisputed control of the sea elsewhere.

Sometimes local command of the sea by a numerically weaker force can inhibit a much superior force from undertaking offensive actions not only in the same theater but also in an adjacent narrow sea. Such a situation prevailed for most of World War I in the North Sea. The German High Seas Fleet's command of the Heligoland Bight kept the Kattegat closed, thereby preventing the Grand Fleet from conducting offensive actions within the Baltic in support of Britain's besieged ally, Imperial Russia.

Local command of a semi-enclosed sea theater is steadily more contested as one moves away from the center of a stronger force's hub of power. The Germans in their invasion of Norway in April 1940 clearly exercised a fair degree of control in the narrow waters of the Skagerrak. However, this control progressively became more uncertain the further north the German forces moved. In the execution of the invasion of Norway, Admiral Erich Raeder pinned all his hopes on achieving complete surprise and the ability of the *Luftwaffe* to control the skies over Norway and the adjacent waters, thereby countering any subsequent Allied action against the German naval forces.

In the archipelago type of coast, local sea control can be obtained by seizing a series of islands of tactical or operational importance. Control of the mainland coast is almost invariably necessary to hold command of the adjacent sea and offshore islands for any length of time. In May 1941 Admiral Raeder considered the Greek mainland, and the islands of Crete, Lemnos (Límnos) and Melos (Mílos) to be of decisive importance to the Axis, as these positions would serve as the starting points for conduct of war in the Eastern Mediterranean. Germany's ultimate objective was to eliminate the British position in the Eastern Mediterranean, including Alexandria and Suez. This objective would be accomplished by the close cooperation of all services, combined with the best use of the newly acquired positions. The end result would be to obtain command of the Eastern Mediterranean and eliminate any threat to the Aegean and North Africa by the British Navy.[15]

Lack of local sea control even for a relatively short time doomed many plans for large-scale invasions. Germany's inability to obtain local control in the Channel in the summer of 1940 was one of the main reasons that the Germans abandoned their plan to invade England (Plan *Seeloewe*). In 1940 not only the RAF but also British flotillas based at Harwich, Dover and Portsmouth swept the narrow seas day after day, night after night, finally forcing the Germans reluctantly to accept that they could not obtain an adequate degree of local sea control to ensure that their slow-moving convoys could survive even a short sea passage. The British strategy, which forced the Germans to abandon their invasion plan in 1940, was identical to the one that had been used against any would-be invader since the defeat of the Spanish Armada in 1588.

Duration of Control

Sea control can be either permanent or temporary. A *permanent* sea control exists when one side in the conflict completely dominates a given theater. This, of course, does not mean that the opponent can do nothing, but rather that he cannot interfere with one's maritime shipping or amphibious landings in such a way as seriously to affect the course of the war. Or, expressed differently, permanent sea control means that your opponent cannot conduct his maritime trade or maritime expeditions except at unacceptably high risk.

A *temporary* sea control is usually the result of the inability of either fleet to obtain a decision. When one of the fleets loses the initiative, for whatever reason, to its opponent, then the abandonment may be permanent or temporary. The weaker force usually falls back on the defensive and keeps its main fleet in bases, avoiding action at sea. However, if a weaker side at sea obtains superiority in the air, then this in itself is often sufficient to allow one's own use of the sea for a specific purpose and for a limited time. This situation would probably have existed for the German *Kriegsmarine* in the English Channel in the summer of 1940 if the *Luftwaffe* had been successful in obtaining command of the sky.

Degree of Control

In practice, sea control can range from absolute to contested; it can also mean the free use of particular types of ships, but not others. *Absolute* sea control means in practice that one's fleet operates without major opposition, while a hostile fleet cannot operate at all. It aims in general to obtain control of the entire maritime theater or area of operations. Then, one side can use its fleet whenever and wherever needed without serious threat from its opponent.

In the modern era absolute control of the sea even in a part of the ocean has been difficult to accomplish. In a narrow sea, however, a superior force could possibly obtain such a control relatively quickly after the beginning of the hostilities, as happened in the Gulf War of 1990–91. On the open ocean permanent and absolute command of the sea cannot be obtained against a relatively strong opponent. Therefore, it is probably more realistic to aim at establishing fluid and temporary zones of control, wherever it is desired to use a particular stretch of salt water for particular purposes.

SEA CONTROL AND SEA DENIAL

In the modern era absolute control on the open ocean has usually meant control of the surface, and to some extent of the air, while control of the subsurface was difficult or impossible to attain. The US Navy held complete control of the surface during the Cuban missile crisis of 1962, but not subsurface (because of the presence of Soviet submarines). Likewise, the Coalition's naval forces attained absolute control of the Gulf during the first few weeks of their war against Iraq in 1991; yet their control of the subsurface in the northern Gulf was disputed because of Iraqi mines.

Absolute but temporary sea control exists when one side enjoys superiority over its opponent in the entire area or theater of operations, but is short in duration. After the battle of Beachy Head, off the southern coast of England, in July 1690, the French held temporary control of the Channel over the combined Anglo-Dutch fleet. However, that command was of little use because the French failed to assemble an army in time to invade England.

Limited sea control, also called a *working* or *conditional* control, exists whenever one fleet has a high degree of freedom of action, while the other operates at high risk. It can also be described as a situation where only one type of ship can operate without undue risks, while the other ships operate at high or unacceptable levels of hazard. Then, regardless of the inequality of naval strength, the weaker force always has a chance to carry out hasty raids against some weak and unprotected points on the enemy coast. British Admiral Reginald Bacon noted that the Royal Navy held command of the North Sea in World War I only as far as battleships were concerned, while the Germans still had the ability to use battle-cruisers, light cruisers, destroyers and U-boats effectively. No matter how tight the blockade was, the British still needed direct defenses of shipping and coastline behind the lines.

A *disputed* or *contested* sea control normally prevails in a war between two strong opponents on the open ocean or in a large narrow sea such as the Mediterranean. It often occurs in the initial phase of a war and is characterized by almost uninterrupted struggle for control of a certain sea area. Once command is obtained, it is usually not maintained for a long time, but is lost from time to time and then obtained again. In coastal or offshore waters, sea control by a stronger fleet can be contested even if the major part of a weaker fleet is destroyed. Besides the Army on the coast and land-based air, defensive minefields and coastal missile and gun batteries can be used to dispute control by a stronger force.

When sea control is in dispute on the open ocean, both sides operate at high risk because their strength is roughly in balance. Then, one side usually controls one or more parts of a maritime theater of operations, while its opponent controls the remaining part of it. Control of a specific sea area for each side is usually limited in time. In some cases, however, one side can obtain more or less permanent control of a specific part of a maritime theater. Disputing or contesting sea control is primarily carried out by one's submarines, aircraft, and mines.

In the modern era control of even a small ocean area has rarely been established for any length of time. Hence, it is far more common that in a war on the open ocean sea control is in dispute. In a narrow sea, however, a fleet too weak to win command by offensive actions may yet succeed in disputing control by assuming a generally defensive attitude. The High Seas Fleet held contested control of the Baltic in World War I, although Russia's naval position was more favorable because of its control of the Gulf of Finland and Livonia.

In World War II, the English Channel was a sea area in which both the British and the Germans constantly fought to obtain full control of the sea, but for most of the war, both sides disputed control of the Channel. The last action of British Coastal Forces took place off Boulogne and Calais on 1–2 October 1944. Afterwards, the focus of activity by British inshore patrols shifted to the Belgian and Dutch coasts. Although maritime traffic from Britain to the Belgian and French Channel ports continued to be heavy, control of narrow seas was disputed only by the occasional German U-boat and unswept mines.[16]

Contested sea control existed in the Mediterranean in 1940–43. The Royal Navy with its forces based at Alexandria and Gibraltar controlled the eastern and the western part of the Mediterranean, while the Italians held general control of the central Mediterranean. Nevertheless, this control was never complete for either side. The Axis' submarines contested Allied control at both ends of the Mediterranean, while British submarines and aircraft based on Malta did the same in the central part of the sea. This did not prevent either the Allies or the Axis using the sea to run convoys when and where they needed. The Axis used the sea more frequently than the Allies did because their communications were so much shorter and Allied maritime routes were longer and more exposed. However, to use these routes, the Allies needed only to obtain a temporary control of a 'moving zone' in which the ships sailed. The British were deprived of using a direct sea route through the Mediterranean except for the passage of occasional military convoys from Alexandria to Malta which they wanted to use as a base to attack Italian convoys to Libya. The British surface forces were unable to be based regularly on Malta. To attack successfully the Axis' communications in the Central Mediterranean, the employment of light forces, submarines, and aircraft was a prerequisite; and that was not achieved. Also, British submarines were too large to work safely and effectively in the Mediterranean.[17]

The state of divided control of the surface prevailed in the Black Sea in 1941–44. Largely because of the inactivity of the Soviets, the Axis' forces controlled the western part of that sea; the Soviets controlled only its eastern part. The Axis' naval forces were small and weak, except for six 280-ton coastal submarines. At the same time, German and Soviet submarines and land-based aircraft contested control. This situation came into being partly by the great lack of support from the air for the Soviet surface ships.[18]

An example of temporary but disputed sea control was the situation at the sea approaches to Guadalcanal in 1942–43. The outcome of the struggle for that island depended upon whether the Americans or the Japanese could establish control of the adjacent seas to supply and reinforce ground troops contending ashore for mastery. After the battle of the Eastern Solomons in September 1942, a virtual exchange of sea mastery took place every 12 hours; Americans dominated waters around the island from sunrise to sundown, but at night the Japanese *Tokyo Express* transported troops by destroyers and light cruisers. The Japanese rarely attempted daytime raids and Americans more often interfered with the night surface express, but any such attempt proved to be expensive.

In the Yom Kippur (Ramadan) War of 1973 Israel sought to deny the area off its Mediterranean coast to the forces of its Arab opponents, while the Egyptian forces tried to deny the same much further south in the Red Sea in the Strait of Bab el-Mandeb. The Egyptians called off operations apparently on the appearance in the area of a US aircraft carrier group. The Egyptian threat to close the Gulf of Aqaba also precipitated the Six-Day War in June 1967. Other examples of area denial were the Korean and Vietnam wars when the US sought denial of passage close to carrier groups. For example, in the Iraq–Iran War both sides had an interest in the use of the sea and had some capacity for conducting sea denial. However, Iran's dependence on the sea was more critical to its survival than it was for Iraq.

Sea Denial

Denying the use of the sea to an opponent has often been regarded as the obverse of sea control, but this is an oversimplification. Sea denial is essentially sporadic warfare at sea. In this way a markedly weaker fleet can successfully thwart a stronger fleet and strike at a time and place of its choosing to achieve maximum surprise. Sea assertion and sea denial are complementary. Sea denial can be a strategic objective at any stage of the war. It does not depend on a complementary need for sea use or control. It may be used as a means of helping to secure sea use, either in the same geographical area or elsewhere. The notion that sea denial is a means of sweeping the seas clear of enemies has not worked well in the past against a determined and resourceful opponent. The failure of offensive anti-submarine warfare in both world wars is a particularly cogent example. If one side loses sea control or does not obtain it, the other side does not necessarily gain it.[19]

A country bordering one or more enclosed seas and marginal seas of one or more oceans might select or be forced by circumstances to accomplish a combination of strategic objectives – full control in enclosed seas, while conducting a sea denial in semi-enclosed seas and parts of the adjacent oceans. This course of action was adopted by the German Navy in both world wars. Likewise, the Imperial Japanese Navy after 1943 focused on maintaining complete control of the Inland Sea, the Sea of Japan, and the Yellow Sea, while disputing control of ocean areas elsewhere in the Pacific.

The strategic objective for a major power bordering several narrow seas, but weaker at sea than the coalition of its enemies, is to obtain full control of these seas, while contesting control of certain areas of the open ocean. During the Cold War era, the Soviet Navy most likely intended to seek sea control in the Baltic Sea, Barents Sea, Black Sea, Sea of Japan, and the Sea of Okhotsk, while conducting area denial operations in other areas of the world's oceans, and specifically the marginal seas such as the Norwegian Sea, the North Sea, and the Sea of Kamchatka.

Choke Point Control

A unique feature of narrow seas is the extraordinary influence that straits and narrows – commonly called *choke points* – play in naval strategy. There are tens of thousands of straits among the world's oceans. Their importance varies depending on their geostrategic position. The total number of straits of some importance worldwide does not, however, exceed 200. Some straits are the sole exits from enclosed seas to the adjacent sea. The majority of international straits are hubs of maritime communications. The most important ocean routes from Europe to North Africa and the Middle East converge in the Strait of Gibraltar and the English Channel. The Turkish Straits and the Danish Straits are vital for the flow of trade to the riparian states of the Black Sea and the Baltic Sea. A large volume of the world's oil trade is carried through the Straits of Hormuz, Malacca, Sunda and Lombok.

By establishing control over straits and narrows in peacetime one creates the main prerequisites for gaining control in the adjacent sea or ocean shortly after the outbreak of hostilities. For a blue-water navy general command of the sea is hardly possible without establishing superiority on the open ocean and control of several passages and narrows of vital importance to the protection of international shipping.

Physical occupation or indirect control of a sea's only exit would make it possible in a short time to execute inter-theater movement, to prevent the transit of warships and merchant ships to other sea or ocean areas, and to institute naval or economic blockade of the riparian states. Artificial canals can also be used for the movements of naval forces from one narrow sea to another. The German Navy in both world conflicts used extensively the 53-mile-long Kiel Canal for the transit of warships and merchant ships from the North Sea to the Baltic and vice versa. The canal shortened the distance between Hamburg and the Baltic by 425 miles, and between London and the Baltic by 240 miles. The Soviets in the 1930s built the 500-mile-long Baltic–White Sea Canal. This waterway linked Leningrad and Belomorsk via Lake Onega. The transit for destroyers took ten to 14 days and for small vessels three to four days to complete.

The straits or narrows through which pass the shortest or the safest and most convenient routes for the movement of shipping have particularly great importance. In wartime, they constitute the most vulnerable sections of sea communications. They can be used primarily to blockade hostile naval forces and merchant shipping.

Control of a sea's only exit can be used to move one's forces from one sea or ocean area to another by the shortest route and rather quickly. This also makes it possible to swing one's fleet forces between theaters with the aim of building up forces in a specific sector of a maritime war theater.

The strait zone is the most suitable part of the seas for crossing one's ground troops. It was control of the Tsushima Strait that allowed the transport of the US 25th Division, followed by the 24th Division, across to Pusan in July 1950. Straits are also a convenient place to defeat an enemy invasion. The Koreans by controlling the Strait of Korea effectively cut off the Japanese armies from their sources of supply in 1588. Faced with starvation, the Japanese were soon in full retreat. They were able to establish themselves in a fortified strip of Korea's south coast whence, after three years of negotiations, they were able to sail home.[20]

In actions aimed to seize an archipelago, control of a strait is often critical for the ultimate success of the enterprise. The Japanese occupation of Singapore and the island of Java in 1942 gave Japan control of the Malacca and Sunda Straits.

For a country bordering several enclosed sea theaters the only way for its fleet forces to reach the open ocean is to obtain full control of the sea's only exit. The Russian home-based fleets seeking to reach the open waters of the ocean must pass through a series of nine choke points – GIUK (Greenland–Iceland–United Kingdom) Gap, Skagerrak, Strait of Gibraltar, Turkish Straits, Suez Canal, La Perouse Strait, Tsugaru Strait,

Korea Strait, Tartar Strait – all of which with the sole exception of the Tartar Straits are controlled by potentially hostile powers.

Control of a major strait or several straits can be used to cut off or isolate enemy forces in the adjacent war theater. The occupation of Denmark and Norway in April 1940 secured Germany's full control over the Baltic Sea's exits and thereby completely cut off British access to the Baltic. The Germans also used the economic resources of the Scandinavian countries for the duration of the conflict.

Sometimes it is not necessary to control both shores of an important strait or narrows but often that was not well understood and the result was a waste of sorely needed resources and more importantly a waste of time. After the capture of the bases in the Huon Gulf, New Guinea, General Douglas MacArthur thought that in order to advance further northwest toward Madang and Wewak on New Guinea's coast a full control not only of the Vitiaz Strait, but also of the Dampier Strait must be obtained. On 26 December 1943 about 12,500 troops of the US First Marine Division landed on Cape Gloucester, and six days later the entire promontory was in Allied hands. After suffering heavy losses the Allies secured a beachhead and perimeter by 16 January 1944. In retrospect, the entire operation was probably superfluous because control of one side of the Vitiaz Strait was adequate.[21]

The maintenance of command in one or another strait zone depends on establishing superiority in the air and at sea. It is almost always necessary to have available naval and air bases nearby and to have reliable communications with them. Control of a choke point is sometimes adequate to ensure control of the surrounding sea area even if negligible fleet forces are available for the task. Yet, superiority in the air is the prerequisite to achieve command of the sea. The Germans secured control of the Aegean after their conquest of Greece in May 1941 by establishing control over the straits linking the Aegean Sea and the Mediterranean. Afterwards, the Aegean Sea became inaccessible to Allied naval forces. Also, by using air bases on the island of Crete and other islands in the Aegean, the Germans paralyzed to a large degree Allied communications in the Eastern Mediterranean. The Japanese followed similar objectives in their invasion of the Netherlands East Indies in 1941–42. The invasion routes led through the straits of Borneo, Celebes, Sumatra and Java. The Japanese used land-based aircraft to obtain air superiority in the strait zone and to destroy enemy ships at sea and in their bases. The offensive rolled simultaneously from the Philippines and Malaya from one island to another until Java, the most important of all, was captured.

One of the principal differences in conducting naval strategy in a narrow sea as opposed to the open ocean is the possibility of obtaining local sea control by using even coastal defense forces. This is especially true in respect to such small bodies of the waters as straits, narrows, or channels. The Germans during their withdrawal from Sicily used their coastal gun batteries to obtain temporary control over the Strait of Messina. Between 3 and 17 August 1943 they transported more than 40,000 German and also some 62,000 Italian troops with large quantities of guns, vehicles, stores and ammunition. This effort was made mostly by day across the Strait of Messina where the Allies possessed full command of the sea and the air.[22]

The consequences of losing control of an important strait or narrows on whose shores a land combat is in progress is often fraught with dangers for one's fleet forces. In late August 1914 the German advance threatened the Channel ports of Ostend, Boulogne and Le Havre. The serious consequences of a possible British defeat at the battle of Mons led Admiral Sir John Jellicoe to warn that the Grand Fleet would shift its bases in case the Germans seized Calais and thereby obtained control of one shore of the Strait of Dover. As a direct result of possible major defeat on land, the British were faced with the prospect of a fundamental redistribution of their fleet. The British Admiralty began to make plans to evacuate all stores not immediately required by the Army. Cherbourg was favored by the War Office as a new base because of the ease with which the Cotentin Peninsula could be made impregnable so long as the British had command of the Channel. However, for really effective command of the Channel it was critical that the Flemish ports did not fall into the German hands. Therefore, the Admiralty, while pushing ahead with all preparations for transferring the base, did not want to abandon the more easterly Channel harbors, specifically Dunkirk (Dunkerque), Calais and Boulogne.[23]

The role and importance of straits and narrows in a war at sea are bound to increase steadily in the future because of the considerable expansion of ocean and sea shipping, especially for the transport of oil and its derivatives and other strategic raw materials. The scale of a coalition war is bound to enhance the importance of an inter-theater maneuver of fleet forces. The steady increases in the range and destructiveness of naval weapons will substantially expand their capabilities in strait warfare.

Control of the Air

In World War II as well as in local conflicts fought since 1945 control of the air has proved to be one of the most decisive factors in ensuring the successful outcome of the war at sea. This has been even more true in wars waged in narrow seas. Because of the increased range, endurance and speeds of modern aircraft the ever-larger sea and ocean areas have become the area of combined employment of naval forces and land-based aircraft. Today, no part of any narrow sea is free from observation and attack from the air. Struggle for air superiority in narrow seas is closely linked to that over the littoral area. Control of the air is a prerequisite to sea control in an enclosed maritime theater. A fleet can conduct sustained actions only when there is control of associated airspace. The British learned in many campaigns fought in World War II – notably in Norway in 1940, Malaya in 1941–42 and in the Mediterranean in 1940–43 – that any attempt to use ground troops where the adversary held air superiority was very costly, and in many cases doomed to failure.

Sometimes control of the air over a part of the sea or ocean can in itself be sufficient to exert a decisive influence on the outcome of the struggle on land. The victory of RAF Fighter Command in the summer of 1940 not only prevented the *Luftwaffe* from gaining mastery of the air over the Channel, but brought the German leadership face to face with the unpleasant fact that their planned invasion could not be carried out as long as the *Kriegsmarine* was totally inadequate to the task of obtaining a sufficient degree of sea control to allow the invasion troops to sail with any prospect of success.

In contrast to the war in the Atlantic, the struggle for control of the Mediterranean in World War II depended primarily on airpower, and therefore on the possession by one side or the other of land bases from which to operate aircraft. Hence, the island of Sicily at once assumed a dominant position in the war. Sardinia also played a major part and Malta's importance as an air base came fully to equal its importance as a naval base.

By May 1941, the Germans were in full control of Greece and the Aegean islands. Crete, however, remained a threat to their southern flank. Although the Allies still held control of the sea, the Germans had practically undisputed control of the air.[24] The plan to seize the island of Crete (Operation *Merkur* – Mercury) was approved by the German Supreme Command on 25 April 1941, the day after Allied evacuation from Greece began. The attack on Crete from the air started on 20 May. The entire area between Crete and the African coast, called 'Bomb Alley', was subsequently dominated by the *Luftwaffe*. Even if the Allies could have held Crete, its resupply would have been virtually impossible. The Germans used their air superiority to overcome Allied resistance ashore and drive away British ships. The ultimate capture of Crete was a landmark in the history of war: never before had a territory only accessible by sea been overrun by a power not possessing even local sea superiority.[25]

After the fall of Crete, the British, although enjoying superior naval strength, did not hold control of the Eastern Mediterranean, except for a small area. The reason for this was German superiority in the air. The

Italians, although superior in surface ships in the central Mediterranean, could not prevent the loss of their merchant ships inflicted by the British submarines and aircraft. The German reoccupation of the island of Léros (Dodecanese Islands) in November 1943 was due to the overwhelming air superiority of the *Luftwaffe*. However, the margin of the German victory was extremely narrow.[26]

To obtain superiority in the air, it is necessary to possess or to seize a number of suitable air bases; otherwise, that objective cannot easily be accomplished if at all. The British in their attempt to seize the Dodecanese Islands in the fall of 1943 were unable to establish satisfactory advanced air bases on either Léros or Cos, since both islands were situated within easy range of the *Luftwaffe*'s aircraft based on the Greek mainland, on Crete, and on Rhodes.[27]

In early 1943, the situation in the central Solomons was similar to that which prevailed around Guadalcanal in 1942. Allied command of the air was so complete that the Japanese did not dare to use their surface forces in daylight, but during the night the waters of New Georgia were hotly contested. The Japanese repeatedly tried to run reinforcements, while Allied cruisers and destroyers tried to intercept them.[28] By mid-1943 Allied forces stood poised and ready to resume the offensive in the Solomons. Although the Allies possessed air superiority in the area, they lacked well-placed and sufficient airfields in the archipelago. For this it was necessary to seize the islands north of Guadalcanal so that fighters could accompany bombers sent to attack key Japanese positions around Rabaul in New Britain. The Japanese, by contrast, possessed an excellent chain of airfields stretching south from New Britain to New Georgia in the Solomons, and west to Lae, Salamaua, Madang and Wewak on the northern coast of New Guinea. It was clear that control of the air over Rabaul could never be gained until the Japanese had been driven from at least some of their positions.[29] Control of the waters around the Bismarck Archipelago passed to Allied aircraft in 1943. The success of Allied neutralization of Rabaul from September 1943 through April 1944 was predicated on obtaining local air superiority.[30]

Control of the air in combination with control of the surface almost surely results in opening up the possibility for success by ground troops on the coast. Allied air superiority combined with a high sorties rate of combat aircraft resulted in such a weakening of the German hold on the Yugoslav coast in the spring of 1944 that a drive up the Adriatic and into central Europe became a real possibility.[31]

Command of the air in itself sometimes cannot ensure success in a war at sea especially if the entire effort must be carried out on a sustained basis. In 1941, Admiral Raeder urged the Italian Navy to increase its activity and to attack Allied communications in the Aegean, and thereby protect the German southern flank. The Germans possessed superiority in the air in the Central Mediterranean, but they had no hope of obtaining command of sea surface. Yet the Italians were quite unwilling to risk their fleet because of their perceived uncertainty of air support.[32]

Strategic objectives in wartime in a typical narrow sea to be accomplished by a blue-water navy or by the navy of a riparian state considerably differ in scope, duration of actions, and means used. Besides fleet forces, land-based aircraft play a critical role in narrow seas for obtaining control not only in the air and surface, but of the subsurface as well. Sometimes, it is possible that land-based aircraft can obtain the principal objectives of naval strategy alone or with relatively small contribution from one's naval forces. Yet the outcome of the war in a narrow sea is largely decided by the combat on the ground. Also, by seizing the enemy bases from the landside, one's ground troops can obtain local or even absolute command of the sea in a respective area or theater of operations. However, that control will be obtained much quicker if both fleet forces and ground troops cooperate closely with each other. Coastal defense forces alone can sometimes achieve local and temporary control of small but critically important bodies of water such as straits or river estuaries.

A prerequisite for obtaining permanent control of a narrow sea by a blue-water navy is first to secure control of the sea's exit to another narrow sea or to the open ocean. In today's terms, this control doesn't necessarily mean physical control of the exit, but can be limited to obtaining control of the airspace above it. Clearly, full control of the exit in all three mediums is much preferable. Otherwise, although a blue-water navy can enjoy control of the adjacent sea or the ocean area, it would be unable to exert the necessary degree of control within a given narrow sea. A coastal navy, in contrast, has an easier task to frustrate a stronger fleet to obtain full and permanent control of a narrow sea. Its main objective will often be defensive and aimed to contest control of the sea.

A stronger navy should initially aim to obtain sea control in a major part of a narrow sea, thereby having freedom of action to carry out its principal tasks. However, a weaker fleet in combination with land-based aircraft and coastal defense forces can sporadically, and possibly even permanently, deny the necessary degree of control to its opponent. This is especially true if a weaker side or a neutral country controls one or both shores of the sea's only exit. Most often a superior fleet must control not only the sea's exit(s) but also several operationally significant positions within the confines of a narrow sea. In modern times, choke point control must often be combined with sortie control to obtain and maintain sufficient degree of control in a narrow sea. Today, no fleet can safely operate for any length of time in a narrow sea without superiority in the air.

NOTES

1. Julian S. Corbett, *Principles of Maritime Strategy* (Annapolis, MD: Naval Institute Press, 1988), p. 91.
2. 'MCM in the Gulf – More Information Comes to Light', *NAVINT: The International Naval Newsletter* (London), 3, 8, 26 April 1991, p. 1.
3. Harold M. Sprout and Margaret T. Sprout, *The Rise of American Naval Power 1776–1918* (Princeton, NJ: Princeton University Press, 1939), pp. 351–2.
4. Juerg Meister, *Der Seekrieg in den osteuropaeischen Gewaessern 1941–45* (Munich: J. F. Lehmans Verlag, 1958), p. 340.
5. Ibid., p. 336.
6. Ibid., p. 265.
7. Ibid., p. 336.
8. Michael Salewski, *Die deutsche Seekriegsleitung 1935–1945*, Vol. 1: *1935–1941* (Frankfurt a.M: Bernard & Graefe, 1970), p. 399.
9. Ibid., *Die deutsche Seekriegsleitung 1935–1945*, Vol. 2: *1942–1945* (Frankfurt a.M: Bernard & Graefe, 1975), p. 449.
10. R. C. Anderson, *Naval Wars in the Baltic during the Sailing Ship Epoch, 1522–1854* (London: Gilbert-Wood, 1910), p. 47.
11. David Woodward, *The Russians at Sea: A History of the Russian Navy* (New York: Frederick A. Praeger, 1966), p. 85.
12. Donald Macintyre, *Sea Power in the Pacific: A History from the 16th Century to the Present Day* (London: Military Book Society, 1972) p. 119.
13. Sprout and Sprout, *The Rise of American Naval Power, 1776–1918*, p. 232.
14. Meister, *Der Seekrieg in den osteuropaeischen Gewaessern 1941–45*, p. 244.
15. Salewski, *Die deutsche Seekriegsleitung 1935–1945*, Vol. 1, p. 336.
16. Stephen W. Roskill, *The War At Sea 1939–1945*, Vol. III: *The Offensive*, pt. 2: '1st June 1944–14th August 1945' (London: Her Majesty's Stationery Office, 1961), p. 137.
17. Stephen W. Roskill, *The War At Sea 1939–1945*, Vol. I: *The Defensive* (London: Her Majesty's Stationery Office, 1954), p. 305.
18. Woodward, *The Russians at Sea*, p. 217.
19. Mitchell B. Simpson, III, ed., *The Development of Naval Thought: Essays by Herbert Rosinski* (Newport, RI: Naval War College Press, 1977), p. xix.
20. Macintyre, *Sea Power in the Pacific*, p. 52.
21. Samuel E. Morison, *The Two-Ocean War: A Short History of the United States Navy in the Second World War* (Boston, MA/Toronto: Little, Brown, 1963), p. 292; Roskill, *The War At Sea 1939–1945*, Vol. III, *The Offensive*, pt. 1, '1st June 1943–31st May 1944' (London: Her Majesty's Stationery Office, 1960), p. 228.
22. Roskill, *The Strategy of Sea Power: Its Development and Application* (1981), p. 203; Walter Baum and Eberhard Weichold, *Der Krieg der Achsenmaechte im Mittelmeer-Raum* (Goettingen: Musterschmidt Verlag, 1973), p. 337; I. C. B. Dear, gen. ed., *The Oxford Companion to World War II* (Oxford/New York: Oxford University Press, 1995), p. 1001.
23. Julian S. Corbett, Vol. 1: *To the Battle of the Falklands, December 1914*, in *Naval Operations: History of the Great War Based on Official Documents* (London: Longmans, Green, 1920), p. 96.
24. S. W. C. Pack, *Sea Power in the Mediterranean: A History from the Seventeenth Century to the Present Day* (London: Arthur Barker, 1971), p. 189.
25. J. R. M. Butler, Vol. 2: *Sept. 1939–June 1941*; John P. W. Ehrman, *Grand Strategy* (London: Her Majesty's Stationery Office, 6 vols, 1956–76), p. 515.

26 Roskill, *The War At Sea 1939–1945*, Vol. III, *The Offensive*, pt. 1, '1st June 1943–31st May 1944', pp. 202–3.
27 Ibid., pp. 192, 205.
28 Roskill, *The War At Sea 1939–1945*, Vol. III, *The Offensive*, pt. 1, '1st June 1943–31st May 1944', p. 229.
29 Ibid., p. 223.
30 Macintyre, *Sea Power in the Pacific*, p. 160.
31 Roskill, *The War At Sea 1939–1945*, Vol. III, *The Offensive*, pt. 1, '1st June 1943–31st May 1944', p. 328.
32 Pack, *Sea Power in the Mediterranean*, p. 187.

8

Methods

Modern methods of combat force employment are the result of a long evolution of warfare. In the nineteenth century, 'decisive' battles were the area of study and practice of *tactics*, while *strategy* was concerned with the conduct of campaigns. In that era, campaigns were conducted primarily by the armies, although there are examples where navies took part as well. In generic terms, the principal methods of combat force employment today are *naval tactical actions, major naval operations* and *maritime campaigns*. The objectives they are intended to accomplish and the corresponding command echelon responsible for their planning and execution differentiate the terms.

Naval Tactical Actions

Naval tactical actions is a cumulative term for the employment of one's naval forces ranging from patrolling and surveillance to a battle. These actions take place in given combat zones/sectors, but in some cases can encompass an area of operations. Naval tactical actions are primarily intended to accomplish a certain major or minor tactical objective. Also, when conducted over time, these actions can cumulatively accomplish operational objective(s). The principal methods of tactical employment of naval forces and aircraft in general are attacks, strikes, engagements, and battles. An *attack* is a combination of tactical maneuver and fire conducted by a single or several platforms aimed to accomplish a *minor tactical* objective. The types of attacks are differentiated in terms of platforms and weapon systems used. Thus, an attack could be conducted by a surface ship, submarine, and aircraft firing missiles, torpedoes, ASW weapons, or dropping bombs. An attack is normally an integral part of an engagement or a naval battle, but can also be fought independently.

A *strike* is today probably the most important method of employing naval forces to accomplish tactical objectives. It has emerged due to the introduction of longer-range, highly precise, and more lethal weapons, notably anti-ship and cruise missiles and 'smart' bombs. A *naval strike* consists of several simultaneous or sequential attacks, conducted by a single or multiple platforms, synchronized in terms of time and place, to accomplish *tactical* and sometimes even *operational* objectives. It can be either defensive or offensive in its purpose. A strike can be conducted by the maneuver of a platform (ship, submarine or aircraft), or a maneuver by fire in which fire is shifted from one set of targets to another. The success of a strike is greatly enhanced by covert and swift maneuver and mutual support among the platforms or force elements taking part in it. A strike is usually conducted by a small number of platforms of a single-type force, e.g., missile craft, submarines, or ground-attack fixed-wing aircraft (helicopters). In a narrow sea and especially in the archipelago-type coast, missile- or torpedo-armed surface combatants can carry out strikes from an ambushing position against a much stronger enemy force.

The main feature of a well-planned and conducted strike is shortness of action – often measured in minutes – swift and skillful maneuvering of platforms and almost simultaneous employment of weapons, specifically missiles and/or torpedoes, against the perceived weaknesses in the target defenses. A strike can be carried out by using missiles, torpedoes, guns, bombs or their combination. In a combined strike, the longer-range weapons are used first to weaken the enemy's defenses, followed by the employment of short-range weapons to finish off the target. A well-prepared and skillfully executed strike could quickly and decisively defeat the opponent. In general, the success of one's strike can be anticipated with a higher degree of certainty than a naval battle. Its importance is likely to increase as the new, ever-longer-range, more precise, and more lethal weapons are introduced. A strike in a typical narrow sea may well replace a naval battle as the principal method of employment of coastal navies to accomplish major tactical objectives.

A broader form of strike is a *naval raid* – conducted by a single or several combat arms to accomplish a tactical objective by temporarily denying some position or capturing or destroying enemy forces or coastal installations/facilities. A naval raid can range in scale from an action by a very small force to a rather large tactical action. Sometimes a naval raid may have many features of a major naval operation.

The inferior side usually conducts raids at sea. Temporary or even local control of the sea is not a prerequisite for the success of these actions. The stronger fleet also can conduct feints to divert the enemy's attention or forces him to move forces in the direction of its own sector of secondary effort.

The most immediate effect of a raid is temporary, but its consequences can be significantly larger than initially intended if the opponent reacts operationally in terms of place or time advantageous to one's own forces.

One purpose of a raid can be to divert the enemy's attention or forces or both from one's own sector of main effort. Besides purely physical objectives, a raid is conducted to enhance one's own morale and diminish that of the enemy. The main purpose of a raid should be directed against an objective considered by the enemy as so valuable that its loss or serious degradation cannot be ignored.

Normally, a raid is conducted in an area relatively far away from the sector of the enemy's main effort. A successful raid should convince the enemy that similar actions are likely in the future thereby forcing him to divert significant forces for defense of a larger area of the coast. Temporary and local control of the sea is not always a prerequisite for the planning and execution of a raid.

The forces participating in a raid should not unduly weaken one's forces operating in the main sector of effort. The chances of success are greatly enhanced if multiple objectives and multiple lines of operations for both advance and retreat are selected. The force conducting a raid should be stronger than the force immediately available to the enemy in the area. One's forces should move, if possible, covertly and rapidly towards their assigned physical objectives. Support of a larger force in the area is often required to ensure the success of a raid.

A larger action than a single raid is a *naval surging action* – a series of surprise, short, fast-moving strikes carried out simultaneously or in quick sequences by surface forces and/or aircraft against enemy bases/ports and important coastal installations/facilities.

An *engagement* consists of a series of related strikes and attacks conducted by a single or several combat arms and usually aimed to accomplish the principal tactical objective in a naval battle. In practical terms, an engagement is planned to destroy or neutralize the main part of the enemy force. An engagement lasts longer than an attack or strike, but is shorter than a naval battle. Sometimes, an engagement can occur without being a part of a naval battle and without a plan.

A *naval battle* consists of a series of related engagements, strikes, and attacks, conducted by several combat arms, synchronized in terms of time and place, to accomplish a *major tactical* (and, sometimes, an *operational*) objective in a given area of operations. It is normally planned, prepared, and executed by a single commander. A naval battle could, however, sometimes be fought without a plan drawn in advance. It is the predominant method of combat employment of naval forces and forces of other services in a typical narrow sea. In the Leyte operation in October 1944, the Allies planned the battle of Surigao Strait, while another three battles, in the Sibuyan Sea, off Samar and off Cape Engano, occurred unplanned.

A naval battle can be tactically won, but may represent an operational (or even strategic) failure nevertheless. The battle of Balikpapan fought on 24 January 1942 was a *tactical* success, but an *operational* failure for the ABDA forces, because it did not stop the Japanese invasion of Borneo. The Battle of the Coral Sea in May 1942 (Operation *MO* for the Japanese) was tactically a draw, but there was no doubt that operationally it was an Allied victory, because the Japanese abandoned their attempt to seize Port Moresby by the assault from the sea.[1] In the battle of Savo Island on 9 August 1942, Japanese Admiral Gunichi Mikawa succeeded in inflicting a *tactical* defeat on the Allied force, but failed to accomplish his major tactical objective – the destruction of US amphibious forces at Lunga Point anchorage – which had operational consequences. Likewise, in the Battle of Cape Esperance on the night of 11–12 October 1942, the Americans achieved a tactical victory, but failed to prevent the Japanese from landing fresh troops ashore from two seaplane tenders. In contrast, the battle off Guadalcanal on 12–15 November 1942 was tactically decisive, because it forced the Japanese troops on the island to be on the defensive. Afterwards, these troops were supplied sporadically by Japanese destroyers, while the Americans were able to bring in two fresh regiments of Marines and one infantry division. One month later the Japanese abandoned the hope of recapturing Guadalcanal and began preparations for withdrawal. Subsequently, the Americans went on the offensive to seize successive positions along the Solomons chain.[2]

In a typical narrow sea, conducting diverse tactical actions and measures ensures basing/deployment area control.[3] They are also aimed to prevent operational or strategic surprise. Another objective of these actions is to put the enemy under constant pressure and tie his forces in a certain sea area, thereby creating conditions to change the operational situation to one's advantage. Additionally, a number of protective measures to enhance security and survivability of one's forces and coastal installations/facilities are prepared and conducted as well. Cumulatively, these actions are intended to accomplish an operational objective over time. They are planned and conducted both in peacetime and in case of national emergency and war. They are also normally an integral part of a plan for a major naval operation. Defensive tactical actions include reconnaissance and patrolling/surveillance in one's coastal waters, air defense, ASW, and defense against enemy combat craft, combat swimmers, and commandos. Defensive mining and both defensive and offensive mine counter measures (MCM) in coastal waters are an important part of these actions. Offensive actions encompass strikes or attacks against enemy surface combatants threatening one's bases and ports, naval/air bases and ports, and other installations/facilities on the coast. Protection of one's basing and deployment areas is significantly enhanced by conducting diverse passive and active measures, specifically, electronic warfare (EW), countering enemy reconnaissance/surveillance, and cover and concealment.

METHODS

Major Naval Operations

In generic terms, a *major naval operation* consists of a series of related battles, engagements, strikes, and other tactical actions conducted concurrently or sequentially by diverse naval and combat arms of other services, coordinated in terms of time and place, to accomplish *operational*, and sometimes *strategic*, objectives in a given maritime theater. Major naval operations emerged over the course of many decades as a result of cumulative advances in naval technology and also, concurrently, changes in the character of war on land. By the end of the nineteenth century, naval technological innovations, specifically the introduction of ever longer-range and more lethal and more accurate guns, combined with the larger size, higher speed, and the greater sailing range, endurance, and survivability of principal warships, made it very difficult for any fleet to destroy a major part of the enemy fleet in a single major battle. These advances also led to the proliferation of small warships and thereby to significant increases in the numbers of ships in the world's foremost navies. Great improvements in propulsion plants made it possible to fit powerful engines even on board very small ships. The introduction of torpedoes and mines led to the design of new, small platforms capable of attacking even large ships, which in turn required the construction of other types of combatants to counter the emerging threats. On the other hand, while steam offered warships higher speed, it also initially resulted in considerably smaller ship ranges and endurance. The tactical mobility of warships increased, but their strategic mobility was greatly reduced. Fleets became progressively more dependent on shore bases for fuel and supplies and on auxiliary ships for carrying coal. By the turn of this century, large fleets consisted of numerous ships of all sizes and carrying various type of weapons.[4] They were also deployed over large areas. The result of all these advances was that it became impossible to destroy a major part of a large fleet in a single battle.

Major naval operations are conducted as an integral part of a maritime, and sometimes a land campaign. With respect to the size and character of the physical environment, major naval operations can be conducted on the open ocean, the marginal seas of an ocean, or in enclosed or semi-enclosed seas. Major naval battles on the open ocean have been rare in practice. During World War II, even in the maritime theaters, most naval actions took place fairly close to some large archipelago or continental landmass. In contrast to small enclosed seas such as the Baltic or the Black Sea, a large number of major naval operations have taken place in semi-enclosed seas, such as the North Sea, or larger enclosed seas, such as the Mediterranean.

Today, because of enormous advances in the range and lethality of weapons, even coastal navies, composed of small surface ships, submarines, and land-based aircraft, can conduct major naval operations. These operations in a typical narrow sea will be more complicated to prepare and execute than those on the open ocean because they will involve diverse forces of all services of the country's armed forces.

Major naval operations in a typical narrow sea differ from those on the open ocean not only because of the more limited operational space but also because of the peculiar features of the physical environment. *Operational space* in which a major naval operation is conducted encompasses not only high seas and coastal waters controlled by one's own or enemy forces, but also includes islands and the mainland coast. Also, a major naval operation in a typical narrow sea would include a larger number of smaller ships and land-based aircraft than would operations conducted on the open ocean.

In contrast to naval battles or engagements, major naval operations are normally planned to accomplish a single *operational objective*, that is, the one that might bring about a drastic change of the *operational situation* in a given maritime theater of operations. Such a change can be brought about, for example, by destroying a major part of the enemy fleet at sea or in its bases, destroying a large enemy convoy or successfully protecting one's own large convoy, seizing or defending a major island or a choke point. Sometimes a major naval operation can also be planned to accomplish a *strategic objective*. This usually occurs when a surprise attack on a major part of the enemy's fleet is carried out at the very outset of the hostilities and aimed at the accomplishment of a strategic objective in a principal maritime theater of operations. A major naval operation could be planned to accomplish a strategic objective in the first phase of a new campaign by opening a new front, as operation *Neptune* did for the Allied campaign in France in 1944–45 (Operation *Overlord*).

In general, a major naval operation is planned and conducted when *decisive* results are to be accomplished, in the *shortest* time and with the least loss of one's forces.[5] Such operations are the most effective ways to avoid attritional warfare at the operational or strategic level. Attritional warfare is also inherently protracted. It ties down one's forces in the course of accomplishing a specific objective in a given sea or ocean area, thereby preventing them from carrying out other operational tasks.

A major naval operation with an offensive purpose is usually planned and conducted when one's forces possess at least temporary sea control in a given maritime area of operations. However, major naval operations to protect one's own or to attack enemy maritime trade will be conducted even if command of the sea is under dispute. There may also be the case, that the situation is so desperate for a weaker side, that a major operation to protect a convoy must be conducted regardless of the enemy's control of the sea and the potential losses. The key to success then is to secure at least control of the sea surface, subsurface, and the air in the proximity of a large convoy. The experiences of World War II show that it is also possible to plan a major naval operation when one's forces possess only control of the air. The fact that the enemy possesses control of the sea and air will not necessarily prevent the inferior side, under certain favorable conditions, from mounting a major offensive naval operation. Sometimes an amphibious landing operation can be carried out when sea control is only local and temporary.

Normally, a major naval operation should not be planned if there is a serious mismatch between ends and means. It should also not be conducted if there are not sufficient forces to conduct a consecutive major operation after the initial operational objective is accomplished, as happened in the Allied landing on Guadalcanal in August 1942.

The success of a major naval operation usually cannot be ensured without an adequate theater-wide infrastructure or operational functions to provide comprehensive support. These functions are command and control, operational intelligence, command and control warfare (C2W), logistical support and sustainments, and protection. Obviously, only some of these functions will require proper sequencing and synchronization during support of a major naval operation.

Operational intelligence perhaps represents one of the most important prerequisites for the success of any major naval operation. The operational commander can make a sound decision only with a comprehensive and realistic picture of the situation. The covertness of the preparations for action of one's forces must be ensured. To deceive the enemy as to the main physical objectives to be accomplished, operational intelligence efforts should appear to be routine.[6]

Another important element for success of a major naval operation is operational protection, particularly of one's own center of gravity. One of the major reasons for the heavy losses suffered by the Japanese First Diversionary Attack Force in the Battle of the Sibuyan Sea on 24 October 1944 was an almost total lack of air cover by land-based aircraft. Basing and deployment areas of naval forces and aircraft taking part in a major naval operation must be fully protected. This in turn can be ensured by conducting offensive and defensive tactical actions, such as surveillance of one's coastal waters, laying down defensive minefields, conducting air defense of naval bases and anchorages, launching attacks against enemy forces threatening one's fleet forces, and so on. The ultimate aim of these actions is to establish and maintain a favorable operational situation in one's naval basing and deployment areas.

Major naval operations are normally planned, prepared and conducted by the *operational commander*. Sometimes, however, an operational-tactical (joint task force) commander and his staff can also plan it. These commanders also have the directive authority for logistical support and sustainment. In contrast to a tactical commander, operational commanders also have the authority to sequence and synchronize actions of all the services and force elements under their command.

In making a decision to conduct a major naval operation the estimate of the situation must take into account not only purely naval but also all other aspects of the situation in a given theater of operations – airspace, topography, oceanography/hydrography, climate/weather, and sometimes even political, economic, technological, and others – for several weeks in advance. Because of the much wider range of uncertainty involved and the larger scope of a major operation, the commander's estimate of the situation necessarily contains a larger number of planning assumptions. Consequently, the more assumptions a given plan contains, the more likely it is to require radical changes or even to be completely abandoned during the execution phase.

The main features of a modern major naval operation are decisiveness of action, large scope, high intensity of combat, and participation of diverse, relatively large forces. Also, modern major naval operations are characterized by extensive use of electronic warfare techniques.[7]

A major naval operation lasts considerably longer than a naval tactical action. While a naval battle or engagement is fought for several hours at most (and often even far less than that), a major naval operation can last several days or even several weeks. A major naval operation lasts longer if the combination of large distances between the base of operations and physical objectives, an unfavorable ratio of forces, and logistical constraints, or a lack of operational thinking leads to indecisiveness. Because of the enormous increase in combat power of naval platforms, major naval operations of the future are likely to be highly intense. This would be especially the case in narrow seas, where the combination of a small area and short distances would result in more intense employment of diverse platforms and weapons. The actions at sea and in the air will occur almost continuously. For that reason in major naval operations today extremely high consumption rates of fuel and ammunition should be expected.

Conducting a major naval operation in a narrow sea is far more complex than on the open ocean. The short distances allow an agile and aggressive enemy actively to hinder or even prevent preparations and deployment of one's forces. A covert deployment or redeployment of naval forces or aircraft is difficult to organize and execute in a typical narrow sea. Major naval operations in a narrow sea are also more complicated to plan and prepare than those on the open ocean because of the often high degree of participation of the combat arms of several services. Not only naval forces, but also ground, air and coastal defense forces and special forces would take part in such operations.

Major naval operations can be differentiated with respect to their main purpose, relative participation by services, timing, and the sea or ocean area in which combat actions take place. As to their principal purpose, major naval operations can be offensive or defensive. An *offensive* major naval operation is designed to expand the area under permanent control of one's forces and thereby reduce the area under enemy control. It is planned and conducted in a place and at a time of one's own choosing. Major naval operations against an enemy fleet at sea or in its bases, enemy maritime trade and amphibious landing operations are inherently offensive. A *defensive* major naval operation is, in terms of its objective, planned and conducted when an uninterrupted offensive is not possible or when one's forces need an operational pause after the completion of an offensive operation. Also, an inferior fleet can take a defensive posture from the very outset of the hostilities. A major naval operation intended to protect one's maritime trade or to defend a part of the coast against an enemy amphibious landing is, by its very nature, defensive in character.

With respect to the degree of service participation, independent, joint (multi-service), and combined (multinational) major naval operations are differentiated. In a typical narrow sea, an *independent* major naval operation is conducted *predominantly* by the navy, although air forces or even ground forces can take part as well. The British carrier attack on the Italian main naval base of Tarent (Taranto) on the night of 11–12 November 1940 (Operation *Judgment*) was an example of an independent major naval operation.

A *joint* major naval operation is planned and conducted by forces of the navy and one or more other services and, often, services of allies or coalition partners. In a theater encompassing a large ocean or sea area, major naval operations will be conducted with significant participation by the air force, while the ground forces may also be involved. In contrast, major naval operations in a typical narrow sea are less naval in their character because of the participation of the combat arms of all three services. Examples of major joint naval operations in a narrow sea are the US amphibious landings on Guadalcanal (Operation *Watchtower*) in August 1942 and the German Operation *Knight's Move* (*Roesselsprung*) against the Allied convoy PQ-17 to Russia in July 1942.

A *combined* major naval operation is conducted with navies or other services of two or more countries. Today, combined major operations constitute a frequent method of employing naval forces in a low-intensity conflict. In fact, because of the downsizing of most navies, such major naval operations might well become routine for major regional contingencies or general conflict. A major amphibious landing operation is usually included in the plans. Examples of such major operations conducted in narrow seas are the Allied invasion of Sicily in July 1943 (Operation *Husky*) and the invasion of Leyte in October 1944 (Operation *King II*). Such operations can also sometimes be both joint and combined (as was the Coalition's air offensive against Iraq in the Gulf War of 1991, in which both land and carrier-based aircraft of several nations were involved).

Sometimes several major operations can take place in a single maritime theater of operations. Then, one operation is the *main* or *principal* and others are *supporting* (*ancillary*) major naval operations. In terms of timing or sequencing, *initial*, *successive* (*consecutive*), and *preliminary* major naval operations can be distinguished. A *successive* major operation normally starts during or shortly after the completion of the *initial* major operation. Its objective is based on what has been accomplished in the preceding major operation.

A major naval operation will be the principal method of accomplishing operational and sometimes even strategic objectives in the case of a major regional conflict or general war. Most of these operations will be joint or combined, and relatively few will be conducted either largely or exclusively with naval forces. In generic terms, the main purposes of a major operation today could be to:

- Destroy enemy fleet at sea or its bases
- Amphibious landing on opposed shore
- Destroy enemy coastal installations/facilities
- Interrupt/cut-off enemy maritime trade
- Protect and defend one's own and friendly maritime trade
- Naval blockade
- Naval counter-blockade
- Destroy enemy sea-based strategic nuclear forces
- Protect sea-based strategic nuclear forces
- Support ground forces on the coast.

Naval forces can be employed for a variety of tasks in a low-intensity conflict, because they are better suited than other forces for crisis prevention and peace operations. They also can be employed for carrying out such less complicated tasks as humanitarian assistance, disaster relief, and search and rescue. In general, naval forces involved in such might conduct following:

- Conflict prevention
- Protection of one's own and neutral shipping
- Freedom of navigation and overflight
- Peace operations.

Any major naval operation encompasses several major stages of force employment, ranging from operational deployment, through combat employment and sustainment, to operational redeployment. In some situations, a major naval operation may include strategic deployment or even mobilization of ready reserves.

Deployment of one's naval forces is normally conducted from the base of operations to the area of concentration. However, sometimes naval forces need to be concentrated first and then deployed to their objective areas. Selection of a base of operation can have a considerable effect on the course and outcome of one's major naval operation. In general it determines the position, the directional orientation, and the length of lines of operations. The direct route between a given base and the assigned physical objective is the shortest and usually the best. However, a circuitous route must often be used to avoid a strong enemy position or to enhance the element of surprise. This is especially true in seas possessing a large number of offshore islands or archipelagoes.

The essence of any deployment is speed of movement, and that is dependent on the distances involved and the mobility of one's forces. In contrast to ground forces, naval forces sometimes carry out their deployment in waters under enemy control. Because of the short distances involved, deployment of naval forces in enclosed or semi-enclosed seas is usually tactical. In contrast, forces taking part in a major naval operation on the open ocean or in a marginal sea deploy from several naval and air bases, spread sometimes over several hundred nautical miles.

To reduce the time required for operational deployment, it is highly advantageous to have already deployed a major part of one's forces in a forward position, as the US Navy has in many areas of potential crisis and conflict. In contrast, a small navy would deploy within its own coastal waters or sometimes to sea areas under the enemy's control. All the forces taking part in a major naval operation in a narrow sea can be deployed from their bases simultaneously or sequentially, and their lines of operations are generally much shorter. These forces would often operate along multiple lines of operations, allowing them literally to 'pop out' and execute their assigned tasks at high speed.

A major naval operation can be conducted along *interior* or *exterior* lines of operation. One's forces, if moving on interior lines, can be interposed between two or more parts of a hostile force, and a part of that enemy force can be contained in a secondary sector. A major part of one's own force would then seek to concentrate as quickly as possible in the sector of main effort to destroy the bulk of the enemy force.

A major naval operation conducted along exterior lines consists of concentric advances from several areas towards common physical objectives. A force operating along exterior lines must not only move at higher speeds than a hostile force operating along interior lines but its adjacent elements must keep within mutual supporting distance if it is to avoid being defeated in detail. Such a major naval operation also requires more precise synchronization of movement and actions of one's forces. This is a most difficult thing to do, especially when the distance between one's base of operations and the physical objective is great. Once the forces start to move, the operational commander has to rely on decentralized command and control.

The key to success in combat lies in rapidly achieving superior combat potential at the decisive time and place. This can be accomplished only by concentrating one's forces in the sector of main effort, especially at the point of main attack (or defense). A force is said to be concentrated when all its elements are within supporting distance of each other, or, in other words, when adjacent elements of the force can apply their full strength in coordination against the appropriate part of the enemy force.[8] Concentration of naval forces is usually conducted in the sea/ocean area open to enemy attack or in areas where control is in dispute.[9] Depending on its purpose and scope, concentration can be tactical, operational, and strategic. *Operational concentration* is planned and conducted as a part of a major naval operation or campaign. Thus, it normally takes place in a given maritime area of operations.[10] Despite its obvious advantages, however, concentration makes one's forces vulnerable to enemy fire, which would in turn require dispersal.[11]

A plan for concentration of a force for a major naval operation should include a scheme of maneuver, sequence of movements, rate of concentration, and control of movements. It should also provide for flexibility, secrecy, and security. The scheme of maneuver is the most important element in planning a concentration of naval forces. Command and control of a force must be flexible during the movement to the concentration area. Accurate, reliable, and above all relevant and timely intelligence is the key to a successful concentration of one's naval forces. Alternative plans for the movement must be prepared in a timely manner. Secrecy is another critical part in planning and preparing naval forces to concentrate in a given area of operations. The principal means of enhancing secrecy in planning and carrying out concentration are operational deception, concealment and cover, and the establishment of air superiority in the part of the theater where concentration is to take place. In general, the larger the force taking part in a major naval operation, the more difficult it is to ensure covertness of movement and secrecy in conducting a concentration.

The rate of concentration is the speed at which the force is assembled, and greatly influences the success of the forthcoming operation. In general, the rate of concentration must be faster than the enemy's ability to counter. Thus, the factor time is critical for success. However, this factor should not lead to the faulty disposition of one's forces in the concentration area. Any such error would necessarily require shifting of one's forces to a new area and thereby result in nullifying any gain of time. A rate of concentration must also ensure that concentration is carried out smoothly and without confusion. Sequence of movement is closely related to rate of concentration. It entails such factors as which forces are to arrive in the area first and what supplies are available or can be brought into the area. The operational commander must not lose sight of the logistical implications of the concentration of forces.

Operational concentration of forces for a major naval operation can take place on or off the 'battlefield'. When force elements are brought within supporting distance *before* arriving in the area of operations, they are said to have been *concentrated off the battlefield*. If they are grouped within a supporting distance at the time the combat action is about to start, or shortly after it has begun, they have *concentrated on the battlefield*. When various elements of forces do not arrive within supporting distances until they actually become engaged with the enemy, this type of concentration is carried out in the form of a maneuver. Concentration on the battlefield enhances the factor of surprise. It also makes it difficult for the enemy to deduce where the next blow will come from until it is too late for him to react. Such a concentration also requires meticulous planning and coordination of the movement of naval forces and aircraft. In addition, the opponent operating from a central position has the opportunity to defeat the converging forces in detail, especially when the advance of some elements of the converging forces is delayed due to error in timing, to bad weather, or to some other reason. Concentration on the battlefield is often used to defeat a large part of the opposing fleet. This was the case in the Japanese invasion of the Netherlands East Indies in 1941–42. The Japanese Combined Fleet also tried to carry out concentration on the battlefield in their ill-fated operations in the defense of the Philippines in October 1944.

In a war at sea, concentration off the battlefield is most often used in conducting amphibious landing operations. The amphibious task force, the forces providing direct support, and those in distant (operational) cover and support sail from their staging areas within mutually supporting distance to the assigned debarkation points in the amphibious objective area (AOA). There are some exceptions to this, as in the amphibious landing operation at Leyte on 20 October 1944, during which Task Force 38 (TF-38) acting as distant cover and support, concentrated independently from the rest of the amphibious forces in the waters east of Luzon.

The key to the success of an offensive major naval operation is the *operational maneuver*. In contrast to maneuver by ground forces, naval forces conduct an operational maneuver by moving almost uninterruptedly along their lines of operations to an area or position from which to strike the enemy's main force. In contrast to land warfare, an operational maneuver at sea is not invariably aimed at obtaining favorable positions for defeating or neutralizing the enemy's operational center of gravity. An operational maneuver is usually an integral part of plans for major naval operations to destroy an enemy fleet at sea or in its bases, to destroy coastal installations or facilities, to execute amphibious landing operations, and in major operations to interdict or cut off enemy maritime trade. It is rarely used in other types of major naval operations, such as destroying the enemy's and protecting one's SSBNs or supporting ground troops on the coast.

Normally the actions of one's fleet forces would be focused on delivering strikes against the enemy's main forces in the *sector of main effort*.[12] It is critical to obtain and maintain the initiative. First strikes should be massed, and then sustained, against enemy forces whose destruction would lead to the accomplishment of one's ultimate operational objective.

Actions in the sector of main effort cannot usually be successful unless they are supported by other forces in a *sector of secondary effort*, where they will be preventing or neutralizing enemy forces trying to interfere with one's actions within the sector of main effort.[13] The principal aim of the supporting actions will be to create favorable conditions for resolving the main task. As noted, the intended sector of main effort should be kept hidden from the enemy.[14]

Any offensive major naval operation should have its *point of main attack* within a sector of main effort and within striking distance of the enemy's operational center of gravity. For an amphibious landing operation, the point of main attack is usually the least defended part of the coast or the beach and at the same time offers the best opportunities to advance inland. For an attack on enemy shipping, the point of main attack is the area with the densest traffic, usually at the approaches to major ports or where several major routes converge or intersect. For a defensive major naval operation, such as protecting maritime trade or in an antiamphibious defense, a *main point of defense* must be determined.

A major naval operation normally encompasses several distinctive *phases*. While a tactical action may also consist of several phases, they usually overlap each other; hence there is no clear break in fighting. In contrast, phases in a major naval operation are usually longer and are marked by clear breaks in the intensity of fighting. Time between consecutive phases, or *operational pauses*, vary in length depending on one's losses and logistical sustainment capabilities. These pauses are used to bring up new forces or to regroup existing ones, to replenish and refuel, and to provide rest to personnel. A series of small, almost continuous tactical actions would be conducted by other forces during these pauses to establish a favorable operational situation in a given area of operations and lay the foundations for the conduct of a successive major naval operation.[15] A major naval operation will be shorter in duration and is more likely to consist of a single phase.[16] Because of the high intensity of action, high consumption of fuel and ammunition is to be expected; hence sustained logistical support is critical for success. In narrow seas there will be considerable difficulty in regenerating combat power, because operational pauses will be few and of much shorter duration than in operations on the open ocean. Another problem is much higher vulnerability of one's own forces to enemy strikes.

Synchronization is the most critical part of a major naval operation. It involves several simultaneous or sequential strikes or attacks against targets in the sector of main effort, the aim being to neutralize in a timely way the enemy's ability to interfere with one's own actions in the main sector. These actions may include sweeping certain sea or ocean areas for mines, deploying submarines, or minelaying to restrict the maneuver of the enemy's forces. Precise synchronization in terms of space and time is needed between strikes by the main forces and supporting actions meant to create favorable conditions for the accomplishment of the main task.[17]

The chief aim of any major naval operation is to accomplish, as quickly and with as little loss as possible, *operational success*: an outcome having *operational* importance and decisive impact on the course or outcome of a major operation. Normally, the defeat of the major enemy force in the sector of main effort through a battle or engagement would represent an operational success. Supporting actions can also bring operational results, thereby decisively affecting the outcome of the entire major naval operation. Sometimes operational or operational-tactical commanders come close to operational success but, either through poor intelligence or sheer lack of will, fail to grasp the opportunity. An example of operational success in the Battle for Leyte was Allied victory over the Japanese forces in the Surigao Strait. This prevented the Japanese from attacking Allied troops and supplies on shore in the Leyte Gulf. Likewise, the Japanese missed a chance to accomplish operational success because Vice-Admiral Takeo Kurita, the Commander of the First Diversionary Attack Force, failed to exploit his tactical success in the Battle off Samar.

After the ultimate objective of a major naval operation is accomplished and operational success is consolidated, one's forces are usually *redeployed* to another area of operations, either in the same or an adjacent theater of operations. The purpose can be to obtain a new, more advantageous base of operations for a subsequent major operation, or to consolidate further the gains of the preceding one.

Reconstitution of one's forces may be necessary if losses in ships, aircraft, and personnel are so high that units or elements are no longer capable of conducting their assigned combat mission. Reconstitution involves two distinctive tasks, *reorganization* and *regeneration of combat power*. Specifically, the operational commander is tasked to restore combat-attrited units under his command to a level of combat effectiveness commensurate with mission requirements and the availability of resources. This includes cross-leveling, or replacing personnel, supplies and equipment; reestablishing command and control; and conducting essential training.

Campaigns

One's naval forces operating in a narrow sea will also conduct actions as part of a land or maritime campaign. In generic terms a *campaign* consists of a series of major operations (land, air/space, naval), synchronized in time and place and aimed to accomplish strategic objectives in a given land or maritime theater. These actions are conducted simultaneously or sequentially and are engaged in according to a common plan that is controlled by a theater commander. A strategic objective, especially in a mature theater, encompasses not only military, but also political, diplomatic, economic, and often ideological, ethnic, religious, and informational elements. Therefore, no single service could normally accomplish all the elements of a strategic objective by its own efforts, and a synchronized employment of several and often, all services is needed. This is not meant to imply that a single service cannot make a greater or even decisive contribution to the campaign's ultimate success than other services.

Campaigns can be *offensive* or *defensive* with regard to their main purpose. *Land campaigns* and *maritime campaigns* are differentiated according to the physical environment in which actions predominantly take place. A campaign is planned and conducted for each strategic threat or objective to be accomplished in a theater both in wartime and sometimes even in peacetime (counterinsurgency, counterterrorist and counterdrug campaigns, etc.). However, a campaign in peacetime would consist mostly of a series of related minor (or sometimes major) tactical actions rather than major operations. These actions would be coordinated in time and place to accomplish strategic objectives within a given part of the theater.

Normally, a blue-water navy or coastal navy bordering a narrow sea would be involved in conducting major operations or tactical actions as a part of land campaign, as was the case with the German and Soviet navies in the Baltic and the Black Sea in World War II. Also, Allied navies took part in a land campaign fought in Italy in 1943–45. Sometimes, narrow seas can be a scene of a maritime campaign, as was the case in the Solomons in 1942–43.

To conclude, because of the small size of the area and the short distances involved, most actions of naval forces and aircraft in a typical narrow sea will be tactical in character. This was the case in the past and will most likely remain so in the future. Most tactical actions will be attacks and strikes, while engagements and especially naval battles will be comparatively rare occurrences. The experiences of the Arab–Israeli wars and wars in the Arabian Gulf seem to imply that missile and air strikes have replaced a naval battle as the principal method of combat employment to accomplish a major tactical and sometimes even an operational objective. The importance of a naval battle seems to be considerably reduced compared with the past. Major naval operations in a typical narrow sea are difficult to plan, prepare, and execute because of the physical characteristics of the area and the involvement of combat arms of several services. A high degree of cooperation is required for such operations to be successful. However, they are, especially for a blue-water navy, the quickest and the most effective way of accomplishing operational and sometimes even strategic objectives in a given maritime theater of operations. They are conducted within a certain sea area and within a time-frame determined by the side that takes the initiative. If successful, a major naval operation will result in a drastic change in a situation in a given maritime theater or area of operations. Another great advantage of conducting a major naval operation lies in avoiding the tying up of one's forces in a certain sea area and thereby allowing the sequence of the objectives or tasks to be accomplished as planned. Naval forces and aircraft operating in a narrow sea will often take part in a major operation conducted by ground troops on the coast. They will also play a significant role in the conduct of a land or maritime campaign.

NOTES

1 Stephen W. Roskill, *The War At Sea 1939–1945*, Vol. II: *The Period of Balance* (London: Her Majesty's Stationery Office, 1956), p. 36.
2 Donald Macintyre, *Sea Power in the Pacific: A History from the 16th Century to the Present Day* (London: Military Book Society, 1972), p. 225.
3 The Soviets/Russians use the term 'systematic combat actions' (*systematichiskiye boyeviye deystviy*) aimed to obtain and maintain a 'favorable operational regime'.
4 By 1900, the Royal Navy consisted of 205 warships: 45 battleships, 126 cruisers, and 34 torpedo gunboats. The French Navy in contrast was composed of 33 battleships, 38 cruisers, and 21 torpedo gunboats; Ronald B. St John, 'European Naval Expansion and Mahan, 1889–1906', *Naval War College Review* (March 1971), p. 74.
5 H. Engelmann, 'Die Sicherstellung von Seeoperationen' (Support of Naval Operations), *Militaerwesen* (East Berlin), 3 (March 1980), p. 69; S. Filonov, 'Morskaya Operatsiya' (A Naval Operation), *Morskoy Sbornik* (Moscow), 10 (October), 1977, p. 24.
6 Engelmann, 'Support of Naval Operations', p. 70.
7 Ibid., p. 71; P. Navoytsev, 'Zakonomernosti, Soderzhaniye I Kharakternye Cherty Sovremennykh Morskikh Operatsii' (Regularities, Content and Characteristics of Modern Naval Operations), *Morskoy Sbornik* (Moscow), 7 (July) 1986, pp. 22–3.
8 R. Ernest Dupuy, Trevor N. Dupuy and Paul F. Braim, *Military Heritage of America*, Vol. I (McLean, VA: The Dupuy Institute, 1992 and Dubuque, IA: Kendall/Hunt Publishing, 3rd edn, 1992), p. 15.
9 Friedrich von Bernhardi, *On War of To-Day*; trans. Karl Donat, Vol. II: *Combat and Conduct of War* (London: Hugh Rees, 1913), pp. 361–2.
10 *Tactical concentration*, by contrast, is aimed to mass one's forces prior to or during a naval battle or engagement and is conducted in a given naval combat zone or sector. *Strategic concentration* is intended to create superior naval strength in a given maritime theater of operations or theater of war. It pertains to concentration of the entire fleet or its major part.

11 Headquarters, US Marine Corps, FMFM 1, *Warfighting* (Washington, DC: Government Printing Office, 1989), p. 31.
12 Englemann, 'Support of Naval Operations', p. 69.
13 Ibid., p. 69; Navoytsev, 'Regularities, Content and Characteristics of Modern Naval Operations', p. 21.
14 Ibid., p. 21; Engelmann, 'Support of Naval Operations', p. 70.
15 Filonov, 'A Naval Operation', p. 24.
16 Tihomir Vilovic', 'Specifičnosti priprema i izvodjenja operacija na uskim morima' (Peculiarities of Preparing and Executing Operations in Narrow Seas), *Mornarički Glasnik* (Belgrade), 2 (March–April 1976), p. 213.
17 Filonov, 'A Naval Operation', p. 25.

9

Securing Control

Sea control in wartime is generally accomplished by destroying, annihilating, or neutralizing the enemy's naval forces at sea and in their bases, and by physically seizing or destroying their basing areas and the key elements of support ashore. In narrow seas these actions can be combined and coordinated with the actions of air and ground forces, and sometimes even coastal defense forces. These objectives are also accomplished by seizing a large stretch of the opponent's coast together with naval and air bases and ports, choke points and the key positions offshore.

Actions Against an Enemy Fleet

The destruction or neutralization of the enemy fleet is usually the immediate objective of one's naval forces and is aimed to gain and exercise control of the sea. Actions intended to destroy or possibly annihilate enemy naval forces at sea at the very outset of the hostilities are the most effective methods of securing control both on the open ocean and in a narrow sea. Historically, the most common method used to destroy an enemy fleet was by seeking a *general fleet action* aimed to destroy, incapacitate, or capture the enemy's main body and thereby accomplish the principal strategic objective of the war at sea. Yet, even in the sailing era it was difficult to bring about a decisive encounter with the opposing fleet. In some cases the maintenance of command of the sea was more important than seeking out decisive action.

    A general fleet action often leads to a major and *decisive battle*, resulting in a drastic change of the naval balance in a given theater. Such a battle sometimes had a decisive impact on the outcome of the war as a whole. British Admirals Phillip Colomb and Herbert Richmond observed that the idea of seeking a decision by battle on the high seas or by a preemptive strike on the enemy coast had been developed in the sixteenth century during England's struggle with Spain. Admiral Sir Francis Drake always sought to seize the initiative from the Spanish by striking them shortly after the outbreak of hostilities and before mobilization of the main body of the Spanish fleet was completed. The past wars at sea show that even in the case of a decisive defeat of the opposing fleet, the victor was rarely relieved from the enemy threat. Even in the Anglo-Dutch Wars, the likelihood of completely finishing off the adversary and thereby exercising absolute command of the sea was not great. The defeated fleet was sometimes able to avoid total destruction simply because of a sudden fog or a change of wind.

In the past, a decisive battle often led to a dramatic change of a situation in a theater. The most important factor was not the immediate losses or damages inflicted on the opposing fleet, but the long range consequences in the aftermath of the battle. After its victory at Tsushima in May 1905, the Japanese Navy was forced to keep close watch on the remaining Russian fleet forces. However, even when a weaker force is kept under observation it does not necessarily mean that a stronger force has secured undisputed control of the sea. An active and energetic enemy, operating from a long coastline endowed with numerous harbors will almost invariably take the opportunity to launch attacks and cause diversion of one's own efforts.[1] In fact, in most naval wars in the past, neither side enjoyed command of the sea.

A decisive victory at sea sometimes allowed the victorious fleet to obtain or regain the initiative and thereby to use the sea more or less freely for its own purposes while preventing the enemy from doing the same. The decisive defeat of the Spanish Armada in August–September 1588 restored the strategic initiative to England and subsequently led England to create a large maritime empire and ultimately acquire the status of a world power. The defeat of the Spanish Armada also led to the rise of Dutch sea power some 80 years later. The British defeat in the 'Four Day Battle' in June 1666 (the Second Anglo-Dutch War) allowed the Dutch to obtain control of the English Channel and close the Thames estuary to trade.[2] In the Battle of Sole Bay in May 1672 (the Third Anglo-Dutch War) and in the two engagements in Schoonveldt Channel in May 1673 the Dutch made great use of their navigatorily dangerous coast for defensive–offensive actions by their fleet. Admiral Michael A. De Ruyter defeated his opponents in all the battles, prevented the landing of an invasion army and broke up England's attempt to blockade the Dutch coast.

In the Battle of Beachy Head in July 1690 (the War of the Grand Alliance against France) the combined French Mediterranean and Atlantic fleets defeated the combined English and Dutch fleet by driving it back into the Thames. However, the French were unable to exploit their temporary command of the Channel because they did not have ready an army to invade England. Moreover, they failed to follow up their success at Beachy Head and inflict a decisive defeat on their opponents at sea. By the end of 1690 and throughout 1691, the Anglo-Dutch fleets in the Channel were so strong that the Comte de Tourville avoided taking up a battle. The entire French effort at sea was ineptly conducted and the French fleet fell away in efficiency, while its main adversaries at sea, England and Holland, improved their fleets.[3]

In the War of 1812 the naval action on Lake Erie had similar strategic importance to the battle of Lake Champlain, because it ended the British invasion of the United States. It also had a highly favorable effect on the peace negotiations at Ghent. The Duke of Wellington observed that the success of military operations in Canada depended on control of the lakes, of which there was evidently little prospect, and he advised a peace settlement without territorial demands.[4]

In the aftermath of the decisive battle off Heligoland on 9 May 1864, the Danish fleet withdrew into its coastal waters, thereby making it possible for the Prussians to open up the Elbe estuary and the German ports in the North Sea for maritime traffic.[5] The Battle of Lissa on 20 July 1866 decided the question of command of the sea in the Adriatic. On the same day an armistice ended the fighting between Austria and Prussia. The Austrians completed their withdrawal to the Isonzo River, leaving Venice in Italian hands. France and Prussia pressured Italy to conclude an armistice on its own with Austria. The Italian Prime Minister, Baron Bettino Ricasoli, refused the call and insisted on obtaining 'natural' frontiers for Italy. These included direct cession of Venice and the South Tirol and a guarantee that Italian interests in Istria were respected. However, Ricasoli had completely ignored the fact that Admiral von Tegetthoff had won command of the sea and that the Austro-Prussian armistice had strengthened Vienna's hand. The Italian fleet was unable to challenge its Austrian counterpart. In the end, the Italians signed the armistice with Austria on unfavorable terms for themselves.[6]

The battle of the Yalu River in September 1894 resulted in incomplete victory for the Japanese fleet, led by Admiral Sukenori Ito, over the Chinese squadron. However, that victory was sufficient to secure control of the Yellow Sea. The Japanese soon afterwards seized Dairen and captured the fortified base of Port Arthur. By February 1895 they also took control of the port of Wei-hai-wei and ultimately accepted the surrender of the Chinese fleet.[7]

Sometimes, what was intended to be a decisive battle resulted in an *inconclusive battle*. After the battle off La Hague (Barfleur) in May–June 1692, the French Navy never recovered and Louis XIV concentrated instead on expansion on land. Another example of a tactically inconclusive, but operationally decisive battle was that in the Yellow Sea between the Japanese and Russian fleets in August 1904. Then, although both sides suffered serious losses, the Russian squadron never emerged to fight again. Five months later the Russian Squadron met its end from the guns and howitzers of the Japanese army besieging the naval base of Port Arthur. Afterwards, the Japanese fleet was free to exercise absolute control of the Yellow Sea and the Sea of Japan, and made certain the eventual Russian defeat in the war.

The Battle of Jutland in 1916 came closest to what was then considered a general fleet action. It in fact consisted of several major battles and smaller encounters between the British Grand Fleet and the German High Seas Fleet. All of these actions were related and controlled by a single commander, and were aimed at accomplishing an operational objective in the North Sea theater of operations. The Battle of Jutland was tactically won by the German High Seas Fleet, but operationally it was clearly a British victory. However, the operational and strategic consequences of the Battle of Jutland were less obvious than the losses in material. If the battle had ended with a clear British victory, the Royal Navy would have obtained control of the North Sea, and would perhaps have even been able to operate within the Baltic in support of the Russian Navy. The difficulties of neutralizing the threat of the German U-boats might also have been considerably eased. On the other hand, had the Germans won the battle, the initiative would have been passed into their hands and they would have been able to blockade England's coast in the North Sea. Moreover, their U-boats and heavy surface ships would have been able to attack allied shipping in the Northern Atlantic. As it turned out, the Battle of Jutland did not affect the British command of the North Sea and Germany did not improve its overall strategic position at sea.

Naval warfare in the 1914–18 conflict showed that the fleets of the major opponents were too large and deployed too widely to be destroyed during a single battle or even a couple of battles. This signaled the final demise of the *decisive battle*, and demonstrated that *operational* objectives in the theater could be accomplished only by a series of related naval battles and engagements, sequenced and synchronized in time and place – a major naval operation – and controlled by a single commander.

The quickest and the most economical method of obtaining sea control in a typical narrow sea is to mount a major naval operation aimed at destroying or even annihilating major parts of the enemy fleet at sea or in its bases. However, the instances of such major naval operations in such waters were few and far between. In practice, most fleet-on-fleet encounters in World War II took place when one fleet provided distant cover and support for a major convoy or amphibious task force or when the stronger fleet used the threat of amphibious landing to lure an inferior force into a 'decisive' battle. The Italian heavy surface forces (one battleship, six heavy and two light cruisers, and 13 destroyers) sailed out on 26 March 1941 to attack British convoys to Greece in the area south of Crete. The entire operation was to be supported by the German Tenth Air Corps. However, the British were able to obtain information on the impending action by decoding German orders to Tenth Air Corps. A strong British force sailed out to intercept the Italian fleet and in the ensuing battle on 28–29 March, three Italian heavy cruisers and two destroyers were sunk, while one battleship, one heavy cruiser, and two destroyers were damaged. The attacks by the German planes on the British ships were unsuccessful.[8] This victory dramatically changed the naval situation in the Mediterranean to Allied advantage.

A small but excellently led navy can obtain relatively quickly local control of the sea even against a stronger combination of opponents by conducting a series of decisive tactical actions. This was accomplished in what may be considered a major naval operation by the Israeli Navy during the 19-day Yom Kippur (Ramadan) War of 1973 with Egypt and Syria. Early on the Israelis secured control of those parts of the Eastern Mediterranean declared by Syria and Egypt to be war zones.[9] From the first day of the hostilities, the Israelis seized the initiative and inflicted heavy losses on its enemies. A force of five Israeli missile craft sank three Syrian missile craft, and one torpedo craft and one minesweeper in a battle off Latakia (Al Lādhiqīyah) in the night of 6–7 October.[10] A naval battle between six Israeli *Saar*-class missile craft and Egyptian *Osa*-class missile craft took place off Damietta-Balṭīm (off the Egyptian coast) on the night of 8–9 October. In the ensuing exchange, the Israelis sank three Egyptian missile craft, while one was heavily damaged and subsequently destroyed by artillery fire.[11] These victories drastically changed the operational situation at sea to Israeli advantage.[12]

Sometimes, owing to lack of aggressive leadership, a superior fleet fails to mount a major operation and achieve decisive victory over its much weaker opponent at sea. This was the case with the Soviet Baltic and Black Sea Fleets in World War II. Both fleets failed to take their chance to obtain command of the sea in their respective operating areas or to provide strong support to the embattled troops on the coast. The actions of the Black Sea Fleet are perhaps more illustrative of the profound lack of aggressiveness than those of the Baltic Fleet, because its margin of superiority over its opponents was so large. By the outbreak of hostilities in June 1941, the Soviet Black Sea Fleet was numerically ten times as strong as the Axis' naval forces deployed in the area, and even during the rest of the war the Soviets retained an at least fourfold superiority over its adversaries. Moreover, the Soviet naval forces occupied a very favorable base of operations. Nevertheless, despite these advantages the Soviets accomplished only modest results. They were unable to cut off the Axis maritime traffic along the Ukrainian coast totally or to blockade the Turkish coast and the Danube estuary until the end of war.

In a more recent example, the US Navy and the navies of other Coalition members destroyed in a series of strikes and attacks most of the Iraqi forces in the northern part of the Gulf between 22 and 24 January

1991. Then, two minelayers, one oiler (acting as a scouting ship), two patrol craft, and one hovercraft were destroyed from the air. On 29 January, in the engagement off Būbiyān Island, the US and British missile-armed helicopters and ground-attack aircraft destroyed four and forced 14 patrol craft, probably carrying commandos to take part in an Iraqi attack on Khafji, to run aground. In a separate incident, a British helicopter destroyed a large patrol craft.[13] A day later, the US and British helicopters and ground attack aircraft attacked a force consisting of one former Kuwaiti patrol craft and three Iraqi amphibious craft and one mine-sweeper, all of which ships suffered various degrees of damage. In another encounter, a force of eight combat craft, including some *Osa*-class missile craft were attacked by US ground-attack aircraft in the northern part of the gulf, four craft being sunk and three damaged.[14]

Another option for a stronger side is to mount a major operation to destroy a substantial part of the coastal navy either at seas or at its bases or both. Historically, there were four main methods of destroying an enemy fleet in its bases: conducting strikes from beyond the range of the enemy base's defense, penetrating the enemy naval base with one's surface forces, raiding by teams of special forces, and a blockade from the sea in support of troops besieging the enemy's naval bases from the land. A prerequisite for success in striking an enemy fleet in its bases is to obtain local, temporary sea control in a given sea or ocean area. In the past, the best results were achieved by striking from beyond the effective range of the enemy's defenses. The British force of one aircraft carrier (*Illustrious*) with 21 aircraft and with accompanying escorts launched an attack (Operation *Judgment*), from a distance of about 180 miles, on the Italian heavy ships at the naval base of Tarent on 11 November 1940. A complete surprise was achieved and, despite strong Italian defenses, the British aircraft heavily damaged one new battleship (*Littorio*) and two older battleships (of the *Giulio Cesare*-class), of which one sank (*Cavour* – later salvaged and put back into service), while one heavy cruiser and one destroyer were slightly damaged. All but two British aircraft returned safely to their carriers.[15] This attack drastically, though only temporarily, had changed the naval situation in the central part of the Mediterranean to Britain's advantage.

In the first phase of their invasion of Soviet Russia, the Germans tried to seize Leningrad from the land side. They intended to destroy heavy surface ships of the Soviet Baltic Fleet bottled up in the Leningrad–Kronstadt area. The First Air Fleet responsible for cooperation with the Army Group North, assigned one squadron of its *Stuka* dive bombers, specially adapted to carry a single 2,200-lb bomb to attack the ships. The first attack between 16 and 19 September 1941 failed, but more success was achieved on 21 September when heavy damages were inflicted on one battleship, which subsequently sank. The same day the German bombers damaged another battleship and one heavy cruiser, while one destroyer was sunk. On 23 September, two heavy cruisers, two destroyers, one submarine tender, one submarine and several torpedo craft suffered various degrees of damage, and one flotilla leader capsized, while one destroyer and one submarine were sunk.[16]

The Germans decided to conduct a major air operation (Operation *Eisstoss*) to eliminate the threat posed by the remnants of the Soviet Baltic Fleet in April 1942. The First Air Fleet received an order on 26 February to prepare an attack aimed at destroying all the remaining Soviet ships in the Gulf of Finland (one battleship, two heavy cruisers, one minelaying cruiser). The plan also envisaged support by the siege artillery of the 18th Army Corps.[17] The operation carried out by the I Air Corps started on 4 April when 95 Ju-87s/-88s bombers attacked the ships, while 37 He-111a aircraft suppressed Soviet air defenses. Fighter protection was provided by about 60 Me-109s. The Germans carried out another attack with 31 He-111s, on the night of 4–5 April. These attacks resulted in heavy damages to one battleship, four cruisers and one destroyer, while one minelaying cruiser, one destroyer and one training ship each suffered light damage.[18] In another series of attacks (Operation *Goetz von Berlichingen*), carried out in the last week of April, the German bombers attacked Soviet ships and shore facilities, resulting in further damage to one cruiser. Thus, in the end, despite many efforts, the Germans failed to achieve their declared objective of destroying the Soviet surface ships bottled up in the Gulf of Finland.[19] Moreover, the Soviet warships continued to provide fire support to the defenders on the Leningrad front and the Oranienbaum beachhead.[20]

Attacks on enemy naval bases have also occurred in many regional conflicts since 1945. Aircraft of the US and Coalition forces attacked Iraqi ships based on Umm Qasr, in the Būbiyān Channel, and in the port of Kuwait between 25 and 28 January 1991. One Iraqi minelayer, two patrol craft, and one transport were sunk in these attacks.[21] In another action on 4 February, the Coalition aircraft attacked the Iraqi naval base at Al Kalia and disabled two missile craft. Helicopters from an American frigate engaged four Iraqi patrol craft off the Maradin Island. One of them was sunk and another damaged. With this action, the threat posed by the Iraqi Navy to the US and Coalition navies operating in the northern part of the Gulf practically ended.[22]

Special forces proved to have a great potential for attacking enemy ships in the their bases in narrow seas. By using submarines, fast attack craft, or aircraft, small groups of commandos can be covertly deployed in the base's proximity and then carry out their tasks. In World War II, the Italians conducted several successful raids by their naval commando teams against the naval bases at La Valletta (Malta), Soúdha Bay (Crete), and Alexandria. The most successful was the attack on Alexandria on 19

December 1941 when three Italian combat swimmers penetrated the harbor and planted limpet mines that eventually sank two British battleships (*Queen Elizabeth*, *Valiant*), and a tanker. This action led to a dramatic, although temporary change in the naval situation in the Mediterranean. The British had only three light cruisers left in the eastern part of the Mediterranean, and because of the entry of Japan into the war, there was no possibility of sending additional battleships or aircraft carriers as none were available. Fortunately, for the British, because their battleships did not capsize, about two months passed before the Italians realized just what their swimmers had accomplished.[23]

In a more recent example, during the Yom Kippur (Ramadan) War of 1973, Israeli frogmen penetrated Port Said, a major Egyptian port. In the Gulf of Suez, patrol craft penetrated two Egyptian anchorages and entered the main port of Adabia. In the northern part of the Red Sea, the Israeli naval commandos penetrated the Port of Hurgada and destroyed at no loss to themselves four Egyptian *Komar*-class missile craft. These attacks forced the Egyptians to remove their remaining missile craft from the base, which in turn strengthened the Israeli navy's control of the Gulf of Suez. The repeated raids on the Port of Hurgada also contributed to the slow down in the flow of oil from the Suez wells to Israel.[24]

Today, land and carrier-based aircraft, submarines and surface ships armed with long-range cruise missiles are the most effective platforms for destroying enemy ships in their bases. Attack submarines armed with land-attack cruise missiles can strike enemy naval bases from distances of several hundred nautical miles. In the Gulf War of 1990–91, US SSNs deployed in the Arabian Gulf repeatedly struck with their *Tomahawk* missiles a variety of land targets, including the Iraqi naval base at Basra.[25]

Air strikes against the enemy naval bases in a narrow sea can be far more effective than those mounted from the open ocean, because of the much shorter distances and the larger number of land-based aircraft that can be employed. They can be conducted at high intensity and repeated at short intervals. In some instances not only fixed-wing aircraft but also missile-armed helicopters can be effectively employed.

Today, only large navies, specifically the US Navy and the Russian Navy, have the capability to mount major operations aimed at destroying the enemy's sea-based strategic nuclear forces. Such operations may be focused on destruction of the enemy SSBNs either at their basing areas, during their transit to or from them, or in their patrol zones. Likewise, a major naval operation to protect one's sea-based, strategic nuclear forces may be conducted as well. Such major operations will most likely be conducted by predominantly naval forces, specifically attack submarines and maritime patrol aircraft in and over the open ocean. However, in areas closer to one's own or the enemy's coast, such a major operation will require the employment of forces of other services.

Attrition

Another method of destroying enemy naval forces at sea and in their bases is to attrite enemy fleet forces and land-based aircraft over time. Also called 'sporadic warfare', this method is practiced by both inferior and stronger fleets operating in a typical narrow sea. The main objective is to avoid or postpone a major clash with the opposing fleet and occasionally to engage the enemy forces whenever favorable opportunities exist. These objectives are accomplished by conducting tactical actions of smaller surface combatants, land-based aircraft, diesel-electric submarines, and by offensive mining. Optimally, the operational objective is accomplished by equalizing naval strength in a given theater, thereby creating favorable conditions for an inferior fleet to engage an initially stronger fleet on its own terms. Attrition warfare was widely practiced by the German High Seas Fleet in the North Sea from the outset of the hostilities in August 1914 until the Battle of Jutland in late May 1916.

The stronger fleet may often combine attritional warfare with major naval operations aimed at destroying or annihilating the enemy fleet. In contrast to the inferior fleet, the stronger fleet can conduct attritional warfare at the tactical level. After the outbreak of hostilities in August 1914, the Royal Navy blockaded both exits from the North Sea, and the Germans relied on their land power to achieve their strategic objectives in the war. They deployed their fleet in a triangle of bases on the rivers Weser, Ems and Jade, from where they conducted tactical forays into other parts of the North Sea. Because the initial ratio of forces was 15 to ten, the Germans tried to balance naval strength by conducting attritional warfare before seeking a general fleet action. However, these actions were not a part of an overall military strategic objective or strategic idea, nor were they an integral part of an operational plan. Despite pressure to go on the offensive, the High Seas Fleet remained on the defensive strategically and acted tactically.[26]

Attritional warfare can also ensue as a result of accomplishing an operational objective with inadequate forces or by the lack of operational thinking. The struggle for control of the Solomons in 1942–43 is a classic case of attritional warfare at sea and in the air. Both the Allies and the Japanese suffered heavy casualties in ferocious fighting in the waters around Guadalcanal and in the air. No fewer than seven naval battles and engagements were fought in the struggle for Guadalcanal alone. Each side lost approximately the same number of ships. Japanese losses totaled two battleships, one light aircraft carrier, three heavy and one light cruisers, 11 destroyers and six submarines. The Allied losses amounted to two aircraft carriers, six heavy crusiers, two light cruisers and 14 destroyers. In addition, Japanese naval air strength was so severely depleted that fleet aircraft carriers could not be properly manned. A more serious problem for the Japanese was the fact that new construction was unable to make up the losses. No more battleships or heavy cruisers were built by the Japanese and no more than half of the lost destroyers were ever replaced.[27]

A new phase in the struggle for Guadalcanal began a few months after the Japanese were forced to evacuate Guadalcanal in January 1943. Both sides started to build up their strength in the southern Pacific in preparation for the next inevitable clash. In the Solomons, the Japanese established bases at Munda on New Georgia and at Vila on Kolombangara. From Rabaul they used the *Tokyo Express* to reinforce their troops in New Guinea.[28] Most naval actions in recent local wars have consisted of a series of sporadically fought engagements, strikes and attacks that were designed to attrite the opponent's strength over a relatively long period. If a stronger fleet possesses overwhelming firepower, then the overall strength of its opponent can be attrited rather quickly, as the actions by the US Navy against the Iranians in the Gulf in 1987–88 clearly demonstrated.

Neutralization

A stronger fleet can neutralize the weaker force in a typical narrow sea by blockading it in its bases or significantly limiting its operating area. The stronger fleet then deploys behind an arbitrarily selected blockading line to watch out for the enemy fleet should it try to sail out from its bases or ports. Sometimes, a weaker force may remain in its bases and ports to avoid annihilation by a stronger force. The effect of any blockade is invariably the same, the weaker fleet is neutralized because the stronger fleet as a whole holds a central position and operates along interior lines with respect to the blockaded force.[29]

A blockade can be aimed against an enemy fleet or overseas maritime trade. However, in practice, naval and commercial blockades are intrinsically linked, and therefore almost indistinguishable. The immediate objective of a naval blockade is to force the enemy fleet to come out of its basing areas. In practice, however, naval blockade has rarely been successful in luring a weaker fleet to come out and fight. Admiral Richmond observed that in no case in all the many wars at sea in which an enemy has been forced to keep under the shelter of the defenses of his ports has a blockade forced him to fight. In their wars with England, neither Spain, Holland, France nor Germany sent their fleets out to fight the stronger British Navy.[30]

Another objective of a naval blockade is to prevent the opposing fleet from interfering in any substantial way with the ability of the stronger fleet to use the sea as it wishes If the weaker force is neutralized, the blockading force is effectively in command of the sea behind the blockading line. The ships not actually involved in the blockading operations are employed to exercise that command.³¹ However, the degree of security from one's forces exercising command of the sea is necessarily less complete than that from wholesale destruction of the enemy fleet. In conducting a blockade, a fleet of adequate strength, suitably disposed geographically, is usually capable of blockading a weaker force from the adjacent seas or oceans, thus providing a cover under which other forces farther out actually exercise command.

Close Blockade

Until the advent of steam, a naval blockade always meant a close blockade normally aimed at preventing the enemy fleet from leaving its bases or ports. The real objective of such a blockade was ultimately to draw the enemy fleet out of its port to be engaged and destroyed or annihilated. A close blockade, even if successful, could obtain only local and temporary control of the sea and not permanent control. The concept of naval blockade as practiced in the era of sail has had significant impact on naval blockades conducted in modern times.

Close blockade of the enemy fleet was conducted by almost all the major navies in the era of sail. The Dutch fleet of 100 ships under Admiral Maarten H. van Tromp blockaded the Spanish fleet of 70 ships in the Downs along the English coast in September–October 1639. The Dutch fleet also blockaded the English fleet based on the island of Elba and in Leghorn (Livorno) for six months between September 1652 and March 1653. Both of these blockades were effective and, in the latter case, the Dutch completely destroyed the English fleet as it tried to escape. A close blockade of the British ports by the Dutch fleet in 1667 forced the British to discuss peace terms with the Dutch in a matter of weeks.³²

A close blockade was conducted by the Royal Navy in the Seven Years War (1756–63). British Prime Minister William Pitt the Elder made a naval blockade an integral part of the national strategy. Britain subsidized its allies on the continent to fight wars for it, while the Royal Navy conducted a blockade of the French fleet and provided support to the British army. The aim was to keep the French fleet divided between the Atlantic and the Mediterranean ports. After the battle of Quiberon Bay in November 1759, the British maintained an unopposed blockade of French ports that eliminated French ships from the high seas. The Royal Navy kept its main forces back at Torbay or even Spithead, and relied on light forces off the French coast to keep an eye on the French ships if they tried to come out. The French merchant marine was devastated and the finances of the country exhausted. The British further restricted French trade by seeking out contraband in neutral shipping. At the end of the Seven Years War in 1763 the British were firmly in command of the sea.³³

During the French Revolutionary Wars, the string of naval bases along England's Channel coast combined with control of Gibraltar, Mallorca (Balearic Islands), and La Maddalena (Sardinia) allowed the Royal Navy to blockade the French and Spanish fleets from Texel to Toulon. In contrast, the French were handicapped because along their Channel's coast there was no harbor large enough to provide shelter for a major part of their fleet. In addition, the prevailing winds blew up the Channel and towards the southern coast of England. The French were unable to obtain permanent control of Antwerp, the only large port within striking distance from the British major naval bases. The British policy was always to have close and friendly relations with the Low Countries and thereby deny the use of Antwerp to the French fleet.

In the Napoleonic Wars, the British did not try to seal all the French or Spanish ports hermetically, but effectively to prevent any large French-led force to sail out to sea. For the French, it was essential that parts of their fleet could sail out when needed using routes that ensured a rapid concentration at a particular point. In the blockade of the French main naval bases of Brest in 1803, the English fleet based at Torbay flanked a possible French advance up the Channel. However, it was not enough for the British to watch only the Channel, as the examples of the Lazare Hoche expedition to Ireland in 1796 and Admiral Eustache Bruix's to the Mediterranean in 1799 showed. Hence, to prevent that from happening, the British blockading squadrons took station off Brest, except during strong southwesterly winds when an enemy sortie out of the port was very difficult if not impossible. The most successful close blockade was conducted by Admiral William Cornwallis in the winter and summer between the outbreak of the war with Napoleonic France and the Battle of Trafalgar in October 1805.

In conducting a close blockade, a stronger fleet was stationed off the enemy bases watching ship movements from a distance. It occupied a central position in regard to the enemy fleet and operated along tactically interior lines, with all the advantages and disadvantages these lines offer. The blockading ships cruised periodically off the enemy bases and hence exerted only a general control of local waters. The objective was to prevent the enemy from concentrating his forces, or in other words, precluding a great combination of ships from all the ports.

For Napoleon I it was essential that parts of his fleet could escape to the open sea, but to do so at a time of his own choosing and by using routes that would ensure a rapid concentration at a particular point.[34] To prevent this from happening, the British conducted a close blockade in 1803 by deploying their ships off minor French bases in the English Channel. In the event of French detachments escaping out of ports, the orders were given for the outlying British squadrons to fall back on their central position off Brest, thereby enabling them to pursue the French either to Ireland or up the Channel. If the French moved up the Channel, they would be met by the British squadron based in the Downs and in the North Sea.[35]

Normally, it was a stronger fleet that conducted a close blockade. Each blockading squadron represented a concentration against the force it blockaded. Sometimes, however, a close blockade played into the hands of a weaker fleet by tying the stronger fleet to a given area and thereby reducing or even preventing it from operating elsewhere. It also required large forces to be effective, and was extremely exhausting and demanding on ships' crews and support services. Despite its high demand for ships, a close blockade was a more economical method of watching the enemy fleet than attempting to find its scattered parts in the vast expanses of an ocean.[36]

One of the prerequisites for success in conducting a close naval blockade was to have a superiority in numbers over the blockaded force. In the sailing era, a blockading force slightly larger or equal to the blockaded fleet was often unsuccessful in preventing a blockaded force from escaping. The Turkish fleet of 76 ships blockaded in the Dardanelles in 1695, by a Venetian fleet of only 26 ships, was able to escape.[37]

In conducting a close blockade, the individual elements of the blockading fleet were within mutually supporting distance. A close naval blockade to be effective required the availability of geographically favorably located bases for the blockading fleet. Therefore, it was critical to have available bases near major ports and in the vicinity of the focal area of shipping. In the eighteenth century, Gibraltar and Minorca provided the Royal Navy opportunities to resupply and refit ships assigned to a long period of blockade duty in the Mediterranean. In 1794, the British lost Minorca but obtained a new base on Corsica used for resupply of the British ships blockading the French Mediterranean ports. However, in 1803, Nelson lacked favorably situated bases for the conduct of a close blockade of Toulon to prevent the French from acting against Sicily or some other part of the Mediterranean. Although Nelson's nine ships-of-the-line were badly in the need of repairs, he kept them constantly on the move. Naples and the ports of Sicily were closed to British ships. Nelson did not have a regular base nearer than Gibraltar or Malta. Gibraltar was inadequate as a fleet base because it was some 900 miles from his station off Toulon and Malta was some 700 miles away, or six to seven weeks of sailing time. However, Malta was the key for control of the western and central Mediterranean. The British ships based there were used to watch the French bases in Apulia and the entrance to the Adriatic, and to protect British commerce in the Levant.[38]

Bases flanking the enemy transit routes are especially valuable in conducting a naval blockade. The British used their naval bases in the Channel to provide support to squadrons blockading the French base at Brest. They also used Maddalena Bay in Sardinia during their blockade of Toulon in 1804 and 1805.

The steady and uninterrupted resupply of the blockading ships was another prerequisite for the success of a close blockade in the era of sail, as was shown by the Dutch in maintaining their blockade of the Spanish fleet at anchor in the Downs in 1639. Admiral Edward Hawke's ships were only able to remain on station in the 1750s because they could be continuously resupplied from Portsmouth.[39]

In the era of sail, the blockaded force was often able to escape from the port if it was composed of shallow draft vessels and the blockading force lacked the same. In July 1790, the Russians blockaded the Swedish fleet in Vyborg Bay. By using a diversionary surprise attack, combined with the covering fire from the larger warships, the Swedes, with their shallow draft galleys and transports, succeeded in slipping out of the ports through the shallow waters.[40] In the opposite scenario, a blockading fleet cannot always be successful if it does not possess ships capable of operating in shallow waters. This was one of the reasons for the rather ineffectual blockade by the British of the Russian fleet in the Baltic in 1854–55.

Sometimes, the use of some new and dramatic technological advances was the key for success of the blockading force. The Dutch used fire ships in their blockade of the Spanish fleet in the Downs in October 1639. During the blockade of Copenhagen in September 1807 the British ships used new incendiary shells to bombard the city, while the Russians used explosive shells in their blockade of the Turkish fleet at Sinope (Sinop) in November 1853.

The weather also played a great role in the conduct of close blockade in the sailing era. The ships were often forced to abandon their patrolling areas off the enemy ports and that gave an opportunity for the blockaded fleet to escape. In blockading the French base at Brest in November 1759, Admiral Hawke was forced by severe weather to abandon his blockading position, thereby allowing the French ships to escape. Nevertheless, Admiral Hawke went in pursuit and destroyed the French fleet in Quiberon Bay. The weather was one of the main factors in deciding whether a close blockade should be conducted.

A close blockade was generally most successful when it was combined with attack on land. A Swedish fleet under Count Karl Gustav Wrangel blockaded Copenhagen by land and by sea in 1658. The Danish fleet, equal in numbers, engaged the Swedes in the Sound. Although the battle was not decisive for either side, it was sufficient that Swedes lifted the blockade of Copenhagen. In 1807, the British blockaded Copenhagen from the sea and landed their troops on Sjælland to blockade the city from the land side.[41]

Disadvantages of a close blockade were increased in the age of steam because the large consumption of fuel by the blockading forces greatly increased the logistical needs of the fleet. Shifts in wind no longer granted respite to the blockading squadron. Coal-burning ships did not have the capacity for almost indefinite cruising and, hence, the task of on-station replenishment became even more essential than it was in the era of sail. However, fleets did become independent of the wind.

The French planned to conduct a blockade of Germany's coast in the Franco-Prussian War of 1870–71, but because of the rapid advances of German armies on land the blockade had to be lifted. This example highlights the importance that operations on land have on the conduct of naval blockade.[42]

In the Spanish–American War, the US Admiral William T. Simpson blockaded the Spanish squadron under Admiral Pasquale Cervera at Santiago Bay on 28 May 1898. Simpson, unable to force the bay's narrow entrance, requested the army's help to suppress shore batteries. However, US troops became bogged down elsewhere and therefore failed to neutralize the shore batteries. The Spanish squadron tried to escape but was subsequently destroyed by superior US forces. Within two weeks Santiago fell into American hands.[43]

A close blockade was conducted for the last time in the Russo-Japanese War. The blockading Japanese fleet in front of Port Arthur suffered disproportionately high losses from Russian mines. Not only did Admiral Togo lose two battleships, or one-third of his battle force, but he was also forced to loosen his pressure on Port Arthur. Togo's worries were increased greatly by the Imperial General Staff's delay in landing an army at Yentoa Bay on the Liaotung Peninsula, only 53 miles from Port Arthur. This action was very risky because the Russian squadron at Port Arthur could possibly have intervened and attacked the Japanese army's flank during one of the absences of the Japanese battleships for coaling or maintenance. Admiral Togo asked for a dozen merchant ships to be used for the blockade of Port Arthur, but that request was denied because of the general shortage of shipping for the transport of troops and supplies.[44]

Distant Blockade

By the late nineteenth century, close blockade had become more costly for a superior fleet because of the defender's use of mines and torpedoes. To these hazards were added submarines and aircraft in the years just prior to the outbreak of World War I. A *distant blockade* emerged as the solution for a stronger fleet to neutralize the weaker fleet. Its primary objectives were confining the enemy fleet to its ports, watching its movements, and bringing it to action if it came out. In this way the stronger fleet tried to occupy a position from which it was able to intercept the weaker fleet and bring it to action.

A distant blockade leads necessarily to a contested command of the sea, a kind of naval no man's land. The inferior force can operate at will, not only with submarines, but also often with surface ships. A distant blockade also made greater demands upon one's own naval strength than a close blockade. The opponent could strike at any of several widely separated areas, thereby forcing the stronger fleet to keep large forces in more than one place. It is less effective than a close blockade because the moment a stronger fleet retreats to a distance, it concedes to the inferior fleet a certain freedom of action. It does not keep an enemy fleet immobile but only threatens the ships with interception and possible destruction if they slip too far away from their bases or ports.

The differences between a close and a distant blockade lie in the proximity of the blockading squadron to the enemy bases and the degree of certainty of contact when the enemy fleet comes out. In contrast to a close blockade, the fleet conducting a distant blockade holds an operationally exterior position in regard to the blockaded force and hence operates along exterior lines of operations.

The Russo-Japanese War of 1904–5 showed how difficult it was to maintain a close blockade of enemy bases because of the threat to one's ships from enemy mines and torpedoes. The advent of submarines and aircraft made this problem even more difficult. However, many navies of the day did not realize that the time to conduct a close blockade of a large fleet had passed forever. The Imperial German Navy assumed that the British Grand Fleet in case of war would sally forth to blockade the Heligoland Bight at close range, which would have offered the German High Seas Fleet the opportunity to counterattack, especially by using its highly trained torpedo boats. The Germans completely ignored the fact that the highly favorable strategic position of the British Isles with respect to its potential opponents bordering upon the North Sea and the Baltic gave the Grand Fleet practically all the advantages of command of the sea, which normally can be acquired only through a decisive battle or close blockade, but without having to run any significant risk.

A distant blockade was instituted by the Royal Navy in the North Sea during World War I. The backbone of this blockade was the Grand Fleet and the objective, in part, was to lure the German High Seas Fleet out of its bases and to destroy it on the open sea. Although the British failed to achieve their primary objective, they substantially succeeded in their other objective, the protection of maritime interests behind the blockading line. Offensively, the control of German shipping which the Grand Fleet exercised from Scapa Flow and the Channel ports was no less complete than if the German North Sea ports had been blockaded at close range. It was probably even more effective in the final result because the blockade included the neutral countries adjoining Germany as well. Defensively, a long-range blockade was sufficient to accomplish its main purpose. As long as the northern and southern exits from the North Sea were firmly closed, German heavy surface ships were unable to attack the Entente's shipping on the Atlantic routes. The German Navy, for all its impressive strength, was literally locked in, and operated only tactically within the confines of the North Sea. However, the Germans did in fact operate offensively there, for their U-boats slipped through the blockade to attack British commerce on the open ocean.

A distant blockade forces a stronger fleet, while exercising overall control over a given maritime theater, to allow a certain degree of tactical freedom of action to the weaker fleet. The reason for this is that no complete and permanent control of the surface, the subsurface, and the air can be obtained, unless the enemy fleet is destroyed. Thus, a stronger fleet may occasionally conduct a major naval operation if the inferior fleet dares to challenge its control of the sea or interfere with maritime trade or amphibious landing. Likewise, the inferior fleet can conduct a major naval operation to accomplish similar objectives.

A distant blockade of a semi-enclosed sea normally does not prevent the weaker force from exercising control of the adjacent enclosed sea theater. The Germans possessed control of the Baltic throughout the whole of World War I, despite the British distant blockade of the German fleet in the North Sea. In World War II, a distant blockade provided the stronger fleet with the best means of containing an inferior though potentially dangerous enemy, and so improved its chances of being able to use the sea in relative security. At the same time, technology and strategic circumstances gave the blockaded navies (those of Germany and Italy) much more freedom of action than they had had in the past.

In the past, a stronger sea power normally declared a naval blockade at the outset of the hostilities, while the inferior fleet responded by a counter-blockade. The stronger fleet usually tried to neutralize or destroy the inferior fleet at sea and in its bases if the opportunity presented itself. Blockade is inherently attritional in its nature and, therefore, the majority of actions conducted by one's naval forces and land-based aircraft in a typical narrow sea are tactical in character.

Since World War II naval blockade has evolved to become even more integrated and indistinguishable from other actions by naval forces. Naval blockade is a traditional term rarely used today, because it signifies that the stronger fleet intends to apply a drastic form of pressure on its opponent. Modern terms used are 'maritime exclusion zone' (MEZ) and 'total exclusion zone' (TEZ). The US Navy used the more benign term 'quarantine' during the Cuban missile crisis of 1962. The main objective of naval blockade today is to prevent any hostile submarine, surface ship, or aircraft from leaving its basing area and thus endangering one's control of the sea area behind the blockading line. At the same time, a naval blockade must keep the enemy's ports closed to traffic.

In the modern era, the success of a distant blockade depends on having much greater strength than the blockaded fleet. Distant blockade is much less successful against a continental country than against an island nation or a peninsular country. It is also closely tied to the progress of war on land. The lack of a formal declaration of war hinders the ability of a stronger fleet to impose a naval blockade on a weaker fleet. Superiority in the air is invariably one of the key prerequisites for the success of a naval blockade in a typical narrow sea. The existence of sheltered passages within a blockade zone is usually a distinct disadvantage for the blockading force in enforcing a naval blockade. Also of great importance is the ability to introduce some new technological advances to counter the technological advances of the inferior fleet. The support of allies or coalition partners and, in many cases, the implicit support of neutrals are other important factors for the success of a naval blockade in a typical narrow sea.

Modern forms of naval blockade are *sortie control* and *choke-point control*. In the first method a weaker fleet is bottled up or itself chooses to take refuge in its bases or ports. Individual submarines or surface ships are destroyed if they try to sail out. A blockade of enemy naval bases is applied primarily by the side on the offensive. In favorable conditions it usually ends by seizing enemy ships by an attack from the land side. The principle of concentration of forces should be applied to obtain superiority over the adversary. Sortie control primarily depends on the effective employment of small surface combatants and mines, as well as submarines, by the stronger fleet. In enclosed seas, land-based aircraft are a critical factor in the success of this form of a naval blockade.

Often, the best place to engage the enemy fleet is in a geographical bottleneck through which it must pass or by conducting choke-point control. A blockade of the exits from a larger narrow sea such as the Mediterranean is essentially a form of a distant blockade. A blockade of the larger gulfs, even if they are called seas, such as the Red Sea or the Adriatic, is essentially a form of close blockade. Blockade of straits, narrows, gulfs, or bays were extensively used in both world conflicts.

The main purpose of blockading a choke point is the disruption of enemy maritime communications and the prevention of deployment and maneuver of his fleet forces. Blockading the only exit of a sea is one of the most effective methods of achieving the main prerequisite for neutralizing and subsequently destroying the enemy fleet. Specifically, the objective is to prevent the enemy fleet or individual submarines and surface ships from transiting straits or narrows to reach the open water of the sea or an ocean. A stronger fleet conducting choke-point control can use smaller units, which normally cannot be used for sortie control.

Control of a sea's only exit is not in itself sufficient to obtain control within the blockaded sea, as the experience of the Entente navies in the Adriatic Sea in 1914–18 illustrates. The French fleet blockaded the Strait of Otranto early in the war, but made only occasional forays further north. The French ships bombarded the outer fortifications of the Bay of Cattaro but never tried to penetrate into the base. They sailed only once into the central part of the Adriatic, near the island of Lissa, which lies about 150 miles south from the Austro-Hungarian main naval base at Pola.[45] This situation left the weaker Austro-Hungarian fleet with almost undisputed control of the Adriatic throughout the war. If the Entente navies had made a strong effort to defeat the Austro-Hungarian forces they could have endangered the flank of the Central Powers and the Dalmatian coast would not have been used by the German and Austro-Hungarian U-boats as a base of operations for their attacks on the Entente shipping in the Mediterranean.[46]

The main prerequisite to ensure a successful blockade of a strait or narrows is to seize (preferably from the land side) all or most of the enemy naval bases in the vicinity. The main obstacle to establishing a permanent blockade of the Adriatic by the Entente and Associated Powers in World War I was the existence of the advanced base at the Bay of Cattaro used by Austro-Hungarian light forces to carry out many raids against Allied ships guarding the Strait of Otranto. On 15 May 1917 an Austro-Hungarian squadron tried to break the blockade of the Strait. Yet, despite sinking one Allied destroyer, two merchant ships, and 14 armed trawlers, this foray did not loosen Allied control of the Strait of Otranto.[47] The capture of the Bay of Cattaro by the Allies was not possible, however, because it would have required more force than they had available.

The Allies tightened their control of the Strait of Otranto at the conference in Rome held in November 1917. They decided to greatly improve the barrage by speeding up delivery of material to establish an anchored-net barrage and net mines between Otranto, Fano, and Corfu. In addition, about 90 destroyers and submarine chasers (SCs), 125 fishing steamers and a number of seaplanes were used in this effort. The British base at Otranto and the French base in Corfu were primarily used in the search for U-boats. These measures came into effect in May 1918 and remained in force until the end of war.[48]

A neutral country guarding a sea's only exit, can, by closing the waterway on both sides, in effect create a more favorable situation for a fleet trying to prevent a weaker fleet within the confines of a narrow sea from operating on its trade routes. This was the case with the Danish decision to close the exits from the Belts and the Sound by laying minefields in 1914. Curiously, this decision was made with the approval of both the British and the German Admiralty. Denmark recalled all personnel for its sea fortresses on 31 July 1914 and for its fleet a day later. After mobilization of the German fleet, the Danes deployed about 16,000 troops east of the Great Belt (Store Baelt) and 2,500 troops west of it. On 4 August, the Danish Government informed all the belligerents about its state of neutrality.[49] The Danes were determined to maintain strict neutrality and to prevent all military operations in the straits. Hence, blockade of the straits was extended to the Danish waters of the Great and Little Belt and the Sund. Between 5 and 10 August about 1,400 mines were laid in these waters.[50] By this action the Germans were unable to move their naval forces out of the Baltic into the North Sea through the Danish Straits.[51] The closure of the Danish Straits deprived the German fleet of the possibility of transiting the Skagerrak and reduced its lines of retreat to one, that from the North Sea to the Heligoland Bight.[52]

An inferior fleet guarding a sea's only exit can in turn be blockaded by a stronger force, as happened to Turkey in World War I. The Entente navies established a combined naval and commercial blockade of the Turkish coast in the Aegean Sea on 2 June 1915. The British Navy was responsible for the Smyrna Patrol designed to watch from the Gulf of Adramyti (Edremit) to Khíos. The French Navy monitored traffic from the southern point of Greece to Marmarica on the Asiatic coast and southward to the island of Samos (Sámi). The British Southern Patrol watched the area between Mudros and the Gulf of Adramyti, while the Northern Patrol guarded the northern part of the Aegean.[53]

A stronger fleet is often unable to break through the blockade of a sea's only exit imposed by a weaker fleet. This was the case in 1915, when the Turks closed the Dardanelles by using mines and coastal guns to enforce a blockade. Allied naval forces alone without troops on land proved unable to seize the control of the straits. The Turkish action had the most momentous consequences because the links between Imperial Russia and its western allies were cut-off for the duration of the war. In other cases, a mere threat by a blue-water navy may be sufficient for a weaker fleet to abandon blockade of the sea's only exit. This happened in the Yom Kippur (Ramadan) War of October 1973. The Egyptian blockade of the Bab el-Mandeb Strait was lifted after a US carrier task force was sent to the area to protect US shipping.[54]

A stronger fleet may be prevented from operating in the main theater by a decision to make one of the sea's exits impassable to any ship. This happened shortly before the outbreak of the Six Day War of 1967, when Egyptian President Gamal Abdel Nasser ordered that the Suez Canal be blocked by sinking barges and a few unusable Egyptian merchant vessels. This action was designed to make it impossible for the Israelis to use the Canal in case they should seize the Sinai Peninsula. The Egyptian action effectively bottled up about 30 per cent of the Israeli navy (two destroyers and six torpedo craft) in the Red Sea, where they were a small threat to the Egyptians.[55] At the same time, this action prevented any redeployment of the Egyptian surface ships and submarines from the Red Sea to the Mediterranean and vice versa.

Offensive Mining

Mines are one of the most effective weapons to blockade enemy naval forces, the enemy fleet's basing and deployment areas and the transit routes of enemy surface ships and submarines. The objective is to make the operational deployment of enemy fleet forces difficult or to destroy enemy surface ships and submarines on their transit routes and in their operating areas. The areas mined by both sides in a conflict in a typical narrow sea considerably change the geography. They should be treated like a land obstacle and thus, whenever possible, avoided. The *suspected* presence of a minefield may achieve results comparable with the existence of a real one. Also, it is often very useful to attempt to convince the enemy of the existence of a fictitious minefield.[56]

Mines can be used operationally or tactically. In the first case, major parts of a narrow sea or a sea's only exit are mined to deny the sea area to the movement of the enemy fleet. Tactical use of mines is aimed at inflicting damage, delaying or hampering enemy naval activities, and closing shipping traffic routes.[57] To remain effective, minefields should be protected by other forces, specifically light forces and coastal anti-ship missile and gun batteries.

Mines played a significant role in the war at sea during the Russo-Turkish War of 1877–78. The Russians neutralized a major part of the Turkish fleet by mining selected parts of the Danube Estuary. Although only one Turkish ship (an ironclad) was sunk, the Russians obtained the initiative in the war.[58]

In the Russo-Japanese War of 1904–5 the Japanese used mines not only to block the movements of the Russian ships out of their Port Arthur base, but also to lure the energetic Russian Admiral S. O. Makarov out of Port Arthur to do battle. Before any engagement had taken place, the main Japanese forces laid a new minefield, thereby sending the Russians back to their port. The Russian battleship *Petropavlovsk* ran onto the mines and sank immediately, with about 600 men, including Admiral Makarov. Control of the Yellow Sea was in Japanese hands and this allowed them to debark their troops undisturbed by the Russian fleet.[59]

In both world wars, mines were used extensively and sometimes with great success to restrict or to prevent completely enemy naval movements through a strait, narrows or large gulf. The Dover Barrage, comprising both deep and shallow minefields with detector loops and flare drifters between Dover and Calais, was established in November 1917. Because the Strait of Dover was closed, the Flanders Flotilla U-boats began in May 1917 to use the northern route, which, though safer, added greatly to their transit time to the operating areas. In July 1917 seven U-boats tried to use Dover again but only one returned. By the end of September 1918 several in a group of 12 U-boats were destroyed while trying to cross the barrage. Altogether, a total of 27 U-boats were lost to mines of the Dover Barrage.[60] This is in contrast to the one U-boat destroyed by mines laid off the Flanders coast.[61] The first British offensive minefields were laid off the Flanders coast in October 1914 and more mines were laid there as the war progressed. The British also mined large parts of the Heligoland Bight, starting in January 1915. These minefields tied down considerable forces of the German High Seas Fleet in sweeping them. They also forced the Germans to use the much longer route through the Skagerrak for their U-boats. This, by increasing their time in transit, reduced their number in the operating areas at any given time.[62]

The Northern Barrage was established primarily on US initiative between March and early June 1918 to restrict the movements of German U-boats from the North Sea into the Atlantic. In its final form, the barrage stretched for 240 miles between the Orkneys and the Hardanger Fiord (Norway). Although the initial requirement called for some 200,000 mines, in the event only about 15,100 British and 56,000 American mines were laid.[63] Yet, despite this effort, the Northern Barrage proved not very effective since the Germans continued their U-boat activities as before. Probably not more than two U-boats were destroyed by the barrage, and they continued to use the same route from the North Sea that they had used before.[64]

The Entente's barrage in the Strait of Otranto was designed to prevent German and Austro-Hungarian submarines from exiting or entering the Adriatic. Initially, the barrage consisted of net-drifter patrols, established in October–November 1915. However, the German and Austro-Hungarian U-boats left the Adriatic at will and returned from their patrols in the Mediterranean to their base at the Bay of Cattaro. The Otranto Barrage was much strengthened in September 1918 by laying successive lines of deep mines to catch the U-boats that dived under the nets, and by laying a shallow minefield between Otranto and Cape Linguetta. However, the overall effectiveness of the Otranto Barrage was not great because the water's depth exceeded 3,000 feet, more than the depth of water in the Northern Barrage.[65]

Throughout World War II, a major part of the Royal Navy and Coastal Air Command was used to maintain a blockade of the North Sea's exits. At the outset of war the Royal Navy instituted a blockade of the English Channel, the Strait of Dover and the northern Atlantic, through which passed the deployment routes of the German U-boats. Some 10,000 mines were laid in the St George's Channel, but no U-boat was lost on any mine laid. The series of minefields laid off England's east coast in 1939–40 (approximately 35,000 mines) also failed to have any appreciable influence on U-boat activities.[66] The British plan of November 1939 was to lay a 250-mile-long mine barrier between Scotland and Norwegian territorial waters. However, Germany's capture of Norway made this plan impracticable. Hence, in May 1940 it was decided to lay a barrage with about 81,000 mines between Scotland, the Faeroes, Iceland and the Denmark Strait. The existence of this mine barrier was declared in July 1940. However, the barrier resulted in only one U-boat being destroyed, and there was no appreciable effect on operations by U-boats.[67] The British also laid an anti-submarine mine barrage in the Dover Strait.[68] After losing three U-boats, the Germans were forced to abandon the Strait of Dover as a deployment route for their submarines.[69]

The British laid a total of 126,000 mines between the Orkneys and Iceland, and the St George's Channel. However, despite this monumental effort, there is no evidence to indicate that mining had any appreciable effect on German naval activities in the affected areas.[70]

The British also blockaded the Strait of Gibraltar in World War II to prevent German U-boats from entering or exiting the Mediterranean. Nevertheless, a total of 62 out of 93 German U-boats were successful in transiting the Strait of Gibraltar.[71]

The most effective method of bottling up the remnants of an enemy fleet in its bases is to establish several successive mine barriers across the gulf or narrows serving as the only possible transit route to reach the open waters of the respective narrow sea. In June 1941 the Germans planned to lay a number of mine barriers intended to restrict the freedom of movement of the Soviet Baltic Fleet, and thereby take away the Soviet freedom of action to conduct offensive warfare at sea.[72]

The Germans also envisaged the laying of a large mine barrier between Memel and Oeland, while small mine barriers were planned to close the exits from the Bay of Riga.[73] The first German mine barrier, *Wartburg*, was laid from the tip of Oeland to the Lithuanian coast between 18 and 21 June 1941. Under German pressure, the Swedes also laid within their territorial waters a mine barrier to supplement the *Wartburg* barrier. On the night of 21–22 June the Germans also mined the approaches to Libau, Windau, Soelosund, and Moonsund, as well as the Irben Passage.

As it turned out, the Germans did too much mining in the Baltic because the Soviets showed no desire to break out from their home bases. Moreover, the *Wartburg* barrier made the movement of German forces much more difficult and by the end of 1941 had caused the loss of at least ten German merchant vessels and two minesweeping craft. This example shows that offensive mining should not be overdone, because too many mines in an area required for the transit of one's own surface ships or submarines may restrict one's own freedom of maneuver and cause unnecessary losses.[74]

The Germans also planned to close the Gulf of Finland completely by laying several mine barriers patrolled with surface craft and defended by coastal gun batteries. By late June 1941 the German Naval Staff believed that it was only a matter of time before the Soviet Baltic Fleet would be bottled up in its bases in the inner part of the Gulf of Finland and ultimately eliminated from the operational area.[75] The first mine barrier in the Gulf of Finland was laid on 12 June 1941 by German ships camouflaged as merchant vessels. Another mine barrier, *Apolda*, was laid down between Oro and Takhona on 21 June 1941 and the *Corbetha* barrier between Kallbadargrund and Pakerort. After the Finns entered the war on 26 June, their ships laid the *Kipinola* and *Kuolemanjarvi* minefields. Between 22 and 28 June 1941 the Finnish submarines laid a number of mine barriers in the area between Hochland and Great Tueters.[76] The Finns were not enthusiastic about massive mining of the Gulf of Finland because of the subsequent restriction on the movement of their own forces.[77] The Germans, however, laid a large mine barrier (*Juminda*) in the Gulf of Finland in late August 1941. The Soviets, seemingly insensitive to losses, sporadically broke through the barrier by using submarines. The mine barrier was then reinforced to prevent the Soviet supply of and evacuation from the fortress Hangö.[78]

In contrast to the Germans, the Soviet use of mines for offensive purposes in World War II was quite ineffective. In the Black Sea, except for the Kerch Strait, the effectiveness of the Soviet offensive mining was negligible, both in the number of mines used and in the results achieved. This was due to the use of technically obsolete mines and the passive measures employed by Soviet surface ships to lay offensive mine barriers.[79]

In the Pacific War, US Army aircraft mined the straits in the Inland Sea to limit Japanese naval movements. In support of the Okinawa landing (Operation *Iceberg*), US aircraft mined the Shimonoseki Straits, the naval bases at Kure and Sasebo, and the military port of embarkation at Hiroshima. The objective was not only to make the movements of the Japanese warships more risky, but, more importantly, to prevent the

Japanese from sending their fleet through the Shimonoseki Straits southward to Okinawa and impeding the Allied landings. The mines forced Japanese ships attempting to sortie from the Inland Sea to use the Bungo Suidō exit, where US submarines aided in the detection and destruction of the *Yamato* task force.[80]

One of the principal objectives of NATO maritime strategy during the Cold War was to prevent the exit of the Warsaw Pact naval forces through the Danish Straits or the Turkish Straits in European waters, or the exit of the Soviet Pacific Fleet through La Perouse Strait and the Korea Strait in the Pacific. Denmark and Turkey by virtue of their geostrategic position were able to effectively block the egress of the respective Soviet fleets to the open ocean.

Seizing Choke Points

General control of a narrow sea can be accomplished in wartime by physically controlling or seizing one or both shores of a sea's exits. The Italians controlled both shores of the Strait of Otranto in both world wars. What they failed to do in World War I, the Germans accomplished in World War II by obtaining control of one shore of the Danish Straits and the Norwegian coast facing the Shetlands. They also obtained a full control of the entire Channel coast in 1940, a success that eluded them in World War I.

In World War I, the German Navy was unable to operate on the open waters of the Atlantic because the Army failed to seize the Channel ports in the first months of the war in 1914. Admiral Alfred von Tirpitz noted that one of the most important German objectives was to seize the Channel ports and thereby prevent transport of the British Expeditionary Force to Cherbourg or Brest and to operate in the Atlantic instead of within the narrow confines of the North Sea. The Germans' steady interference with the sea traffic at the approaches to the Thames estuary would inflict serious economic losses on England. He also thought that the capture of the coast of Flanders was inadequate to threaten immediately and seriously the French Channel ports. By using the Dutch and Belgian ports the Germans carried out sporadic attacks against British sea traffic in the southern part of the North Sea, but without having a decisive impact on the war at sea as a whole. To complicate matters, the German operational planning prior to the war was almost exclusively focused on a general fleet action with the Grand Fleet in the North Sea. Their fleet held a very favorable position in the southern part of that sea, because of their control of the ports of Zeebrugge, Ostend and Antwerp which could be used as bases for attacks against British warships and transports at the approaches to Dunkirk,

Calais and Boulogne.[81] Likewise, the Austro-Hungarian army failed to capture the Albanian coast all the way up to Valona, the best natural harbor along the Albanian coast. The Austrians reached the port of Durazzo in 1916, but afterwards the front stabilized and remained stationary until almost the end of war.[82]

Control of a sea's exits by itself cannot usually secure absolute control within an enclosed or semi-enclosed sea. To obtain such control it is necessary to seize or neutralize some operationally important positions within a narrow sea. During World War II, the Royal Navy was in full control of the two Mediterranean exits: the Strait of Gibraltar and the Suez Canal. Yet, control of the central Mediterranean was in dispute in 1940–42. The island of Malta barely functioned as a naval and air base because of almost constant attacks by the German and Italian aircraft based on Sicily and Sardinia. The British did not control either the western or the eastern part of the Mediterranean, because all operationally significant positions were in Axis hands until the summer of 1943. German control of the Danish Straits in World War II, however, did not ensure them full control of other parts of the Baltic, specifically the Gulf of Finland. Plan *Barbarossa* anticipated that the Finns would seize key positions dominating the western end of the Gulf of Finland, the Aaland Islands and Hangö, at the outset of hostilities with Soviet Russia. Afterwards, German S-Boats and Finnish craft would be employed up to longitude 29°E. Any Soviet attempt to break through the blockade would be prevented. Finnish submarines would operate in the inner part of the Gulf of Finland against Soviet battleships and cruisers, while German U-boats would do the same in the eastern part of the Baltic.[83]

The Iranians in their war with Iraq (1980–88) held control of one shore of the Strait of Hormuz. The Strait of Hormuz waterway was within effective range of the Iranian coastal anti-ship missile batteries and was controlled by Iranian surface warships. However, this in itself proved to be insufficient to obtain absolute control over the Arabian Gulf, because other riparian states either controlled or shared control of some operationally significant positions within the Gulf. The Iranians also seized the Al Fāw Peninsula (in the northern Gulf) in 1986. The Iranians fired *Silkworm* missiles from their positions on the Fāw Peninsula, starting in September 1987 against ships using the port of Kuwait City. However, the intensity of their attacks was low (on average one *Silkworm* missile launch per day). The Iraqi army recaptured Al Fāw in April 1988 and with that ended Iranian attacks on Kuwait.[84] To secure control of a larger part of the eastern part of the Gulf, the Iranians deployed small detachments of the Revolutionary Guards (*Pasdaran*) on Lārak, Hengion and the Sirrī Islands. Field gun and AA-gun positions were built up on the Qeshm and Greater Tunb Islands. In addition, the Iranians reinforced their forces on Fārsī Island.[85]

Securing Control by Land-Based Aircraft

Land-based air power plays a key role in securing and maintaining command of a narrow sea or parts of it. Because of the growing range, endurance and speeds of modern aircraft ever-larger sea and ocean areas become the areas of combined employment of naval forces and land-based aircraft. Today, virtually no part of a typical narrow sea is free from observation and attack from the air.

Struggle for air superiority in narrow seas cannot be separated from one fought in the airspace over the adjacent coastal area. Short distances increase the effectiveness of air strikes against enemy ships and targets on the coast considerably more highly in a narrow sea than on the open ocean. Land-based aircraft in a typical narrow sea can fly more sorties within a given time frame than on the open ocean. In a sea with many offshore islands and islets, land-based aircraft can strike from bases flanking the transit routes of enemy ships. The aircraft can be quickly deployed and redeployed from one airfield to another or from one part of the sea to another. The arrival of the German X Air Corps in Sicily from Norway in January 1941 changed the entire operational situation in the Central Mediterranean practically overnight. The Germans possessed enough airfields on land, and combined with bases on the Dodecanese, they were able to cover all the coasts of Libya and the Eastern Mediterranean. Their aircraft could also relatively easily be serviced and maintained, thereby allowing a large number of them to be kept operational at any one time.

Sometimes air power alone can obtain control of a major part of a narrow sea as shown in the German example in the Mediterranean in 1941–42. The Germans strengthened their hold on the Aegean after seizing mainland Greece by mounting a major airborne operation on 20 May 1941 aimed at capturing the island of Crete. Crete was defended by about 28,600 British, Australian and New Zealand troops, plus two Greek divisions.[86] The German forces that landed on the island were composed of the XI Air Corps with the 7th Parachute Division, the 5th Mountain Division, and a part of the 6th Mountain Division, and some other units which were transported by about 500 Ju-52 aircraft.[87] The entire operation was supported by the *Luftwaffe*'s VIII Air Corps. The Germans dropped about 3,000 troops of the 7th Parachute Division on the first day of the operation, 20 May.

Crete's defenders offered unexpectedly strong resistance, and the Germans failed to make substantial progress in the first days of their invasion. The Germans' initial plan was to bring troop reinforcements and heavy weapons by sea. However, lack of large ships forced them to improvise by organizing two transport groups (convoys) of Greek *caiques* (powered sailing craft) carrying on board two battalions of the 5th

Mountain Divisions and heavy weapons, including some tanks. The first convoy of 25 *caiques* almost reached Crete on 21 May, but the Royal Navy ships, after some strong resistance by the Italian destroyers sank most of the ships carrying troop reinforcements. A day later, the British turned back the second attempt by the Germans to bring in reinforcements to Crete. During their invasion of Crete, the Germans inflicted heavy losses on the Royal Navy in the first major air-sea battle of World War II, sinking two cruisers and one destroyer, and damaging several other ships.[88] The Germans eventually brought in some 22,000 troops to seize the island after 12 days of heavy fighting.[89]

The Germans had had a plan to seize Cyprus, and then the Suez Canal from the air. While reportedly sympathetic to the idea, Hitler did not endorse it, probably because his mind was too preoccupied with the pending invasion of Soviet Russia. A year later, Hitler initially approved the plan to seize Malta from the air (Operation *Herkules*). However, he changed his mind, and canceled the entire operation because of his fears that if the Royal Navy appeared on the scene, the Italians would go home and leave the German troops stranded on the island.[90] This decision in retrospect seemed to have momentous consequences for the future course of the war in Northern Africa, because Malta being in Allied hands was crucial to control of the central Mediterranean.

In some cases, a multitude of offshore islands spread over a relatively large area allows one's own forces to attain command of the sea by using predominantly land-based air supported with either surface naval forces or ground troops or both. After seizing Crete, the Germans rapidly extended their hold over the Aegean by occupying the key islands in the archipelago. They also reinforced the Italian-held garrisons in the area. Not only was the Aegean virtually closed to Allied shipping, but also good use was made by the Germans and Italians of island bases for carrying out air attacks against British convoys in the eastern Mediterranean. German and Italian submarines made regular use of their harbors and the light forces and aircraft stationed there were able to command the Aegean sufficiently to protect their own coastwise and inter-shipping traffic.

In another example, the German occupation of Cos in the Dodecanese Island group in October 1943 was carried out mainly by transporting troops by air, while destroyers were used to transport troops to the island of Léros, whence some troops were ferried north to Sámos in small craft.[91] The German success was in part due to their ability rapidly to reinforce their air assets in the area. By early September 1943, the *Luftwaffe* had only 285 aircraft in the theater, and there were no long-range bombers. However, within a week of the British occupation of the Aegean Islands, German reinforcements started to arrive. By 1 October the *Luftwaffe*'s strength had risen to about 360 aircraft, including a substantial number of long-range bombers. The most important factor was the proximity of the airfields used by the *Luftwaffe* to the decisive points in the area. Cyprus was only 350 miles from Cos, while the Germans had two airfields on Rhodes, some 70 miles from Cyprus, and two more on Crete, which was only 150 miles away.[92]

Seizing Basing Areas from the Land

Strategic objectives in a typical narrow sea can be accomplished by employing one's ground troops to capture the choke points, a large part of the mainland coast, and key offshore islands with their naval/air bases and ports. As one's troops advance along the coast, the enemy's naval position is also steadily reduced. Full sea control is obtained after the enemy's entire coast or archipelago has been seized by one's troops. With this, the enemy fleet is either destroyed or interned in neutral ports. These objectives can be accomplished even if the advancing army on the coast has little or no support from its own fleet forces, as the German and Soviet actions in the Baltic in World War II clearly demonstrate. However, there is no question that it is usually preferable to use fleet forces and land-based aircraft in support of the advancing troops on the coast.

By their rapid advances along the coast German ground troops almost obtained full control of the Baltic and the Black Sea in 1941–42. The initial operational plan for German Army Group North envisaged blockading the Soviet ships based in Libau, Windau and Riga by using mines, and finally seizing the Baltic ports from the land side. As it turned out, Army Group North advanced quickly along the Baltic coast in the first few weeks of the war. By 1 July, German troops seized Windau, and Riga fell into their hands three days later. The southern shore of the Gulf of Riga including the port of Ust'–Dvinsk was also seized, and by 8 July the Germans had entered Pernau (Estonia). However, by mid-July the German advance had slowed down because of the stiffening of Soviet resistance and the front became stabilized northeast of Riga. The Soviets held off the German attacks along the coast and thereby kept Reval in their hands. However, the German 18th Army moved further east and by 7 August had reached the coast of the Gulf of Finland at Kunda and Juminda Cape, about 65 and 30 miles east of Reval respectively. By this, the Germans cut off the land connection between Reval and Narva.

On 27 August the German 42nd Army Corps reached the suburbs of Reval. Still Soviet troops held off German attacks on the Baltic islands of Moon (Muhu), Oesel (Saaremaa), and Dago (Hiiumaa). They also held the fortress of Hangö in the southeastern part of Finland. By 21 October 1941 the Germans had occupied the entire Baltic coast. The Supreme

Command of the *Wehrmacht* thought briefly that the Soviets might decide to send their heavy ships to internment in Sweden. To prevent this, a 'Baltic Fleet' was formed (including among other ships the 42,000-ton battleship *Tirpitz*). This 'fleet' was in existence for only a few days before it was dissolved, because it became clear that the Soviets intended to concentrate their remaining battleships and cruisers in the Kronstadt–Leningrad area.

By 3 December 1941 the Soviets evacuated their base at Hangö, thereby allowing the Germans to obtain control of the eastern part of the Baltic.[93] The German Army, supported by the *Luftwaffe* played the principal role in reducing the Soviet naval position to only the small stretch of coast in the inner part of the Gulf of Finland. Besides Kronstadt and Leningrad, the Soviets retained control through the entire war of two islands in the inner part of the Gulf, Lavansaari and Seiskari. The *Kriegsmarine*'s activities were limited mainly to minelaying and escorting convoys. However, the German large ships were held back, allowing the Soviet Baltic Fleet almost unrestricted actions in the eastern part of the Baltic, specifically in the Gulf of Riga and the Gulf of Finland, until these positions were overrun by German troops.[94]

By early 1943 the stalemate on the Leningrad front prompted Army Group North and the German Navy's High Command to agree on a plan to clean up the Baltic by finally seizing Leningrad and Kronstadt. However, by the end of February the commander of Army Group North decided to shorten considerably the front around Leningrad by withdrawing to the new line at Narva. The German Navy's High Command protested to the High Command of the Army that in case of Army Group North's withdrawal, the Soviet Baltic Fleet would be able to operate freely in the Baltic. This would put at great danger German imports of iron ore by sea from Sweden and oil from Estonia. At that time, approximately 80 per cent of Germany's needs in iron ore came from Sweden, but because of requirements in other maritime theaters, the Germans were unable to reinforce their naval forces in the Baltic. The main argument of the German Navy's High Command was that the loss of control of the Baltic would mean losing secure training areas for the U-boats, which would have disastrous consequences for the conduct of the U-boat offensive in the Atlantic. Hence, the Navy's High Command instead proposed that Army troops should try to seize the Schepel-Oranienbaum area as the most effective way to obtain control of the entire Baltic area.[95] This proposal was never carried out, and the front around Leningrad remained stable until early 1944.

Soviet ground forces were the principal factor in the Soviet reoccupation of the Baltic coast in 1944–45. The Soviet offensive against Army Group North started in January 1944 when the German 18th Army was thrown back from its positions around Leningrad. By late June the Soviets after some very heavy fighting broke through the fortifications of the Finnish *Mannerheim* Line and captured the city of Vyborg. In July the Soviets resumed their offensive against Narva and forced the Germans to fall back. Reval was cut off, and soon Estonia was lost to the Germans. By September 1944 the situation for the Germans became even more precarious after the Finns signed an armistice with the Soviets. Soon afterwards the Soviets began the reconquest of the Baltic Islands. The Germans offered a stubborn defense but were forced to evacuate their last remaining troops from the Sorve Peninsula on 23 November.

In the Black Sea, a similar situation developed to that in the Baltic in 1941. The German Army Group South advanced through the southern part of the Ukraine with the two objectives of seizing the Crimea Peninsula with its large naval base of Sevastopol' and other Ukrainian ports, and of reaching the Soviet oil fields in the Caucasus. German and Romanian troops advanced quickly along the coast and seized the Soviet naval bases and ports on their way, but, as in the Baltic, the Germans encountered heavy Soviet resistance. The large port of Odessa was stubbornly defended by the Soviets. Finally, the Soviets in mid-October evacuated their best troops after realizing that the port was too far in the rear of the advancing German troops. Because of unexpectedly strong Soviet resistance, German troops did not reach the shore of the Sea of Azov until the end of September 1941. A month later, the Germans occupied most of the Crimea including the Kerch Peninsula. The Sevastopol' fortress, however, did not fall into German hands until early July 1942. Because of the almost negligible naval strength available to the Axis in the Black Sea, the Soviets conducted a number of amphibious landings in December 1941–January 1942. However, the Soviets were unable to regain command of the Black Sea, the Axis' maritime traffic was not stopped, and the Soviets did not succeed in forcing the Germans to leave the Crimea Peninsula.

The German offensive in the summer of 1942 almost led to the capture of all the remaining Soviet naval bases and ports in the Black Sea. By mid-July German troops crossed the Kerch Strait and their advance units entered Krasnodar on the Kuban' River in early August. They reached Novorossiysk, but failed to drive Soviet troops from their positions overlooking that port. The naval bases of Tuapse and P'ot'i and other smaller bases along the southern Caucasian coast of the Black Sea were never captured. The Germans failed to eliminate the Soviet naval forces and thereby obtain full control of the Black Sea. The activity of Soviet naval forces remained a constant nuisance to German supply traffic.[96]

Soviet fortunes improved after the end of the battle for Stalingrad in January 1943. The Soviet offensive forced the Germans and their allies to fall back and the front stabilized along the Mius River, that is, where it was in the winter of 1941–42. The Germans established the Kuban' bridgehead to supply by sea the troops on the southernmost part of their front. The Soviets gained access to the entire Sea of Azov, except for the Kerch Strait, which remained in German hands. Novorossiysk was liberated by advancing Soviet troops in September 1943. Shortly after abandoning the Kuban' bridgehead, the Germans were forced to withdraw to the Crimea. However, the Soviet advance continued along the coast and by the late October 1943 Soviet troops stood on the Dniester River. This forced the Germans to evacuate the port of Kherson, and the front became stabilized until the end of February 1944.

After the Soviets resumed their offensive in southern Ukraine in February 1944, their troops recaptured several smaller ports. By late March, the large port of Nikolayev was taken by the Soviets, thereby cutting-off a large number of German troops on the Crimea. The Soviet advance threatened other Ukrainian ports behind the German lines. The Germans conducted a large-scale evacuation of Odessa in April 1944, followed by the evacuation of troops from the Crimea in May. With this, the end of the Axis' position in the Black Sea was near. Soviet successes on land had other consequences, because in the summer of 1944 the Turks closed the Straits to Axis ships, and in August Romania and Bulgaria capitulated. With this the German naval forces lost their base of operations in the Black Sea.[97]

Securing control of a typical narrow sea by a blue-water navy can be accomplished by a combination of methods. The first and the most critical operational task is to obtain air superiority in the area. Simultaneously, or in short sequence, the enemy fleet must be destroyed at sea or in its bases by conducting a single or a series of major naval operations. Another effective but more time-consuming method for neutralizing the threat posed by a small navy is to conduct a distant blockade by seizing control of the sea's only exit and some operationally significant positions within a given narrow sea. This method will invariably lead to a war of attrition. A blue-water navy will eventually obtain sea control, but at higher losses and more time than if a decision is sought by conducting a major naval operation.

For a small coastal navy the problem of obtaining control of a narrow sea or even part of it is much more complicated. However, this task can be resolved in close cooperation with land-based aircraft and troops on the ground. The deficiencies in naval strength can often be overcome with the strength in the air and on the ground. A small navy that controls one or preferably both shores of a sea's only exit enjoys an excellent position to secure control of the major part of a given narrow sea. Because of the need to employ combat arms of several services, securing command of the sea by conducting major joint operations in a typical narrow sea is much more difficult and complicated to plan and prepare, but above all to execute, than it is in the open ocean.

NOTES

1 M. G. Cook, 'Naval Strategy', 2 March 1931, Air Corps Tactical School, Langley Field, VA, 1930–31, Strategic Plans Division Records, Series Box 003, Washington, DC: Naval Operational Archives, p. 11.
2 Alfred T. Mahan, *The Influence of Sea Power upon History, 1660–1783* (Boston: Little, Brown, 12th edn, 1918), p. 117–31.
3 Paul M. Kennedy, *The Rise and Fall of British Naval Mastery* (London: Macmillan, 1983), p. 77.
4 Allan Westcott, ed., *American Sea Power since 1775* (Chicago: J. B. Lippincott, 1947), p. 87.
5 Hans Hugo Sokol, *Des Kaisers Seemacht. Die k.k. oesterreichische Kriegsmarine 1848 bis 1914* (Vienna/Munich: Amalthea Verlag, 1980), p. 69.
6 Lawrence Sondhaus, *The Habsburg Empire and the Sea: Austrian Naval Policy 1797–1866* (West Lafayette, IN: Purdue University Press, 1989), p. 257.
7 Donald Macintyre, *Sea Power in the Pacific: A History from the 16th Century to the Present Day* (London: Military Book Society, 1972), pp. 122–3.
8 Juergen Rohwer and Gerhard Huemmelchen, *Chronology of the War at Sea 1939–1945: The Naval History of World War II* (Annapolis, MD: Naval Institute Press, 2nd revised edn, 1992), p. 56.
9 Syria declared part of the Mediterranean south of 33°N and east of 34°E a war zone, while Egypt's war zone encompassed the sea area south of 33°N and east of 29.5°E; Walter Jablonsky, 'Die Seekriegfuehrung im vierten Nahostkrieg', *Marine Rundschau*, 11 (November 1974), p. 653.
10 Ibid., p. 654.
11 Benyamin Telem, 'Die israelischen FK-Schnellboote im Yom-Kippur-Krieg', *Marine Rundschau*, 10 (October 1978), p. 640.
12 Between 6 and 22 October the Israeli missile craft sank seven Syrian and five Egyptian missile craft, one Syrian torpedo craft and minesweeper each, and two Egyptian patrol craft; Jablonsky, 'Die Seekriegfuehrung im vierten Nahostkrieg', pp. 654, 662–3; in the Red Sea, the Egyptian Navy lost 23 ships (two missile craft, three patrol craft, and 14 armed fishing craft), while four combat craft (two torpedo craft, and two light patrol craft) were captured in the port of Adabia.
13 Hartmut Zehrer, ed., *Der Golfkonflikt: Dokumentation, Analyse und Bewertung aus militaerischer Sicht* (Herford/Bonn: Verlag E. S. Mittler & Sohn, 1992), p. 197.
14 Ibid., pp. 197–8.
15 List of four main methods of destroying an enemy fleet in its bases adapted from A. Kalinin and G. Morozov, 'Evolutsiya Sposobov Unichtozheniya Sil Flota v Punktakh', *Moskoy Sbornik* 6 (June 1988), p. 25. For attack on Tarent, see Stephen W. Roskill, *The War At Sea 1939–1945*, Vol. I: *The Defensive* (London: Her Majesty's Stationery Office, 1954), pp. 300–1; Rohwer and Huemmelchen, *Chronology of the War at Sea 1939–1945*, p. 41.
16 Juergen Rohwer, 'Der Minenkrieg im Finnischen Meerbusen, Part II: September–November 1941', *Marine Rundschau*, 2 (February 1967), pp. 97–100.
17 Gerhard Huemmelchen 'Unternehmen "Eisstoss": Der Angriff der Luftflotte 1 gegen die russische Ostseeflotte im April 1942', *Marine Rundschau*, 4 (April 1959), pp. 226, 229.

18 The plan of attack envisaged the following priority of targets: battleship, *Kirov*-class cruisers, half-completed former German heavy cruiser *Luetzow*, and mine-laying cruiser; Rohwer *et al.*, *Chronology of the War at Sea 1939–1945*, p. 134.
19 Between 4 and 30 April 1942, about 600 aircraft, including 162 *Stukas* attacked a variety of ships in the Leningrad–Kronstadt area that resulted in the loss of 29 German aircraft; Huemmelchen 'Unternehmen "Eisstoss": Der Angriff der Luftflotte 1 gegen die russische Ostseeflotte im April 1942', pp. 231–2.
20 Friedrich Ruge, *The Soviets as Naval Opponents 1941–1945* (Annapolis, MD: Naval Institute Press, 1979), p. 24.
21 Zehrer, *Der Golfkonflikt*, p. 197.
22 Ibid., p. 198.
23 S. W. C. Pack, *Sea Power in the Mediterranean: A History from the Seventeenth Century to the Present Day* (London: Arthur Barker, 1971), p. 194.
24 Ze'ev Almog, 'Israel's Navy Beat the Odds', *Proceedings*, 3 (March 1997), p. 107.
25 Albert Lord and Klaus Tappeser, 'Rolle und Beitrag der Seestreitkraefte', in Zehrer, *Der Golfkonflikt*, p. 197.
26 Hans Fuchs, 'Die Diversion als strategisches Mittel zur Erzielung eines Kraefteaus-gleiches, dargelegt an geschichtlichen Beispielen', *Marine Rundschau*, 4 (April 1938), p. 240.
27 Roskill, *The War At Sea 1939–1945*, Vol. III: *The Offensive*, pt. 1: 1st June 1943–31st May 1944 (London: Her Majesty's Stationery Office, 1960), p. 231; Samuel E. Morison, *The Struggle for Guadalcanal*, Vol. V, *History of United States Operations in World War II* (Boston, MA: Little, Brown, 1949, reprinted 1975), p. 372.
28 Macintyre, *Sea Power in the Pacific*, p. 226.
29 Sondhaus, *The Habsburg Empire and the Sea*, p. 238.
30 Geoffrey Till, ed., *Maritime Strategy and the Nuclear Age* (New York: St Martin's Press, 1984, 2nd edn), p. 123.
31 Ibid., p. 121.
32 David T. Cunningham, 'The Naval Blockade: A Study of Factors Necessary for Effective Utilization' (Fort Leavenworth, KS: US Army Command and General Staff College, unpubl. MA Thesis, June 1987), pp. 19–20, 26.
33 Ibid., pp. 20–3; Till, ed., *Maritime Strategy and the Nuclear Age*, p. 122.
34 Till, *Maritime Strategy and the Nuclear Age*, pp. 122, 124–5.
35 G. J. Marcus, *A Naval History of England*, Vol. II: *The Age of Nelson: The Royal Navy 1793–1815* (New York: Viking Press, 1971), pp. 242–3.
36 Cunningham, 'The Naval Blockade', pp. 16–17.
37 Ibid., p. 28.
38 Marcus, *A Naval History of England*, Vol. II, pp. 239, 241.
39 Cunningham, 'The Naval Blockade, pp. 33, 31.
40 Ibid., p. 38.
41 Ibid., pp. 28, 36.
42 Ibid., p. 68.
43 Ibid., p. 70.
44 Herbert Rosinski, *The Development of Naval Thought* (Newport, RI: Naval War College Press, 1977), pp. 10–11; Macintyre, *Sea Power in the Pacific*, p. 142.
45 Anthony E. Sokol, 'Naval Strategy in the Adriatic Sea During the World War', *Proceedings*, 8 (August 1937), p. 1084.
46 Ibid., p. 1083.
47 Cunningham, 'The Naval Blockade', p. 73.

48 Editors, 'Die Sperre der Otranto-Strasse 1917–18', *Marine Rundschau*, 10 (October 1933), p. 451.
49 D. Welsch, 'Die Sperrung der daenischen Ostsee-Zugaenge: Eine wenig bekannte Episode aus den ersten Augustagen 1914', *Marine Rundschau*, 7 (July 1934), pp. 309–10.
50 Specifically, these reinforcements included 40 British destroyers (six Australian already based in Brindisi, six from England, and an additional 28 that were to be taken away from their escort duties), 12 French destroyers, eight sloops (taken away form escort duties in the Mediterranean), 48 fishing steamers (18 already in the Adriatic, 18 from England, and 12 from escort duties), 76 fishing steamers already in the Adriatic and 36 British SCs; seaplanes operated from the Italian seaplane stations at Varano, Valona, Brindisi and Otranto; ibid., p. 313.
51 Welsch, 'Die Sperrung der daenischen Ostsee-Zugaenge: Eine wenig bekannte Episode aus den ersten Augusttagen 1914', p. 310.
52 Thomas H. Robbins, *The Employment of Naval Mines* (Newport, RI: Naval War College, 20 March 1939), p. 12.
53 Corbett, Vol. 3: *The Dardanelles Campaign: Naval Operations* (London: Longmans, Green, 1923), p. 79.
54 Cunningham, 'The Naval Blockade', p. 85.
55 Randolph and Winston S. Churchill, *The Six Day War* (Boston, MA: Houghton Mifflin, 1967), p. 98.
56 Robbins, *The Employment of Naval Mines*, p. 24.
57 Ibid., p. 17.
58 Andrew Patterson, Jr, 'Mining: A Naval Strategy', *Naval War College Review* (May 1971), p. 54.
59 Roy F. Hoffman, 'Offensive Mine Warfare: A Forgotten Strategy', *Naval Review* 1977, p. 145; Andrew Patterson Jr, 'Mining: A Naval Strategy', p. 55; the Russians lost six warships from Japanese mines.
60 James A. Meacham, 'Four Mining Campaigns: An Historical Analysis of the Decisions of the Commanders', *Naval College Review* (June 1962), p. 88; Eric J. Grove, ed., *The Defeat of the Enemy Attack on Shipping 1939–1945*, revised edition of the Naval Staff History Volumes 1A (Text and Appendices) and 1B (Plans and Tables) (Aldershot: Ashgate, Navy Records Society, 1997), p. 150.
61 Ibid., p. 148.
62 Ibid., pp. 148–9.
63 Henry Newbolt, Vol. 5: *From April To The End of the War, Naval Operations* (London: Longmans, Green, 1931), p. 207; Arthur J. Marder, Vol. V: *Victory and Aftermath (January 1918–June 1919) From the Dreadnought to Scapa Flow: The Royal Navy in the Fisher Era, 1914–1919* (London: Oxford University Press, 1970), p. 66; The British estimated that because of the great depth of water in the area about 400,000 mines would be required to prevent transit of U-boats; however, the invention of the Mk 6 antennae mines by the US allowed the reduction of the total number of mines for the barrage; these mines were laid in the deep section of the barrages, while the British chemical horn H mines were used at both ends of the barrage; J. Meacham, 'Four Mining Campaigns: An Historical Analysis of the Decisions of the Commanders', p. 89.
64 Grove, *The Defeat of the Enemy Attack on Shipping 1939–1945*, p. 149; Some sources claim that three U-boats were sunk and three others damaged, while others claim six confirmed sinkings; J. Meacham, 'Four Mining Campaigns: An Historical Analysis of the Decisions of the Commanders', pp. 90–1.
65 Marder, Vol. II: *The War Years: To the Eve of Jutland* (London: Oxford University Press, 1965), p. 331; ibid., Vol. V: *Victory and Aftermath (January 1918–June*

1919), p. 35; Robert C. Duncan, *America's Use of Mines* (White Oak, MD: US Naval Ordnance Laboratory, January 1962), p. 70.
66 Grove, *The Defeat of the Enemy Attack on Shipping 1939–1945*, pp. 151–2.
67 Ibid., p. 152.
68 This barrier was laid beginning on 11 September 1939 and completed within five days. Initially, the barrier consisted of about 3,000 mines laid in five lines east of the Goodwin Sands and the Dyck Shoal. A secret channel was left off the Goodwins for use by British ships. Approximately 3,640 mines were laid in a deep mine barrier between Folkestone and Cape Gris-Nez in October 1939. The third and final stage included laying a double line of indicator loops between the two minefields to detect any U-boats that might try to pass through the barrage; Roskill, *The War At Sea 1939–1945*, Vol. I, p. 96; Peter C. Smith, *Hold the Narrow Sea: Naval Warfare in the English Channel 1939–1945* (Ashbourne, Derbyshire: Moorland Publishing, and Annapolis, MD: Naval Institute Press, 1984), pp. 11–12.
69 Grove, *The Defeat of the Enemy Attack on Shipping 1939–1945*, p. 153.
70 Ibid., p. 153.
71 A total of 95 U-boats were directed to enter the Mediterranean; sailing of three boats was canceled in September 1943; eight other boats did not sortie for one reason or another; one was damaged and two boats were sunk on their way to the Strait; during the crossing of the Strait, the Allies sank six and damaged another six boats; two boats were deterred from crossing and two other were directed not to try to force the Strait; Grove, *The Defeat of the Enemy Attack on Shipping 1939–1945*, p. 144.
72 Another objective was to secure protection of the shipping traffic between Sweden and Germany; Juergen Rohwer, 'Der Minenkrieg im Finnischen Meerbusen, Part I: June–August 1941', *Marine Rundschau*, 1 (January 1967), p. 17.
73 Ibid., p. 17.
74 Salewski, *Die deutsche Seekriegsleitung 1935–1945*, Vol. I: *1935–1941*, p. 418; David Woodward, *The Russians at Sea: A History of the Russian Navy* (New York: Frederick A. Praeger, 1966), p. 210; Juerg Meister, *Der Seekrieg in den osteuropaeischen Gewaessern 1941–45* (Munich: J. F. Lehmans Verlag, 1958), p. 11.
75 Salewski, *Die deutsche Seekriegsleitung 1935–1945*, Vol. I: *1935–1941*, p. 418.
76 Rohwer, 'Der Minenkrieg im Finnischen Meerbusen, Part I: June–August 1941', p. 21
77 Meister, *Der Seekrieg in den osteuropaeischen Gewaessern 1941–45*, pp. 13–14.
78 Ibid., p. 16.
79 Salewski, *Die deutsche Seekriegsleitung 1935–1945*, Vol. I: *1935–1941*, p. 304.
80 Meister, *Der Seekrieg in den osteuropaeischen Gewaessern 1941–45*, pp. 14–15, 25.
81 Otto Groos, 'Die strategische Bedeutung des Aermelkanals in der Geschichte und Gegenwart', *Marine Rundschau*, 7 (July 1940), p. 376.
82 Sokol, 'Naval Strategy in the Adriatic Sea during the World War', p. 1087.
83 Salewski, *Die deutsche Seekriegsleitung 1935–1945*, Vol. I, p. 370.
84 Frank Uhlig, Jr, *How Navies Fight: The US Navy and Its Allies* (Annapolis, MD: Naval Institute Press, 1994), pp. 380, 382.
85 Anthony Cordesman, *The Iran–Iraq War and Western Security 1984–87: Strategic Implications and Policy Options* (London: Jane's Publishing, 1987), p. 174.
86 B. H. Liddell Hart, 'How Crete Was Lost – Yet with Profit', *Military Review*, 10 (October 1951), p. 10.
87 Detlef Vogel, 'Das Eingreifen Deutschlands auf dem Balkan', in Gerhard Schreiber, Bernd Stegemann and Detlef Vogel, *Der Mittelmerraum und Suedosteuropa: Von*

der *'non belligeranza' Italiens bis zum Kriegseintritt der Vereinigten Staaten, Das Deutsche Reich und der Zweite Weltkrieg*, Vol. 3 (Stuttgart: Deutsche Verlags-Anstalt, 1984), p. 493.
88 Ibid., pp. 504–5.
89 In the process the Germans suffered high casualties; some 6,000 men killed, wounded or missing. By the end of fighting, Eleventh Air Corps had only 185 transport aircraft available, 150 aircraft were a total loss, and an additional 165 were damaged; ibid., p. 509.
90 Hart, 'How Crete Was Lost – Yet with Profit', p. 13.
91 Roskill, *The War At Sea 1939–1945*, Vol. III: *The Offensive*, pt. 1, p. 192.
92 Ibid., p. 193.
93 Ruge, *The Soviets as Naval Opponents 1939–1945*, pp. 20–2; Meister, *Der Seekrieg in den osteuropaeischen Gewaessern 1941–45*, p. 22.
94 Ruge, *The Soviets as Naval Opponents 1939–1945*, p. 23.
95 Salewski, *Die deutsche Seekriegsleitung 1935–1945*, Vol. II *1942–1945*, p. 449.
96 Ibid., p. 384.
97 Meister, *Der Seekrieg in den osteuropaeischen Gewaessern 1941–45*, p. 303.

10

Exercising Control

After obtaining command of the sea, a blue-water navy or a coastal navy must maintain that control. This encompasses a number of strategic and operational tasks that must be carried out both sequentially and/or simultaneously. Exercising control includes, among other things, broad strategic tasks such as projecting power ashore, preventing the enemy's invasion from across the sea, weakening of the enemy and protecting one's maritime trade. There is a great difference in exercising control of the sea in the case of a typical narrow sea in terms of size/mix of force, extent of the area, and duration compared with similar tasks carried out on the open ocean. One of the most important features of narrow seas is that the conduct of major naval operations and even tactical actions require the highest degree of cooperation between all services of the armed forces.

Projecting Power Ashore

In general, projecting power ashore is probably one of the most complicated tasks for any navy. The importance of projecting power ashore for a blue-water navy is shown by the fact that about 70 per cent of the world's population live within 200 miles of the coastline and some 80 per cent of the world's capitals lie within 300 miles of the sea.[1] Some 60 per cent of politically significant urban areas around the world are located within 25 miles of the coastline or 75 per cent of these areas are located within 150 miles.[2] For a blue-water navy it is critical to operate close to the enemy coast and to project its power far inland.[3] For a stronger fleet, the littoral is the area of the sea or ocean that must be controlled to support operations ashore, while for a weaker fleet, the littoral encompasses the area inland from the coast that can be supported and defended directly from the sea. For a coastal navy, projecting power on the opposite shore might well be one of the principal tasks, especially in a strategic offensive.

For a blue-water navy, projecting power ashore within a narrow sea is not the principal but only one of its many tasks in its overall capabilities. To operate successfully in littorals, a blue-water navy must significantly reduce the threat posed by the enemy land-based aircraft, anti-ship missiles, submarines, and mines. This in practice means that such a navy must possess considerable capabilities in anti-air, anti-subsurface, anti-surface, and anti-mine warfare areas.[4] A particularly acute problem is effective defense against the enemy long-range coastal anti-ship missiles and theater ballistic missiles while operating in a typical narrow sea.

In contrast to war on the open ocean, projecting power ashore in a narrow sea can sometimes bring more decisive results than actions against the enemy fleet. Actions against the enemy shore are the ultimate objective of any war at sea, while victory over the enemy fleet creates only one of the prerequisites to accomplish the ultimate objectives of war. Projecting power ashore essentially consists in freedom of action for one's forces to destroy targets on the coast at a place and time not expected by the enemy. This capability is the greatest asset a sea power can possess. Without it, the utility of sea power is considerably diminished. Actions of one's naval forces against the enemy shore in a narrow sea can range from large-scale amphibious assault landings, bombardments of naval bases and ports, destruction of facilities/installations and troop concentration on the coast, to small-scale raids.

Amphibious Landings

Seaborne landings in a typical narrow sea, are usually conducted across much shorter distances and are smaller in size than those mounted against a coast bordering an open ocean. Today, the hazards of operating a large amphibious force in confined waters and under constant threat from the air, submarines, and mines are too great in some narrow seas, specifically the Baltic and the Black Sea.

In general, amphibious landings are aimed at seizing an area of an enemy-controlled coast giving access to an operational objective inland, at speeding the advance of one's troops along the coast, or in some cases, helping to shorten the war itself. These objectives could be accomplished by conducting amphibious landings aimed at eliminating or taking control of a large naval base or port, at preventing the adversary from seizing a base or large port of one's own, at cutting off an enemy army's avenue of escape, and in countering enemy evacuation efforts across the sea. A credible amphibious capability may also help to tie down a sizable enemy force in the defense of a large stretch of its mainland coast or offshore islands. Sometimes an amphibious threat can force the enemy to spend most of its scarce resources on defense of the coast, and thereby reduce the threat at sea to one's own forces.

Prior to the advent of aircraft, lack of sea control made an amphibious landing an extremely risky undertaking. A major operation of war across the sea normally should not be attempted unless naval superiority for an adequate period is probable. Today, the main prerequisite for the success of an amphibious landing in a typical narrow sea is local control of the sea and in the air. Other requirements include: surprise, secure base of operations near the scene of action, highly integrated command and control, smooth inter-service cooperation, and high levels of training and preparation on the part of all participating forces. The establishment of the effective control of the approaches to the landing objective area is generally a prerequisite for any amphibious assault to succeed.

In the era of sail and steam, lack of control of the sea usually doomed maritime expeditions. England did not obtain command of the Channel in the 1376–85 period of the Hundred Years War, which ultimately led to its failure to achieve its stated objective in the war as a whole. For the same reason, Spain was unable to prosecute the war successfully during the Dutch Revolt in 1576–85 because the northern and southern provinces of the Netherlands were divided by water. The Spaniards could not control the coast without having command of the sea, and this they were unable to obtain. In another example, the Japanese invasion of Korea in 1592 failed largely because of the loss of command of the Korea Strait. Likewise, the French in their repeated attempts to land troops in Ireland in the eighteenth century tried unsuccessfully to obtain command of the sea simultaneously with such landings.

Sweden's failure to obtain command of the sea allowed its enemies to roam freely along its coast in 1808 during the war with Napoleonic France. A combined Russian and French force of about 1,600 men had landed on Gotland in May 1808, and then seized Visby, the island's principal port. In the end, a truce was reached; the Franco-Russian force evacuated Gotland, while the Russians agreed to withdraw from the Aaland (Åland) Islands, which they had seized in the winter of 1807 by marching across the ice. A Russian advance to Stockholm was checked by the Swedish and British fleets, although the Swedes had been weakened by the loss of their light coastal forces in the Gulf of Finland.[5]

In the Austro-Italian War of 1866, the Italians assumed wrongly that the Austrian army had crumbled. The moment for decisive Italian action seemed propitious, because about three-quarters of the Austrian Army had retreated from Venetia Giulia for duty in the north. The Italian plan was to lay siege to Trieste, and then cross the Julian Alps for a possible march on Vienna. The Navy would gain command of the Adriatic and blockade the port of Trieste to ensure its capture by Italian troops. Admiral Persano was ordered to take possession of the island of Lissa in the central part of the Adriatic to serve as an advanced base for Italian fleet raids against the Dalmatian coast.[6] He left Ancona on 16 July and planned to land troops on the island three days later. However, the landing was postponed for a day and that allowed the Austrian fleet under Admiral Tegetthoff to intervene.⁷ The entire expedition ended as a costly failure for the Italians because they lost the naval battle off Lissa decisively on 20 July, and with it any hope of obtaining command of the Adriatic. Because of their lack of command of the sea, the Italians also abandoned plans to land troops in Istria.

In the American–Spanish War of 1898, a US landing on Cuba was possible only if local command of the Caribbean was obtained by the US Navy. The objective was to drive Spanish forces from Cuba and the adjacent waters. A naval blockade of the island was eventually to starve out the Spanish garrisons, and a US invasion was to greatly speed up the collapse of the Spanish power there. The US Navy was able to obtain command of the sea only by destroying or immobilizing the Spanish naval forces. Until that was achieved, it was too risky for the US to send its transports to sea at all.

The Entente's amphibious landing at Gallipoli in 1915 was conducted over a 'commanded' sea, but, among other things, the lack of adequate cooperation between the services doomed the entire enterprise. The main objectives of the Gallipoli landing were to take Turkey out of the war, to open a direct link with the Entente's embattled Russian ally, to force the Germans to shift troops from the Russian front, and to influence Greece to side openly with the Entente Powers.⁸ The landing eventually failed because the Turks, fully alerted, improved their deployments and reinforced the troop levels on the peninsula to about 60,000 men.

Far more successful was the German landing in the Gulf of Riga (Operation *Albion*) in October 1917 because of the excellent cooperation between the army and the fleet. The Germans obtained local sea control in the eastern Baltic before the landing. The operational objective of the landing was to open the Gulf of Riga and thereby threaten the rear of the Russian Twelfth Army defending the Baltic coast. The Germans committed one infantry division and assorted troops totaling 25,000 men, 5,000 horses, and only three field gun batteries. These forces were embarked on 19 transports and 21 supply ships.⁹ The first two amphibious landings took place on 12 October, as planned and by 19 October the Germans had seized the islands of Oesel, Dago, and Moon. The Germans achieved a complete surprise and the Russian resistance ashore was weak. The German heavy surface ships successfully fought Russian coastal gun batteries and prevented the Russian fleet from interfering with the landing. The entire operation was a resounding success for the Germans.¹⁰ The capture of the Baltic islands was a necessary step for the Germans to complete their control of the eastern part of the Baltic and for the future operations against Estonia and Finland.¹¹

Sometimes, the extent of sea control needed to conduct amphibious landing has been exaggerated. Several major landings have taken place and been successful with partially disputed control of the sea. The Japanese in their war with Russia in 1904–5, went ahead with their landings in the Yellow Sea despite the existence of the Russian squadrons in Port Arthur and Vladivostok. Likewise, the British and the French shipped their troops to Europe in 1914 despite the threat posed by the German and Austro-Hungarian fleets in the North Sea and the Mediterranean, respectively. Many amphibious landings in World War II took place despite the presence of enemy submarines, aircraft or light forces.[12] In another example, the Allies conducted an amphibious raid at Dieppe (Operation *Jubilee*) on 19 August 1942 although they did not have control in the Channel or control of the air above it. In this significant military effort some 6,000 Allied troops supported by a large number of ships and aircraft landed on a 10-mile wide front near Dieppe. The main purpose of the Dieppe raid was to test German coastal defenses. After some heavy fighting in which the Allies suffered heavy casualties (some 3,400 Canadians died), and the entire operation ended in disaster.[13] However, some valuable lessons were learned that helped the Allies in their invasion of Normandy in June 1944.

In the modern era, it has often not been possible to conduct a successful amphibious landing by obtaining only local sea control and without some degree of control in the air. By the end of the 1930s, the growing capabilities of aircraft in terms of speed, effective range, and weapons payload made it possible to land troops on the opposed shore by obtaining only local and temporary control in the air. The first and most dramatic example of this was the German invasion of Norway in April 1940 (Operation *Weseruebung*). The German plan was to transport troops on board warships and some merchantmen and to seize the main Norwegian harbors to create bridgeheads until the sufficient troops and heavy arms could be sent as reinforcements. The western approaches to the Skagerrak were mined. The operation was successful because, among other things, the *Luftwaffe* obtained and maintained control of the air over the Skagerrak and southern Norway.

Likewise, the American landing on Guadalcanal in August 1942 (Operation *Watchtower*) was carried out despite the fact that the Japanese held command of the surrounding seas. The initial phase of the US landing was highly successful because local command of the air was obtained from the fighters of three US aircraft carriers temporarily deployed in the area.

To be successful an amphibious landing also requires reliable air cover for the naval forces and troops during their transit to the objective area and once the landing has taken place, otherwise the entire landing might end in disaster. The US aircraft carriers were temporarily withdrawn from their operating area during the unloading of troops and *matériel* in the amphibious landing on Guadalcanal on 7 August 1942. The landing force commander, Admiral R. K. Turner, protested vehemently to Admiral Frank J. Fletcher, arguing that the carriers should stay within supporting distance of transports for more than two days. He also explained that the transports could not possibly be unloaded in less than four days, and that they would need air cover during that entire time. In the end, the Americans were fortunate that the Japanese did not take advantage of the lack of air cover for the transports.[14]

In World War II, the British learned, at no small cost, in Norway, Greece, Malaya, and indeed in all other theaters, that control of the air was an essential prerequisite for success in amphibious landings. The speedy Japanese seizure of the Philippines, Malaya, and the Netherlands East Indies would not have been possible without strong and reliable and effective support from the air. The Japanese selected invasion targets rich in raw materials that could provide airfields to cover their future advances. The Japanese Eastern Invasion Force and Central Invasion Force were moved so as to be within supporting distance of each other. The Japanese used their powerful fleet forces to escort the transports to the landing beaches and provide cover against possible offensive thrusts by combined fleets of the Australians, British, Dutch and Americans. At the same time, each of the Japanese fleet detachments was sufficiently strong to present a serious threat to the ABDA forces.[15]

The Allies in their advance through the Solomons in 1941–43 used a similar island-hopping technique that was used by the Japanese in their conquest of the Netherlands East Indies. The opening Allied move was to obtain control of the island of Guadalcanal on which the Japanese had begun to build an airstrip in the summer of 1942. This had forced the Americans hastily to revise their plans in favor of landing about 19,000 Marines on the island. The principal objectives were to expel the Japanese from the island and occupy the uncompleted air strip. Air support for the landing was provided by three US aircraft carriers because the Solomons were beyond the range of shore-based fighter aircraft. After encountering weak resistance, American troops captured the uncompleted airstrip and renamed it Henderson Field. Afterwards, Henderson Field became a base for a strong force of fighters and dive-bombers. The Japanese with their nearest airfield at Rabaul, some 640 miles away, were not in a position to wrest control of the air from the Americans.[16]

The capture of Tunisia in May 1943 enabled the Allies to move their aircraft to a position from which they not only dominated the Sicilian Narrows, but also threatened both Sardinia and Sicily. By June 1943 about 600 first-line Allied aircraft were based on Malta, supported by the latest fighter control and radar installations. However, the Allies were unable to provide fighter cover for the prospective landing area on the island of Sicily. To obtain additional bases for fighter aircraft, they turned their attention to the Axis-held islands of Pantelleria and Lampedusa.[17]

In the winter of 1941–42 the Soviets conducted several amphibious landings aimed at cutting off the southern flank of the German Army

Group South. Feodosiya was selected as the place for the main landing because it allowed entrance to large ships and an easy approach to the nearby central plain of the Crimean Peninsula. Supporting landings north and south of Feodosiya were aimed to tie German forces in the area. The Soviets gave the commander 30 days to plan the entire operation.[18] Although the Soviets did not have any opposition at sea, they failed in their stated objective. Organization of troop transports was inadequate and there was a lack of unity of command. Reconnaissance too proved to have been inadequate, and the Soviets were excessively rigid in the execution of their plan. The operation showed that an effective control of the sea without at least temporary control of the air cannot usually be successful.[19]

Sometimes, the initial amphibious landing was highly successful, but lack of control of the adjacent sea's surface greatly complicated the efforts to expand the bridgehead and consolidate the assigned operational success. This happened even when the attacker possessed local control of the air. The Allies were highly successful in landing troops on Guadalcanal in August 1942, but afterwards they became progressively involved in a series of small but costly actions with Japanese forces on land, at sea, and in the air. The Allies generally had control of the air in the Lower Solomons, but a disputed control of the sea existed each day between sunset and sunrise. The Japanese, who were more skillful in night fighting and torpedo tactics than the Americans, were able to bring new troop reinforcements and supplies to their embattled troops on the island. This attritional phase lasted for almost eight months before the Japanese decided to give up their attempts to regain control of Guadalcanal. By then, this protracted struggle had considerably slowed down the Allied operational tempo, because no further major landings up the Solomons chain were possible until Guadalcanal was made secure. A good argument can be made that had the Allies conducted consecutive major naval or air operations, the struggle for Guadalcanal would have ended much earlier and at far lower cost in ships, aircraft, and personnel.

If used skillfully and energetically much smaller fleet forces can sometimes defeat a much larger enemy landing force inadequately supported by its own fleet. This happened to the Soviets in their landing at El'tigen in October 1943. During the day, German coastal and field guns on both sides of the eight-mile-wide Kerch Strait effectively prevented Soviet ships from reaching the bridgehead at El'tigen. At night, the Germans employed skillfully and almost continuously their MTBs, motor minesweepers, and artillery carriers in what amounted to a virtual blockade of the Kerch Strait lasting for over a month. There were almost nightly encounters between German and Soviet light forces in the strait, and losses on both sides were heavy. In the end, however, the Germans were successful in preventing the Soviets from expanding their beachhead at El'tigen. German and Romanian troops attacked the Soviet positions on 4 December 1943, and seven days later they seized the entire beachhead. The Soviet losses were heavy – approximately 10,000 dead and 3,000 captured, while only about 800 Soviet troops broke through and escaped. In addition, the Soviets lost about 100 miscellaneous craft. The German success was due to many factors, but probably the most decisive was the blockade of the Kerch Strait. The Soviets failed to use their cruisers and destroyers to break up the German blockade. The reason for this was probably Soviet fear from attacks by the *Luftwaffe*.[20]

The most common problem in amphibious landings, which has often led to their failure, was misunderstanding the responsibility of each service. This problem occurred between the general in charge of troops and the admiral responsible for the covering force, and between their subordinates. All too often, the assault miscarried or fizzled out, because of wrangling between the services, which, through incompatibility of views and loyalties, could unintentionally become the enemy's best allies.[21]

To be successful, an amphibious assault landing should surprise the defender. Otherwise, it is almost certain that the entire enterprise will fail rather quickly. The Gallipoli operation of 1915 failed partly because it did not surprise the Turkish defenders. The British left the Turks with their German ally for four weeks before actually landing troops on the peninsula.[22] The disaster at Gallipoli had far-reaching consequences for the Entente powers, for they were unable to establish a direct secure link with their embattled Russia's ally, and thereby tear Turkey away from the Central Powers. If the landing had been a success it is most unlikely that Bulgaria would have abandoned its neutrality and joined the Central Powers. A successful landing also would have removed at one stroke the threat to the British position in Egypt. The wavering respect of the Arabs probably would have been won for the Entente, and Italy's reluctance to join the Western allies would have ended.[23]

The plan for an amphibious landing should be based on the realistic assessment of the operational situation in the theater. This also includes sound appreciation of the forces on both sides and the geographic characteristics of the area in which the landing is to be conducted. One's forces should not be endangered by enemy action to cut them off from their home bases, as apparently was the case with the scheme for an amphibious landing conceived in the opening stage of World War I by Winston Churchill and Admiral John Fisher. Both believed that the Royal Navy should obtain access to the Baltic. Afterwards, the German flank in the west would be turned by carrying out a large-scale landing to seize Schleswig-Holstein, while the Russians would land on the unprotected Pomeranian coast. Known as Lord Fisher's 'Baltic Scheme', the plan envisaged deploying about 600 ships and landing craft of all types which would enter the Baltic, embark the Russian troops, and then land them on the Pomeranian coast. The Germans believed that such an operation was possible, but only after the British Grand Fleet had destroyed their High Seas Fleet, and the Russian army had won a great victory on land. The Germans were right in thinking that such a combination of events was very remote.[24]

In the Korean War, the influence of sea power, and its amphibious element in particular, was most clearly demonstrated when the entire military situation was transformed and the hitherto triumphant North Korean Army found itself encircled and trapped as a result of a brilliantly executed amphibious landing at Inchon, some 150 miles in their rear. After capturing offshore islands on 15 September 1950, about 70,000 troops embarked on 170 transports and amphibious ships and landed at Inchon (Inch'ŏn) (Operation *Chromite*). The UN forces then moved inland, cutting off the road running south from Seoul along which most of the enemy's supplies were transported. At the same time, the only other supply route down the Korean east coast was interdicted by UN naval forces. Seoul was recaptured two weeks later. By 16 September UN forces deployed within the Pusan perimeter had launched a counter offensive. The North Koreans, cut off in their rear from their sources of supply, crumbled in the face of this assault. By 26 September UN forces advancing northwards had joined hands with those landed at Inchon.[25]

Threat of Amphibious Landings

Amphibious capability is one of the greatest assets any navy can have because it forces a land power to devote large forces and a lot of attention on defense of the parts of the coast suitable for amphibious landings. A great benefit of possessing a credible amphibious threat is to tie up enemy forces to defend the coast and offshore islands. In this way, the success of a large-scale attack by one's own troops on the coast can be greatly enhanced. The threat of Allied attack in the Baltic during the Crimean War tied down some 170,000 Russian troops who might have been useful in the Crimea. In another example, the perceived threat of Allied amphibious landings forced the Germans to deploy along the 8,000-mile-long European coastline 53 divisions or 27 per cent of their total strength on the eve of their invasion of Russia in June 1941. As a consequence, the Germans used only 120 divisions or 60 per cent of the total strength in that invasion. In 1943, the Germans committed 18 divisions to Italy against the landing of 15 Allied divisions. They deployed another 15 divisions, or twice the previous strength level, to the Balkans because of their widely held view that the Allies would land there.[26]

In June 1950 the North Koreans would have been able to overrun the entire peninsula and occupy the key south-coast port of Pusan had there not been readily available a large US naval and amphibious force to transport troops of the occupation army from Japan to reinforce the South Korean army. By 5 July 1950 the South Koreans, taken by surprise and outnumbered, had been driven southwards. On that day, advance elements of US 24th Division transported by air came into action and were joined two days later by the remainder of the division transported by sea. By 9 July US 1st Cavalry division had been landed at the east coast port of Pohang (P'ohang) by a US amphibious force just in time to stop the North Korean advance down the east coast toward Pusan.[27]

The value of possessing the capability of a credible amphibious threat was shown most recently in the Gulf War of 1991. General Norman Schwarzkopf kept the 13th Marine Expeditionary Unit (MEU) to deceive the Iraqis and pin down their forces in Kuwait. Reportedly, the Iraqis were forced to deploy at least three infantry divisions to defend the Kuwaiti coast from Kuwait City south to the Saudi–Kuwaiti border.[28] US marines prepared conspicuously for an amphibious landing along the Kuwaiti coast, conducting highly publicized exercises with ominous names like 'Imminent Thunder'. Meanwhile, the US-led coalition sought to convince Iraqi President Saddam Hussein that a major land assault would be launched against Iraq's most heavily defended areas along the Saudi border.[29]

Attack on the Coast

Attacks on coastal installations and facilities are usually conducted by a stronger fleet either to tie up or to divert enemy strength from the main sector of effort. These actions can have a significant but temporary psychological effect by raising morale among one's own forces and population and depressing the opponent. The chief methods of combat employment are attacks and raids. A major naval operation aimed to *destroy enemy coastal installations/facilities* can also be conducted as a preliminary to an amphibious landing or as an integral part of a naval blockade. Such major operations can be carried out either with naval or air forces or by these two services jointly. Major navies today have a much greater ability than in the past to attack a wide range of targets deep in the enemy's operational and even strategic depth by using shipboard guns, carrier-based attack aircraft, and cruise missiles.

In the age of sail and steam, fleets operating in a narrow sea often carried out bombardments of the enemy coast by attacking ports, shipyards, railroads, and road communications. The material damages were usually slight and the results were largely psychological. Nevertheless, such actions often forced the opponent to redistribute its forces to enhance the security of the coastal settlements.

In some cases, bombardments of the coast were planned to provoke the opponent to send out a detachment of larger ships which could then be destroyed in battle. This was the German intent in their bombardment of Lowestoft in April 1916. Admiral Jellicoe reacted with a futile attempt to entice Admiral Reinhard Scheer to take up battle. The next German objective was to bombard the shipbuilding center of Sunderland at the estuary of the Tyne River in order to provoke the Royal Navy into sending out a fleet detachment which would then have been attacked by German surface forces, submarines, and *Zeppelin* airships. However, delays in the repairs of the German ships that were to take part in the action combined with bad weather led to the cancellation of the operation.[30]

The Austro-Hungarian Navy's bombardment of Italy's eastern coast on 24 May 1915 had the character of a *surging naval action*. One day after Italy declared war on Austria-Hungary and Germany, almost the entire Austro-Hungarian fleet led by Admiral Anton Haus bombarded various coastal installations along Italy's eastern coast between Rimini and the estuary of the Potenza River, while three groups of light forces attacked coastal targets on the Monte Gargano Peninsula.[31] The main targets of attack were the coastal railroads and tunnels. The stated objective was to disrupt and delay the movement by rail of mobilized Italian troops along the Adriatic's western coast. Another objective was to cause fear and possibly panic among the Italian populace of the coastal area. As a result, the Italian troop transports were slowed down as the Italians expected that an amphibious landing would follow these raids. This delay in Italian mobilization provided sufficient time for the Austrians to prepare their defensive positions along the country's border with Italy.[32] In retaliation, the Italians sent their cruisers to bombard the Dalmatian coast, but two of them were sunk by Austro-Hungarian U-boats. These raids soon ended and no Italian capital ship even approached the Austro-Hungarian coast until the end of war.[33]

Sometimes, heavy ships were used to bombard heavily protected gun positions on the coast, but such actions invariably proved to be more difficult than anticipated and the results achieved were usually far less than the costs. This was illustrated by Allied bombardment of the Turkish forts in the Dardanelles on 3 November 1914. In this action, two modern 15-inch gun battleships and 16 French and British pre-dreadnoughts bombarded the outer forts for about ten minutes. Only one fort was slightly damaged. The attack proved to be costly in the long run because it alerted the Turks, who from then on began to pay much greater attention to the defenses of the Straits. Hence, there existed no possibility of achieving surprise when the Allies began bombardments of the Straits in February–March 1915 as a prelude to their landing at Gallipoli. A total of 16 battleships plus numerous craft took part in this action. By then, the Turks had deployed more than 100 medium- and heavy-caliber guns in the Straits. In addition, they laid down 11 mine barriers and one anti-submarine net barrier in the Narrows. The Entente's first long-range bombardment was carried out on 19 February, followed by a second bombardment six days later. After preliminary minesweeping, Allied ships bombarded the batteries in the Turkish Narrows on 18 March and succeeded in silencing them, but Allied losses due to the Turkish mines were heavy, three old battleships being sunk, while three others were damaged. This prompted the Allies to call off the attack and hastily leave the Straits. Unbeknown to them, the Allies were on the verge of success. The Turks were at the end of their resources, all gun fire control having broken down and there was almost no ammunition left. The naval effort then ceased, to await the arrival of Allied troops, which did not come until 25 April 1915.[34] One lesson of the Dardanelles action was that fleet bombardment of coastal forts should never be undertaken until the troops are ready shortly thereafter to land and capture the landing objective area. The only way to take the offensive against a land power is to carry out a combined operation. The Allied error was in thinking that a fleet alone could deal with the situation and force the Straits without troops.[35]

One of the largest bombardments of enemy coastal installations was carried out by the British Force H from Gibraltar against the Italian coast. The force consisted of two battleships (*Renown*, *Malaya*), one aircraft carrier (*Ark Royal*), one cruiser (*Sheffield*), and ten destroyers. The attack took place between 31 January and 4 February 1941, when aircraft from *Ark Royal* attacked the dam at Lake Tirso (Omodeo) in Sardinia. However, poor weather forced Force H's commander, Admiral James Somerville, to retire to Gibraltar without striking the ports along the coast of the Upper Tyrrhenian Sea. On 6 February Force H again got underway. During the night of 8–9 February, it separated into two groups: one centered on *Ark Royal*, with three destroyers bombarding Leghorn, while the rest of the force shelled the installations in the port of Genoa. Damage on shore was heavy, but one Italian battleship (*Duilio*) at Leghorn suffered no hits. The British lost only one aircraft. The Italians were unable to mount a counterattack from the air on the retiring British ships because of their ponderous organization and the lack of inter-service cooperation.[36]

In a more recent example, the Israeli Navy, after obtaining sea control in the first few days after the start of hostilities in 1973, carried out the second phase of its plan to exercise sea control. Starting on 10–11 October, the Israeli missile craft and gunboats attacked oil shipping and port facilities at the ports of Latakia, Banias (Bāniyās), and Tartūs. Between then and the end of hostilities on 17 October, the Israelis conducted almost continuous nightly attacks against the Syrian and Egyptian ports of Damietta, Port Said, Rosetta and Alexandria. Additionally, the Israelis attacked many military installations, notably radar stations and SAM sites along the Egyptian-controlled coast, especially in the northern part of Sinai. These attacks were predominantly conducted with missile craft, but commandos took part as well.[37]

By the start of the offensive phase of the Gulf War of 1990–91 (Operation *Desert Storm*), the US Navy carried out massive strikes and attacks against selected Iraqi targets by carrier-based aircraft and surface ships and submarines armed with the *Tomahawk* land-attack missiles (TLAMs). The size of the area and the range of attack aircraft limited US aircraft carriers deployed in the Red Sea to strike only those targets in the western part of Iraq and the Baghdad area. The aircraft from the carriers deployed in the Arabian Gulf concentrated on targets in southeastern part of Iraq, specifically bridges near Basra and naval facilities near Umm Qasr.[38] Nine US guided missile cruisers and destroyers launched TLAMs against selected targets in Iraq from their operating areas in the Red Sea and the Arabian Gulf at the start of the *Desert Storm* phase of campaign on 17 January 1991.[39] TLAM-Cs/Ds missiles were used against well-defended targets that required precision warheads. The US Navy also used extensively its two battleships deployed in the Arabian Gulf to strike a range of Iraqi targets on the coast. The battleship *Missouri* fired 16-inch shells on prefabricated concrete bunkers along the Kuwaiti coast on 3 February. Three days later another US battleship, *Wisconsin*, relieved *Missouri* on the battleline.[40] The US battleships fired more than 1,000 16-inch shells against the Iraqi positions after the start of the ground offensive on 24 February 1991, forcing the Iraqis to abandon their positions at the coastal port of Ras Al Qualayah.[41]

Naval Operations other than War

A blue-water navy often operates in narrow seas either alone or as a part of a multi-national force under the mandate of the United Nations or a regional security organization. It exercises control of a certain sea area in peacetime and in time of crisis by virtue of its size and deterrent capability. The US Navy is forwardly deployed in many areas of the world, including many narrow seas, specifically the Mediterranean, the Caribbean, the Arabian Gulf, the Sea of Japan and the marginal seas of the western Pacific. The navies operating in narrow seas have been in the past, and will be in the future involved in activities of crisis prevention or crisis management/resolution. *Conflict prevention* includes diverse military activities conducted either unilaterally or collectively under Chapter VI of the UN Charter and aimed at preventing the escalation of disputes into armed conflict or at facilitating the resolution of armed violence. These actions range from diplomatic initiatives to preventive deployment of naval forces. Naval forces deployed in forward areas should be of sufficient size and combat power to defeat opposing forces quickly and decisively.

Optimally, conflict prevention under the UN Charter should be conducted with strict impartiality, because all sides in a dispute have to agree to involve other countries as mediators. Naval forces can be deployed in the proximity of a country where hostilities threaten to break out. Normally, carrier forces and amphibious forces have more chance of success in disputes between nation-states than in ethnic conflict or civil war. To be effective the actions aimed to prevent the hostilities must be accompanied by a clear willingness on the part of the international community to use overwhelming force if necessary. Otherwise, preventive deployment of naval forces, regardless of their size and capability, would rarely produce the desired effect.

A major operation to *protect shipping* would require the execution of diverse tasks to protect merchant vessels from unlawful attack in international waters. This is accomplished through, among other things, the escort of merchant ships (sometimes of individual ships, for a specific purpose), coastal sea control, harbor defense, and MCM. In 1987–88 the US Navy protected 11 Kuwaiti ships from possible Iranian attacks in the Arabian Gulf (Operation *Earnest Will*). The Kuwaiti ships were allowed to fly the American flag, which entitled them to protection by the US Navy's ships. Beginning in July 1987 these ships were escorted from Kuwait to a point outside the Strait of Hormuz.[42]

A maritime state, ruled by a radical regime, especially if situated close to maritime-trade choke points, might attempt to harass shipping, requiring the response of naval forces. Protection of shipping requires coordinated employment of surface, air, and subsurface forces, as well as a suitable C3I structure both ashore and afloat. Protection of shipping might include pre-emptive or retaliatory strikes or raids against selected targets at sea or ashore.

A major naval operation to ensure *freedom of navigation (FON) and overflight* is aimed at asserting rights and ensuring an unhindered use of sea or air routes. Normally, a riparian state may exercise jurisdiction and control within its territorial sea. International law, however, establishes the right of 'innocent passage' by ships of other nations through a state's territorial waters. Passage is considered 'innocent' as long as it is not prejudicial to the peace, good order, or security of the coastal nation. Also, freedom of navigation by aircraft through international airspace is a well-established principle of international law. Threats to aircraft through the extension, whether by nations or groups, of airspace control zones beyond international norms can be expected to result in measures, acceptable under international law, to rectify the situation.[43] The US Navy and US Air Force conducted such a major operation (*El Dorado Canyon*) in the Gulf of Sidra in April 1986.

Naval forces are probably most extensively used in various peace operations (PO) – military operations to support diplomatic efforts to reach a long-term political settlement. Peace operations are tailored to each situation and may be initiated in support of diplomatic activities either before, during, or after the conflict.[44]

Peacekeeping and peace-enforcement are the principal types of peace operations. *Peacekeeping operations* (PKO) are intended to contain, moderate, or terminate hostilities between states or within a state, using international military forces to complement political conflict-resolution efforts and to restore and maintain peace. These actions are normally conducted after the sides in a conflict agree to cease hostilities. Impartial observers are usually sent to verify implementation of the cease-fire or to monitor the separation of forces.

Peace-enforcement operations (PEO) involve diverse tasks as authorized by Chapter VII of the UN Charter, specifically implementation of sanctions, establishment and supervision of exclusion zones, intervention to restore order, and forcible separation of belligerents. The objective is to establish an environment for a truce or cease-fire. In contrast to peacekeeping, peace-enforcement operations do not require the consent of the sides or warring factions in a conflict.[45] Maritime forces, when used for peace-enforcement, should be ready to engage in combat if necessary.

During the conflict in the former Yugoslavia, NATO/WEU navies operated in the Adriatic Sea to enforce an arms embargo against the successor states in the area. For the first time NATO navies took part in UN peacekeeping and sanctions-enforcement operations relating to the conflict in the former Yugoslavia in the summer of 1992. This, in effect, punished the Bosnian Muslims and Croats far more than it did the Serbs, who had ample stocks of weapons and ammunition. The NATO foreign ministers also made a similar offer to monitor the arms embargo with their ships.[46]

That part of the activities in the Adriatic by NATO ships was dubbed operation *Maritime Guard*, and the part controlled by WEU as operation *Sharp Fence*. Italy was entrusted with the task of coordinating WEU and NATO forces' operations in the Adriatic. Initially, the activity of WEU and NATO's ships in the Adriatic was limited to monitoring only rather than the blockade often described by the Western media.

Initially, the WEU force operated south of a line from Bari (Italy) to Bar (Montenegro), while the Standing Naval Force Mediterranean (STANAVFORMED) operated north of it. During August 1992 each force reversed its operating area. This pattern has continued since then. Each force received intelligence and surveillance data from the maritime patrol aircraft.[47] By late August the NATO Headquarters in Brussels decided to send a joint Standing Naval Force Atlantic (STANAVFORLANT) to relieve the STANAVFORMED to take over monitoring duties in the Adriatic.

The NATO ministers tightened the trade embargo against the rump state of Yugoslavia at their meeting in Paris on 18 November 1992. In accordance with UN Resolution 787, all merchant ships would be stopped and searched to find out whether they violated the terms of the UN embargo. Ships that tried to break the blockade would be seized. Two days later the WEU foreign and defense ministers at their meeting held in Rome introduced stricter measures in enforcing the UN embargo against Serbia and Montenegro. By the UN Security Council decision, WEU ministers also decided to control the shipping traffic on the Danube River with the help of Bulgaria, Hungary, and Romania.[48] By early June 1993, the NATO and WEU councils had decided to combine their naval activities in the Adriatic and one month later the operations *Maritime Guard* and *Sharp Fence* merged into a single operation called *Sharp Guard*. Allied Naval Forces Southern Europe (NAVSOUTH) took over operational control of this operation.[49]

Sharp Guard was conducted to monitor and enforce compliance with UN sanctions in accordance with UN Security Council Resolutions (UNSCR). Its maritime forces were organized in Combined Task Force 440 (CTF 440). The aim was to prevent all unauthorized shipping from entering the territorial waters of the rump state of Yugoslavia and all arms from entering the former Yugoslavia.[50] A total of 14 navies took part in this operation. A large number of merchant vessels suspected of carrying weapons and equipment were boarded.[51] The Operation *Sharp Guard* was suspended on 1 October 1996 in accordance with UNSCR 1074, which terminated all sanctions on the former Yugoslavia.

Maritime forces continued to operate in the Adriatic as a part of the multinational Peace Implementation Force (IFOR) as part of Operation *Joint Endeavour* formally established on 20 December 1995 in the aftermath of the Dayton Peace Accords (formally signed in Paris on 14 December) that ended hostilities in Bosnia and Herzegovina. The embargo on the delivery of weapons and military equipment to former Yugoslavia (with the exception of some categories of heavy weapons) was formally terminated on 14 March 1996.[52]

Exclusion zones, in the air, at sea, or on land, are often established in peacekeeping operations.[53] The objective is to persuade a country or group of countries to a change in behavior so as to meet the desires of the sanctioning body, rather than face continued sanctions or force. The sanctions are intended to create economic, political, military, or other conditions to that end. The UN or some other international body usually imposes an exclusion zone, but even individual countries may apply it.

NATO air forces started to implement operation *Sky Monitor* in response to UNSCR 781, which requested member states to assist UN Protection Forces (UNPROFOR) to monitor the ban on military flights in Bosnia and Herzegovina. By 31 March 1993 UNSCR 816 extended that ban to cover all flights not authorized by UNPROFOR. It authorized member states to take all necessary measures, in event of further violations, to ensure compliance with the ban. NATO approved the plan to enforce the ban on 8 April and as result Operation *Deny Flight* started on 12 April 1993. The aim of this operation was aerial monitoring and enforcement of compliance with UNSCR 816, which banned flights by fixed-wing and rotary-wing aircraft in the airspace of Bosnia and Herzegovina, provided close air support (CAS) to UN troops on the ground as requested by UN commanders, and carried out sorties on request, all in coordination with UN air strikes against designated targets threatening the security of the UN-declared safe areas. The planning was conducted jointly by AFSOUTH (Allied Forces South) and SHAPE in Mons, Belgium. Tactical day-to-day control of all flights was exercised by the 5th Allied Tactical Air Force (ATAF) with headquarters in Vicenza, Italy. Coordination between NATO and UN was arranged through liaison officers at 5th ATAF and the UN Headquarters in Zagreb and Sarajevo. In the summer of 1993 about 100 NATO multinational tactical strike aircraft and dozens more tanker and command/control aircraft were deployed in the area, with about 4,500 personnel from 12 NATO countries taking part in the operation. These aircraft were based primarily in Italy and on board three carriers (one US, one British, and one French) in the Adriatic. Normally, one US Navy aircraft carrier was deployed in the Adriatic, as were from time to time, one each from France and Britain.[54] During the operation, NATO aircraft carried out a highly successful series of strikes against Bosnian Serb positions (Operation *Deliberate Force*) between 30 August and 20 September 1995.[55] This operation was carried out to force the withdrawal of Bosnian Serb heavy weapons from the total exclusion zone around Sarajevo and to restore complete freedom of movement for UN forces and personnel and non-governmental organizations, and unrestricted use of Sarajevo airport. The operation *Deny Flight* formally ended on 21 December 1995.[56]

After IFOR ended its mission, NATO was directed to organize Stabilization Force (SFOR) to deter the renewal of hostilities and stabilize the situation in Bosnia and Herzegovina. This force, half the size of its predecessor, was formally established on 20 December 1996. SFOR has the same command structure, rules of engagement, and enforcement authority as IFOR. SFOR is a joint operation led by NATO, but it also includes the participation of about 20 non-NATO countries. One of the main missions of SFOR is to deter or prevent a resumption of hostilities or new threats to peace in Bosnia and Herzegovina. The air component of IFOR was transferred to SFOR. The latter's activities, dubbed operation *Decisive Endeavour*, are part of NATO's overall operation *Joint Endeavour*. About 5,500 air force personnel from 12 NATO countries take part in this operation. The SFOR maritime component (Operation *Determined Guard*) was organized into Task Force 436 and was under the control of the Commander Naval Forces Southern Europe. This force, together with

STANAVFORMED, was available and could be called upon to support the SFOR mission. In addition, other maritime forces currently deployed in the Mediterranean, including the US Sixth Fleet, can be made available if necessary to support SFOR.

Maritime intercept operations (MIO) for sanction enforcement employ naval forces, often with the participation of land-based aircraft. These actions normally use coercive measures to interdict the movement of designated items into or out of a nation, specific sea, or ocean area. The political objective is to compel a country or group of countries to conform to the objectives of the initiating body. The military objective is to establish a selective barrier – that is, one which allows only authorized goods to pass. Enforcement of sanctions normally requires action by some combination of surface and air forces.[57] Based on the UN Security Council's Resolutions 661 of 6 August and 665 of 25 August 1990, the US Navy and its Coalition partners initiated MIO against Iraq in the Arabian Gulf and the Red Sea. The key checkpoints to be monitored were the Strait of Hormuz, the Strait of Bab el-Mandeb and the Gulf of Aqaba. The multinational intercept force (MIF) eventually consisted of some 80 surface ships. Of these, 30 ships were non-American, coming from the British, French, Australian, Greek, Belgian, Canadian, Argentinian, Saudi Arabian, Bahraini, Qatari and United Arab Emirates (UAE) navies. In addition, US P-3 patrol aircraft and British Nimrods were used in support of the operation. This effort was quite successful because it reportedly prevented about 90 per cent of Iraqi imports and 100 per cent of Iraqi exports.

Peacekeeping and peace-enforcement operations are difficult to plan separately, and the practice so far shows, that these operations are usually combined, as the example of NATO and WEU forces in support of forces in Bosnia prior to the Dayton Accords of 1995 clearly shows.[58]

Naval forces may also be involved in *expanded peacekeeping and peace-enforcement operations*. These operations are larger in size than peace-keeping, involving perhaps 20,000 men or more. The consent of the sides in conflict is usually nominal, incomplete, or simply non-existent. These operations include more assertive mandates and ROE, including the use of force under Chapter VIII of the UN Charter. Expanded PKO/PEO are conducted with strictly limited objectives, such as protecting safe-flight or no-fly zones, or relief deliveries. If too intrusive, they are likely to draw multinational forces into open hostilities, the naval forces having then either to be pulled out or be committed to full-scale combat. Since December 1995, NATO forces, including maritime elements, have been involved in expanded PKO/PEO in Bosnia and Herzegovina (Operation *Joint Endeavour*).

To sum up, exercising command of the sea requires that one's navy be capable of carrying out simultaneously or sequentially a number of strategic and operational tasks. Because of the generally small area and shorter distances involved, a blue-water navy can exercise command of a typical narrow sea by deploying a fraction of its forces. However, it must be able to project power ashore even in those areas of a narrow seas distant from the sea's exit. For a coastal navy, exercising control over even a portion of a narrow sea requires continuous efforts with the employment of almost all its forces.

Naval forces will conduct in the future diverse tasks in peacetime in narrow seas ranging from forward presence to a more assertive use of combat power, such as may be necessary to establish freedom of navigation and the enforcement of maritime exclusion zones and no-flight zones. They may also provide support to troops ashore in conducting PKO/PEO. These activities may be conducted by multi-national naval forces. Hence, not only large but also medium and small navies may be involved in carrying out these tasks.

NOTES

1 Scott Bowden, *Forward Presence, Power Projection, and the Navy's Littoral Strategy* (Arlington, VA: IRIS Corporation, July 1997), pp. 3, 7; Carl E. Mundy, 'Thunder and Lightning: Joint Littoral Warfare', *Joint Force Quarterly* (Spring 1994), p. 47.
2 Frank J. Murphy, Jr, 'Littoral Warfare: Adapting to Brown-Water Operations', *Marine Corps Gazette* (September 1993), pp. 65–73.
3 Officially, the littoral strategy calls for the US Navy to operate up to 650 miles from the coastline or the furthest striking range of US naval forces today; Bowden, *Forward Presence, Power Projection, and the Navy's Littoral Strategy*, p. 3; however, in practice this distance in US terms varies from 370 miles to 510 miles, that is the effective range of the F/A-18 versions of the Navy's most capable attack aircraft; Bowden, *Forward Presence*, p. 10.
4 Bowden, *Forward Presence*, p. 5.
5 Woodward, *The Russians at Sea: A History of the Russian Navy* (New York: Frederick A. Praeger, 1966), p. 77.
6 Lawrence Sondhaus, *The Habsburg Empire and the Sea: Austrian Naval Policy 1797–1866* (West Lafayette, IN: Purdue University Press, 1989), p. 254.
7 Peter Handel-Mazzetti, 'Kuestenangriff und Kuestenverteidigung', *Marine Rundschau*, 5 (May 1936), p. 218.
8 James B. Agnew, 'From Where Did Our Amphibious Doctrine Come?' *Marine Corps Gazette*, 8 (August 1979), p. 53.
9 Woodward, *The Russians at Sea: A History of the Russian Navy*, p. 101; Elmer B. Potter and Chester W. Nimitz, eds, *Seemacht: Eine Seekriegsgeschichte von der Antike bis zur Gegenwart* (Herrsching: Manfred Pawlak, rev. edn, 1986), p. 406. Significant naval forces provided support: ten battleships, eight light cruisers, six U-boats (for minelaying), ten large destroyers, 11 torpedo craft, one minelaying steamer, one aircraft tender, six airships, and 95 aircraft; Otto Groos, *Seekriegslehren im Lichte des Weltktrieges: Ein Buch fuer den Seemann, Soldat, und Staatsmann* (Berlin: Verlag von E. S. Mittler & Sohn, 1929), p. 214.
10 Randolf Kugler, *Das Landungswesen in Deutschland seit 1900* (Berlin: Oberbaum

Verlag, 1989), p. 35.
11 Groos, *Seekriegslehren im Lichte des Weltkrieges*, p. 217; Mueffling von Ditten, 'Kriegerische Landungen', *Wissen und Wehr*, 3 (March 1929), p. 175.
12 Geoffrey Till, ed., *Maritime Strategy and the Nuclear Age* (New York: St Martin's Press, 2nd edn, 1984), p. 117.
13 About 5,000 Canadian troops, 1,075 British personnel, and 50 US Rangers organized in 13 groups and transported in 237 ships and landing craft, including eight destroyers; 'Dieppe Raid', in I. C. B. Dear, ed., *The Oxford Companion to World War II* (Oxford/New York: Oxford University Press, 1995), pp. 298–9.
14 Samuel E. Morison, *The Struggle For Guadalcanal, August 1942–February 1943*, Vol. V of *History of United States Naval Operations in World War II* (Boston, MA: Little, Brown, reprint 1984), p. 27.
15 Paul S. Dull, *A Battle History of the Imperial Japanese Navy (1941–1945)* (Annapolis, MD: Naval Institute Press, 1978), p. 49.
16 Donald Macintyre, *Sea Power in the Pacific: A History from the 16th Century to the Present Day* (London: Military Book Society, 1972), pp. 212–13.
17 Gilbert A. Shepperd, *The Italian Campaign 1943–45* (New York: Frederick A. Praeger, 1968), pp. 22–3.
18 Claude Huan, 'Die sowjetische Landungs-Operationen auf der Krim 1941–42', *Marine Rundschau*, 6 (June 1962), p. 341.
19 Ibid., p. 355.
20 Meister, *Der Seekrieg in den osteuropaeischen Gewaessern 1941–45*, pp. 272–3; Friedrich Ruge, *The Soviets as Naval Opponents 1941–45* (Annapolis, MD: Naval Institute Press, 1979), p. 355.
21 B. H. Liddell Hart, 'The Value of Amphibious Flexibility and Forces', *Journal of the Royal Services Institution*, Vol. CV, 620 (November 1960), p. 485.
22 B. H. Liddell Hart, *The Real War* (Boston, MA: Little, Brown, 1st edn 1930, repr. 1964), p. 160.
23 Julian S. Corbett, Vol. 2: *From the Battle of the Falklands to the Entry of Italy into the War in May 1915, Naval Operations* from *History of the Great War based on Official Documents* (London: Longmans, Green, 1921), p. 68; ibid., Vol. 3: *The Dardanelles Campaign* (London: Longmans, Green, 1923), p. 106.
24 Arthur J. Marder, *From the Dreadnought to Scapa Flow*, Vol. II: *The War Years: To the Eve of Jutland* (London: Oxford University Press, 1965), p. 191; Woodward, *The Russians at Sea: A History of the Russian Navy*, pp. 167–8.
25 Macintyre, *Sea Power in the Pacific*, p. 254.
26 Hart, 'The Value of Amphibious Flexibility and Forces', pp. 483–5.
27 Macintyre, *Sea Power in the Pacific*, pp. 252–3.
28 Department of Defense, *Conduct of the Persian Gulf War: Final Report to Congress* (Washington, DC: US Government Printing Office, April 1992), p. 294.
29 *The New York Times*, 28 February 1991, pp. A8–9.
30 Potter and Nimitz, eds, *Seemacht*, p. 382.
31 Twelve battleships, six cruisers, 14 destroyers, 30 torpedo craft, three submarines, and six seaplanes; ibid., p. 359.
32 Anthony Sokol, 'Naval Strategy in the Adriatic Sea during the World War', *Proceedings*, 8 (August 1937), p. 1087.
33 Ibid., p. 1087; Potter and Nimitz, eds, *Seemacht*, p. 358.
34 Marder, *From the Dreadnought to Scapa Flow*, Vol. II, p. 201; S. W. C. Pack, *Sea Power in the Mediterranean: A History from the Seventeenth Century to the Present Day* (London: Arthur Barker, 1971), pp. 160–1.
35 Marder, *From the Dreadnought to Scapa Flow*, Vol. II, p. 261.
36 Raymond De Belot, *The Struggle for the Mediterranean 1939–1945*, trans. James

A. Field Jr (Princeton, NJ: Princeton University Press, 1951), pp. 97–8.
37 Walter Jablonsky, 'Die Seekriegsfuehrung im vierten Nahostkrieg', *Marine Rundschau*, 11 (November 1974), p. 656.
38 Michael A. Palmer, 'The Navy Did Its Job', *Proceedings/Naval Review 1991*, p. 88.
39 Stanley R. Arthur and Marvin Pokrant, 'Desert Storm at Sea', *Proceedings/Naval Review 1991*, p. 83; Palmer, 'The Navy Did Its Job', p. 88; Hartmut Zehrer, *Der Golfkonflikt: Dokumentation, Analyse und Bewertung aus militaerischer Sicht* (Herford/Bonn: Verlag E. S. Mittler & Sohn, 1992), p. 198.
40 Between 8 and 9 February *Wisconsin* fired 112 16-inch shells against the Iraqi defenses in southern Kuwait; Palmer, 'The Navy Did Its Job', pp. 89–90.
41 Arthur and Pokrant, 'Desert Storm at Sea', pp. 85–6.
42 Frank Uhlig Jr, *How Navies Fight: The US Navy and Its Allies* (Annapolis, MD: Naval Institute Press, 1994), p. 380.
43 Joint Pub 3-07, *Joint Doctrine for Military Operations Other Than War*, 16 June 1995, p. III-4.
44 Ibid., p. III-12.
45 Ibid., p. III-13.
46 The NATO and WEU representatives at their meeting in Helsinki agreed on 10 July 1992 to conduct a naval action to oversee the carrying out of the UN Security Council's Resolutions 713 and 757, which had stipulated that a naval blockade of the Adriatic was required to ensure the arrival of humanitarian aid to Sarajevo. The WEU Council of Ministers' decision was also aimed at monitoring the arms embargo on all the republics of the former Yugoslavia; 'Further on Arrangements' *Daily Report East Europe* FBIS-EEUR 13 July 1992, pp. 2–3; 'French Vessel Joins Operation' *Daily Report East Europe* FBIS-EEUR-92-135, 14 July 1992, p. 2; 'WEU Chief on NATO Coordination' *Daily Report East Europe* FBIS-EEUR, 14 July 1992, pp. 2–3; 'WEU Head on Ships' Duties' *Daily Report East Europe* FBIS-EEUR-92-135, 14 July 1992, p. 2.
47 The WEU force was supported by three *Atlantic 2*, two P-3C *Orion*, and several other patrol aircraft based in Sicily and Sardinia. The NATO force was supported by US, UK, and Portuguese maritime patrol aircraft based in Italy. In addition the USA and France have deployed E-3A *Sentry* aircraft in Greece and Italy to provide surveillance and intelligence information to both WEU and NATO forces deployed in the southern Adriatic; Michael Chichester, 'Maastricht, the Adriatic and the Future of European Defence', *Navy International*, November–December 1992, p. 361.
48 Guenter Marten, 'Die Westeuropaeische Union und der Balkan-Konflikt', *Marineforum*, March 1993, p. 57; 'NATO Agrees to Use Warships to Enforce Yugoslav Blockade', *The New York Times*, 19 November 1992, pp. A1, A13.
49 'Die Uneinigkeit der NATO belastet das Treffen der Aussenminister in Athen', *Frankfurter Allgemeine Zeitung*, 9 June 1993, p. 4; 'NATO, WEU Combine Navies in Blockade Against Ex-FRY', *Daily Report West Europe*, FBIS-WEU-93-109, 9 June 1993, p. 1.
50 The CTF 440 consisted of the Standing Naval Force Mediterranean (STANAVFORMED) together with the WEU Contingency Maritime Force. CTF 440 was organized into several task groups: two task groups operated in the Adriatic Sea, while a third task group conducted port visits. Maritime patrol aircraft (MPA) operated under CTF 440 through the Commander of CTF 431.
51 Between November 1992 and June 1996 about 74,200 ships were challenged and of these 5,950 were boarded and inspected at sea, while 1,480 were diverted and inspected in port.

52 The naval component of IFOR consisted of: Allied Naval Forces Southern Europe (COMNAVSOUTH), responsible for protection of sea lines of communications in the Adriatic and used for reinforcement and resupply of IFOR forces ashore; and Allied Striking Forces Southern Europe (COMSTRIKEFORSOUTH), available for power projection in case of non-compliance with the terms of the Dayton Peace Accords by the parties. These forces included both carrier-based air and amphibious forces.
53 Joint Pub 3-07, p. III-4.
54 Hans Binnendejk, ed., *Strategic Assessment 1996* (Washington, DC: National Defense University, Institute for National Strategic Studies, 1996), p. 135; 'Further on Arrangements', *Daily Report East Europe* (FBIS-EEUR), 13 July 1992, pp. 2–3.
55 During the same operation, the USS *Normandy* launched 13 Tomahawk (TLAMs) cruise missiles against Bosnian Serb air-defense sites in northwestern Bosnia.
56 Between 12 April 1993 and 20 December 1995 a total of 100,420 sorties were flown in enforcement of the non-fly zone in Bosnia and Herzegovina (23,021 fighter sorties, 27,077 CAS sorties, 29,158 sorties by SEAD, NAEW, tanker, reconnaissance and support aircraft, and 21,164 training sorties).
57 Zehrer, ed., *Der Golfkonflikt: Dokumentation, Analyse und Bewertung aus militaerischer Sicht*, pp. 188–90.
58 About 8,500 ships were stopped, 1,100 ships were searched, and about 60 were diverted from their intended destination in Iraq; ibid., p. 138.

11

Disputing Control

On the open ocean, a weaker fleet will usually contest command of the sea by denying the free use of the sea to the stronger fleet. This situation can also exist in a narrow sea when a stronger navy establishes a blockade or tries to prevent movement of the weaker navy through the sea's only exit. Because of the characteristics of the physical environment in a typical narrow sea, the superiority of a stronger fleet does not prevent offensive tactical actions by the weaker fleet. The threat of submarines, land-based aircraft, coastal missile batteries, and mines makes the presence of large surface combatants very risky. The multitude of islands and islets and a highly indented coastline allows a small navy to challenge the dominance of a stronger fleet operating in a narrow sea. So long as the enemy fleet is in existence and uncontained, control of at least a part of a narrow sea will always be contested or in dispute. The Entente navies secured early on in World War I a working control of the sea, but the Germans disputed that control. The Germans in fact came at one stage close to rendering the Royal Navy's control of the North Sea useless. The Royal Navy never gained control of the Baltic, which remained to all intents and purposes a German lake. The Germans also effectively controlled a part of the North Sea. Control of the Adriatic was also successfully contested by the Austro-Hungarian Navy throughout the entire war.[1]

Within the confines of a typical narrow sea a stronger fleet can sometimes find its command of the sea being contested by a much weaker force. This situation can arise when a weaker fleet operates from a favorable naval position and blockades a sea's only exit or the exit from the basing area of a stronger fleet. The Turkish Navy or Danish Navy, although inferior in strength, may perhaps be able to block the movements of much stronger opponents through the Dardanelles and Bosporus or Skagerrak and Kattegat respectively. Likewise, the Japanese Navy can prevent a hostile navy from moving its forces through La Perouse and the Korea Straits.

The tasks of a navy contesting command in a narrow sea against a stronger force may range from a 'fleet in being', countering naval blockade, strategic diversion, defensive mining and defense of the coast. Additionally, attack on enemy maritime trade is also one of the principal tasks of the weaker fleet in contesting command of the sea. In a typical narrow sea, both the weaker and the stronger fleet carry out the task of defending the coast uninterruptedly, regardless of the strategic situation in the theater.

Fleet in Being

In the past, the usual response to the enemy naval blockade was the 'fleet in being'. The well-known British naval writer J. R. Thursfield defined a fleet in being as a fleet strategically at large, not in itself in command of the sea, but strong enough to deny that command to the adversary by strategic and tactical dispositions adapted to the circumstances of the case.[2] A stronger fleet should watch over enemy bases and the main force should sustain a high state of combat readiness so as to attack the inferior fleet when it comes out. As long as the inferior fleet is undefeated its very existence constitutes a perpetual menace to control of the sea by a stronger fleet. A stronger fleet also must devote a considerable amount of energy and effort to watch the inferior fleet and to make sure that it does not elude the blockading force if it chooses to come out.[3]

Fleet in being was practiced by the British admiral, Arthur Herbert, Earl of Torrington, in 1690. The Royal Navy was dispersed in several detachments, each inferior to the French fleet, under Admiral de Tourville, hovering off the Isle of Wight. Torrington intended to avoid a decisive battle until his reinforcements arrived. However, the important thing was that the French were unable to do any great mischief as long as Torrington's fleet existed. In the end, the Government ordered Admiral Torrington to engage the French fleet and the result was the British defeat in the battle of Beachy Head on 30 June 1690.[4]

In general, an inferior fleet exercises a restraining influence on the stronger fleet. While it avoids a decisive encounter with the superior force, the fleet in being retains the maximum possible threat value. Its threat can be countered only by keeping a superior force constantly ready to intercept and engage it, should it move too far from its bases. A fleet in being, skillfully handled, can prevent the superior fleet from taking full advantage of its strength. Then, success will not depend on who is on the offensive or who is on the defensive, but on who has obtained and maintained the strategic initiative, and the stronger fleet will not necessarily exercise this.

The purpose of a fleet in being can be to prevent an enemy's large-scale seaborne invasion or an enemy attack on one's coast, or it can be to disperse enemy naval strength and to force the strategic redistribution of his superior fleet. Large amphibious landings are usually avoided if the adversary can effectively challenge the superior fleet at sea. The presence of the High Seas Fleet deterred the Royal Navy from landing an army on the Pomeranian coast in the Baltic in World War I; it may also have been the reason why the British did not try to seize one of the Frisian Islands. In a typical narrow sea, the presence of strong land-based air even without a fleet in being can deter the enemy from undertaking any large seaborne expedition.

In the past, a weaker navy has been able to accomplish some positive strategic results by attacking the enemy coast or elements of maritime trade. Some weaker fleets, in contrast, have tried to deny a stronger adversary the benefits of his superiority by harassing actions and avoiding any large encounter at sea. However, there have also been cases where a weaker fleet did nothing except ensure its own survival in the face of a much superior force. Therefore, the variants of fleet in being range from conducting a tactical offensive to passive defense.

An active fleet in being exists when one side avoids actions likely to result in heavy losses, but seizes every tactical favorable opportunity to attack and inflict damage on its adversary. Expressed differently, an active fleet in being exists when the adversary considers general command of the sea as a prerequisite to the projection of power ashore; but one might be able to deny him such command by refusing what Admiral Horatio Nelson called a 'regular battle', while seizing every opportunity for a counter strike.[5]

In the past, the use of the active fleet in being concept was aimed at postponing a general fleet action until such a time when the correlation of forces was at least equal for an inferior fleet. In the Peloponnesian War, 412–11 BC, the Athenians and Peloponnesian fleets watched each other from adjacent ports (Sámos and Miletus) about 20 nautical miles apart and postponed battle for several months. The British used a similar method to defeat the Spanish Armada in 1588. If Alonzo Pérez de Guzmán, Duke of Medina Sidonia, had attacked at once as he advanced up the Channel, he might have compelled the British admiral, Lord Howard of Effingham, to accept a close action and thus have achieved a Spanish victory by virtue of superior numbers. However, the British, fearful of a general action, conducted small-scale and swift attacks at close range, and at the places and times of their own choosing. The result was that the Armada's movement up the Channel was considerably slowed down.

In the past, an inferior fleet has withdrawn to its base and to be protected by fortifications. Then, it was sometimes practicable to land an army with the help of the fleet to capture a base, as the Japanese did in seizing Port Arthur. In such a case, the weaker fleet may share the fate of the army, as happened with the Russian squadron at Port Arthur, or the inferior fleet may be forced to sortie out, as the Spanish squadron did in Santiago de Cuba, to be destroyed by the superior force waiting outside of port.

A superior fleet may also use the active fleet in being concept through either overcaution or timidity, as was shown by Admiral Togo's actions in the initial phase of the Russo-Japanese War of 1904–5. The strategic position then presented to Japan appeared to be simple – the Russians had divided their battle fleet between European and Asiatic waters, and Russian fleet detachments in Port Arthur and Vladivostok were barely adequate for their stated purpose. Consequently, the Japanese had been presented with a godsent opportunity to destroy the Russian fleet in the Far East early on in the war, a chance, however, which Admiral Togo refused to take. The purported reason was that, although it was most likely that the Japanese fleet would have been successful, it would have suffered some losses, making it, in Togo's judgement, unable to meet the expected arrival of the Russian Baltic Fleet. While this course of action seems to have been justified by events, Admiral Togo's critics were perhaps correct in claiming that the fleet actions preceding the Battle of Tsushima cost the Japanese two of their six battleships and that a vigorous offensive in the first few days of the war would probably have won a decisive victory at no greater cost.[6]

Employment of a modern active fleet in being includes tactical-sized actions against enemy ships at sea and in their bases, bombardment of the enemy coast, scouting actions, and offensive mining. In World War I the German Navy harassed the British and conducted minor actions to reduce the British margin of superiority to a level that might eventually allow their High Seas Fleet to take the offensive.[7] The Germans also hoped that successful attacks on the Entente's trade routes might force the British into diverting some of their naval strength into direct trade protection, and thus make the Grand Fleet more vulnerable to sudden ambushes. Taken together, the two courses of action could have forced, and to a certain extent did force, sufficient strategic dispersal on the British to deny them the full rewards of their naval superiority.

The containment of the German High Seas Fleet required the presence of British ships that could otherwise have been away doing something else. The Royal Navy stayed far north and did not dare to attack the German coast to stamp out the U-boat threat at its source. Escorts that might have been valuable in the North Atlantic had to be deployed in the North Sea. Decisive actions against the U-boats, such as laying an effective mine barrier off the German coast, could not be taken throughout the war because of the presence of the German battle fleet in the area.

Attritional warfare is also applied when control of the sea has been contested, that is, where both sides have approximately equal strength in numbers and in quality, or where one side is quantitatively superior but qualitatively inferior or vice versa. Then one or both sides can employ their naval forces and aircraft for a specific and limited purpose. In World War I, the High Seas Fleet tried to inflict as many losses as possible on the Grand Fleet by tactical forays aimed against British forces blockading the Heligoland Bight. The Germans also used mining, and whenever possible their U-boats, for offensive actions off the British coast.

One of such actions was bombardment by German cruisers off the British coast near Gorleston on 3 November 1914. The actual purpose of the bombardment was to provide a cover for a minelaying operation.[8] In a second raid on 16 December, German battle cruisers bombarded Scarborough and Hartlepool on Yorkshire's coast, while one light cruiser laid mines off Filey. The Germans hoped to lure a part of the Grand Fleet into the newly laid minefield.[9] In both raids, British material losses were light, but many civilians were killed and wounded. The High Seas Fleet commander, Admiral Scheer believed that the fleet must strike when the circumstances are favorable. Therefore, it must seek a battle with the Grand Fleet only when a state of equality has been achieved by the methods of guerrilla warfare. The High Seas Fleet must hold back and avoid actions, which might lead to heavy losses. Yet, this did not mean to prevent favorable opportunities from being made to inflict losses on the British fleet.[10] This course of action did not work too well for the Germans because the Grand Fleet conducted a distant not close blockade of the Heligoland Bight. Nevertheless, the very existence of a powerful although weaker German fleet in the North Sea forced the Royal Navy to change its naval dispositions worldwide and the main battle force, the Grand Fleet, had to be deployed in the North Sea for the entire war. The German strategy of fleet in being in fact forced sufficient strategic dispersal of the Royal Navy to deny the British full rewards of their overall naval superiority.[11]

In World War II, Admiral Raeder's plan was to tie the British down with his small fleet of older battleships in the North Sea and then exploit their dependence on overseas supplies with attacks by U-boats and raiders.[12] Although the Germans did not have any hope of obtaining control of the sea surface, their efforts could have been of signal value to their Italian ally. By tying down a good number of British ships, opportunities were created for the Italian Navy (which, however, failed to take advantage of them) in the Mediterranean.

Sometimes the presence of a single but powerful ship and its escorts represents a major threat to control of the sea by a stronger fleet. The German battleship *Tirpitz* was deployed in northern Norway and tied down significant forces of the British Home Fleet. Winston Churchill commented that *Tirpitz* 'exercises a vague general fear and menaces all points at once. It appears and disappears causing immediate reactions and perturbations in the other side.'[13]

A passive fleet in being exists when a weaker force not only holds back but also does little if any damage to its opponent. In the past, passive fleets in being have tried everything to avoid a general fleet action. Such a strategy was futile in the end because it only temporarily prevented a superior force from seeking a decisive encounter. The morale of a weaker force suffers if the concept of passive defense is pursued for too long, as in the French example preceding the Battle of Trafalgar. Not by accident but by a conscious decision, Napoleon I ordered the French fleet to avoid a major battle with the British fleet in 1803–5, because he doubted the possibility of a French victory. His main and only objective was an invasion across the Channel to destroy British army concentrations. Without control in narrow seas, however, this objective eluded Napoleon and, when the decisive battle came, the combined Franco-Spanish fleet proved to be no match for the British fleet under Admiral Lord Nelson. The French often stayed on the defensive even when no longer necessary. French Admiral and strategist Raoul Castex argued that passive defense must be avoided and continuous attacks must be carried out against enemy shipping and the coast. The enemy must be forced to disperse his fleet.[14] However, this is not easily done in practice.

The Russian Navy during the Crimean War of 1853–56 practiced a typical version of a passive fleet in being. The Russians had two battle fleets, one in the Black Sea and one in the Baltic. Both fleets shut themselves up in harbors as soon as the Franco-British fleets arrived on the scene. The Russian fleet at Sevastopol' did not even try to attack French warships used as troop transports.

The Russian battle squadron based at Port Arthur in 1904–5 was another example of a passive fleet in being. At the beginning of hostilities, the Russians had withdrawn their ships into the inner harbor and showed no sign that they were about to venture out to sea. The Japanese for their part did little to destroy the Russian squadron in Port Arthur. They did not realize that at that time the combat readiness of the Russian squadron was too low owing to lack of training, low morale, and the poor state of maintenance. The Japanese feared for the safety of the troop and supply convoys upon which their entire military strategy was based. Therefore, they postponed indefinitely the movement of their troops into Manchuria and, as long as the Russian battle squadron in Port Arthur was intact and able to come out, the Japanese were forced to retain their entire modern fleet in the area. This left the Russian battle squadron at Vladivostok unguarded and free to threaten Japan's northern coasts. In reality, the unpreparedness of the Port Arthur garrison was such that had the Japanese taken immediate advantage of their temporary naval superiority to land an army and take the defense in the rear, the base would quickly have fallen, but this was beyond the limited resources of the Japanese who were still in the process of mobilizing their main armies. Moreover, it called for skillful improvisation of the overall plan, which the Japanese Imperial General Staff was incapable of doing.

In retrospect, the Japanese were perhaps overly concerned with the threat posed by the Russian fleet in Port Arthur. Moreover, the Japanese General Staff seemed to be greatly influenced by the action of the Russian squadron from Vladivostok, which had made a foray into the Strait of Korea at just about the time that the Port Arthur squadron was making a sortie. After sinking two Japanese troop transports and damaging another, the Vladivostok squadron cruised up the Japanese coast and captured and sank several other merchant ships. It eventually escaped unharmed, aided by thick fog, and returned to Vladivostok.[15] After the first indecisive action, the Japanese fleet obtained temporary local control of the Yellow Sea and thereby secured safe passage for the troops to Korea. The Japanese, on 9 February 1904, within a month of the outbreak of the war, landed troops at Chinnanpo (Namp'o) and three months later landed right at the head of Korea Bay. All this took place despite the presence of Russian fleets, one 200 miles away, and the other only 60 miles distant.[16] Yet, Admiral Vilgelm K. Vitgeft, Chief of Staff of the Russian squadron in Port Arthur, asserted that a landing of Japanese troops either on the Liaotung or the Korean Peninsula was impossible as long as his fleet had not been destroyed. The Russians' unquestioned acceptance of the fleet in being theory was largely responsible for the subsequent Russian disaster in Manchuria.

In 1940–43 in the Mediterranean the Italian Navy essentially adopted a passive fleet in being posture, avoiding risk to their battleships unless there was a great chance for success. As these circumstances did not occur, the war in the Mediterranean became a war of attrition. The continued existence of the Italian fleet on both flanks of the supply routes to Malta played an important role in bringing that fortress within sight of starvation and capitulation. Consequently, it also hampered severely the capacity of the British to interfere with the supply traffic to North Africa. However, the Italian refusal to risk their fleet, even when the situation was in their favor, was a prime example of a spiritless strategy.

The concept of a fleet in being can also be practiced in peacetime, as shown by the German 'risk theory' conceived by Admiral Tirpitz and put into practice by their Naval Act of 1900. The idea was that the German fleet should be strong enough to threaten a superior British fleet with such damage that Britain would feel dangerously exposed in consequence. Britain would therefore be deterred from using its naval superiority to pursue policies inimical to German interests. The Germans wanted to have such a fleet, which would force the first-ranking sea power to think twice before attacking the German coast.[17] However, this concept ended in a disaster for Germany. Britain, instead of appeasing Germany, took the initiative by resolving its differences with Russia and France.

Counter-Blockade

A state bordering a narrow sea usually resorts to counter-blockade to neutralize or even nullify the effects of a blockade imposed by a stronger power. In general, the actions of a blockaded country are aimed at lifting either a naval or a commercial blockade. Success against the enemy naval blockade means the resolution of both tasks. Some theoreticians consider counter-blockade to be a form of 'sporadic' warfare at sea. A weaker force is usually unable to obtain the necessary superiority at sea. This is a more difficult problem for a fleet based within the confines of a narrow sea, even if it is numerically stronger than the blockading force, because a disadvantageous maritime position does not allow the full use of all available forces.

Actions to break through a naval blockade are primarily directed towards allowing one's naval forces access to the open ocean so that they can accomplish their assigned tasks. Another objective is to re-establish the country's uninterrupted links with overseas. One's fleet is usually employed in cooperation with the other services of the country. The stated objective is accomplished by destroying or seriously weakening the hostile forces conducting the blockade.

A blockaded fleet can choose to contest an enemy naval blockade by breaking through the blockading line with part of its surface forces or submarines. A weaker fleet can also hope under favorable circumstances to force the straits or narrows defended by a blockading fleet. However, this task is usually difficult to accomplish, as shown by the repeated attempts by German surface forces against Allied forces defending the Dover Barrage in World War I. The Austro-Hungarian light forces based in the Bay of Cattaro fought a number of engagements with Allied ships in the Strait of Otranto, but despite some success they were unable to lift the blockade of the Adriatic Sea's only exit.

The repeated attempts by Soviet submarines in 1941–44 to break through the German–Finnish mine barriers in the Gulf of Finland were only partially successful. The first wave of eight Soviet submarines succeeded in breaking through the mine barriers and coastal patrols in mid-June 1942. Although these submarines subsequently operated in the western part of the Gulf of Finland, off the coast of Sweden and Pomerania, they sank only four ships of about 8,000 tons. The second wave came with groups of seven and three boats in mid-August and early September 1942 respectively. During the breakthrough they were energetically supported by Soviet land-based aircraft and patrol craft. Three submarines ran on to mines and one was sunk, while two others were damaged but managed to return to base. The submarines, which reached the open sea, torpedoed seven ships totaling 14,000 tons, of which five ships totaling 10,000 tons were sunk. The third wave of 18 submarines crossed the

German–Finnish mine barriers in two waves in late September, in October and in early November 1942. Four outward-bound Soviet submarines were lost on mines, while three other boats were sunk by Finnish submarines. Additionally, at least two Soviet submarines were damaged by mines while returning to their bases. Those submarines that reached the open waters of the Baltic destroyed only six ships with 12,000 tons, while three other ships of 12,000 tons were lost on mines presumably laid by Soviet submarines. In 1942 a total of ten Soviet submarines were lost, and seven damaged while trying to break through the German–Finnish blockade of the Gulf of Finland. In addition, the Soviets lost at least 18 patrol craft, MTBs, submarine chasers, and minesweepers. Soviet submarines made repeated attempts to break through the German–Finnish blockade in 1943 as well as in the spring and summer of 1944. However, not a single one of them crossed the successive mine and net barriers in the Gulf of Finland.[18] For the Germans the price of a successful Soviet submarine offensive in the Baltic would have been high. It would have cut off the iron ore traffic from Sweden and shale oil from Estonia (this was important because of declining amounts of oil received from Romania), cut off maritime supply routes to Finland and Army Group North, and interrupted U-boat training in the Baltic.[19]

An example of a successful operation to break through an enemy naval blockade and operational redeployment was the escape of the German battle cruisers *Scharnhorst* and *Gneisenau*, as well as the heavy cruiser *Prinz Eugen* and escort ships, from Brest through the English Channel to their North Sea bases on 11–13 February 1942 (Operation *Cerberus*) before eventually being redeployed to Norway. This operation was conducted on Hitler's wishes. He wanted to bring the battle squadron from Brest to Germany because of his obsession with a possible Allied invasion of Norway. Despite some excellent information obtained from reading the German codes (the *Ultra* system), the British were fooled as to the exact timing of the passage of the German ships through the Strait of Dover. Other factors in the British failure were poor inter-service cooperation and lack of coordination among combat arms. Although the Germans accomplished a great tactical victory, their success was also a significant operational loss, since having their battle squadron at Brest was a far greater menace to Allied convoys on the Atlantic than it was while based at home or in Norway.

Strategic Diversion

When there is great disparity of strength, the task of an inferior fleet is to bring about a balancing of forces. This is especially critical when a stronger fleet exerts too much pressure in home waters. Then, an inferior force should try to disperse the enemy naval strength by conducting a *strategic diversion* in some other sea or ocean area far away from the enemy's home waters.

Concentration and diversion of forces have different objectives, but common for both is that their results are not permanent. Concentration does not mean that all available forces must be concentrated in a given sea or ocean area. It aims at assembling one's own forces against the enemy's forces with the objective of eliminating the remainder of the enemy force and possibly attaining a force superiority. In other words, force concentration is not the sum of available forces. It is one-time concentration.

Strategic diversion is designed to stretch out enemy naval strength and obtain a favorable ratio of forces or even superiority in the main theater of war or operations. It does not necessarily aim to achieve the qualitative balance of naval strength. Like strategic concentration, its objective is temporary. The fundamental problem facing an inferior fleet in its attempt to achieve final victory cannot be resolved by conducting a diversion.[20] Nevertheless, a successful diversion can greatly facilitate the task of a weaker fleet in home waters in achieving decisive results.

Because of a close interrelationship of maritime interests with the total life of a country or group of countries, several options exist to apply strategic diversion in a war at sea. The objective may be purely military, as for example to partially divert enemy forces from exerting strong pressure in home waters, to inflict losses on the enemy fleet, and to weaken the enemy fleet as a preliminary to achieving a decision in the main theater.[21] In the Seven Years War the French used a deception plan to cover their strategic diversion. In 1756 strong troop contingents embarked on transports in Brittany's ports tied down the British fleet while, at the same time, the French secretly prepared and armed a second transport contingent in Toulon. The French deception was highly successful. They achieved complete surprise, seized the Balearics with its fort at Port Mahón and thereby gained control of the western Mediterranean.[22]

During the French Revolutionary Wars, the French had several strategic options to choose from: invade England, fight a decisive battle to defeat the Royal Navy, cut off the British lifelines with the outside world, or attack the British overseas possession to undermine England's source of power and its position in the world.[23] Napoleon I tried all four options and then applied a strategic diversion by threatening England in India, which was the key to England's world position. As a preparatory measure, the French secured routes in the Mediterranean by agreement with Sardinia, disarming Venice, and seizing control of the Ionian Islands. The French Mediterranean fleet was preparing to transport an army from Toulon to Egypt. The rumors of the pending invasion of England across the Channel were spread to preserve the element of surprise, provide cover for the real plans, and tie down strong British forces in the Channel. The subsequent French expedition to Egypt, which landed in May 1798, was a surprise for the British. Napoleon I became a master of Egypt within six weeks. Malta was seized by the French in early June 1798. However, Nelson's victory at Aboukir (Abūqir) on 25 July destroyed all French hopes of obtaining mastery of the Mediterranean. Sea routes cannot be maintained without mastery of the sea. The ultimate failure, however, cannot diminish the cleverness of the French in planning these actions.[24]

One example of an improvised strategic diversion was the escape of the German squadron under Admiral Wilhelm von Souchon, composed of the battle cruiser *Goeben* and the cruiser *Breslau*, to the Dardanelles in early August 1914. These ships were in the Adriatic at the outbreak of war. The British forces in the area under the command of Admiral Archibald Berkeley Milne consisted of three battle cruisers, four armored cruisers, four light cruisers, and 14 destroyers. Admiral Milne received a telegraphic message from the Admiralty on 30 July directing him to support the transport of the French army troops from Algeria to southern France. On 3 August he received an additional message to monitor the exit from the Adriatic and watch for movements of the German squadron. To complicate things, the Admiralty ordered Milne to cover Gibraltar and prevent the German ships from reaching the Dardanelles. The French Toulon squadron under Admiral Boué de Lapeyrère also had a double task, to observe the German battle cruiser *Goeben* and cover the transport of army troops from Algeria. There was no joint command and no arrangements were made to coordinate the actions of the two navies in the Mediterranean. Admiral Milne also did not know that the French had delayed the transport of troops from Algeria because of the presence of the German squadron.[25]

Because of the confusion among the British and French, Admiral Souchon escaped from the Adriatic and then sailed into the central Mediterranean to bombard the Algerian ports of Bône and Phillippeville on 4 August. This action in retrospect was unnecessary and might have sealed the fate of the German ships if the British and the French had acted with more energy and dispatch. After this action, the German ships sailed to the Italian port of Messina to take on coal before reaching their ultimate destination, Constantinople. They encountered two British battle cruisers which were on their way to Gibraltar. However, because Britain had not officially declared war on Germany, the British ships were not allowed to open fire on the German ships. The higher speed of the German ships allowed them to elude shadowing by the British battle cruisers. Since Italy had declared its neutrality the German ships were allowed to stay at Messina only 24 hours. Admiral Milne was forced to divide his forces, leaving most of his ships deployed northwest of Sicily to prevent German ships from interfering with French troop transports. Only a single light cruiser watched the southern exit of the Strait of Messina. The German ships left Messina in the late afternoon of 6 August and made a feint toward the Strait of Otranto, but during the night changed course toward the passage between Crete and mainland Greece. The British Rear Admiral E. C. T. Troubridge, who with four armored cruisers monitored the exit from the Adriatic, intended to engage the German ships, but abandoned the idea in the early morning on 7 August. The same day Admiral Milne with three battle cruisers left Malta and sailed towards the eastern Mediterranean. The Admiralty compounded its errors by sending a premature message to Milne on the state of war with Austria-Hungary. This led Admiral Milne to proceed towards the Strait of Otranto to provide support to Admiral Troubridge in case the Austro-Hungarian fleet tried to exit the Adriatic. The Admiralty on 8 August corrected its previous message and ordered Admiral Milne to resume his pursuit of the German squadron. However, it was too late and the Germans safely reached the Dardanelles. This unintended strategic diversion had the most momentous consequences for the Entente's cause. By 13 August Turkey declared the purchase of *Goeben* and *Breslau* for its navy, but Admiral Souchon remained in command. In late October 1914, the German–Turkish squadron bombarded the Russian ports of Odessa and Novorossiysk, laid mines off the main naval base of Sevastopol', and sank one Russian gunboat. This action led directly to war between Turkey and Russia.[26]

Despite difficult military and political circumstances, Admiral Souchon tied down strong British and French forces in the Mediterranean for days. The eventual escape of the German squadron to Constantinople led to a dramatic change of strategic situation. Turkey joined the Central Powers and eventually Bulgaria did the same. This action resulted in the most momentous consequences for Russia, which became cut off from its allies in the Mediterranean. Among other things, the principal causes of Allied failure were the Admiralty's blunder in assigning multiple objectives to the British fleet, inaccurate assessment of the enemy's intentions, lack of coordination between the British and French navies, and last but not least the lack of the initiative on the part of Admirals Milne and Troubridge.

The Germans did not adequately use diversion in World War I as for example in paralyzing British trade overseas. Another missed opportunity for the Central Powers was the offensive used by the Austro-Hungarian fleet to tie down the Entente fleets in the Mediterranean. In the Atlantic, the German position would have been much improved if they had seized the Channel ports in the early days of the war. The British Isles would have been enveloped from the south forcing the Grand Fleet to redeploy to the Channel, thereby enhancing the opportunity for German heavy surface forces to operate in the northern part of the North Sea and possibly even in the northern Atlantic.

Sometimes, the main purpose of strategic diversion may be predominantly economic, to weaken or reduce the enemy's economic pressure on a country. A strategic diversion can be used to reduce unbearable political or moral or other conditions. Regardless of the purpose, however, the methods of conducting a strategic diversion are essentially the same, the weakening of pressure on one's own vulnerable places by applying pressure at vulnerable or sensitive places of the enemy.[27]

Strategic diversion is potentially a strong card to play in the hands of an inferior but aggressively led fleet. Normally, a weaker fleet will be under pressure from a superior fleet. The inferiority in *matériel* can also lead to the lowering of morale and self-confidence and, hence, have the most profound effect on the fleet's overall combat power. Thus, one of the key problems for a weaker force is to find how it can act offensively even when the opponent's strength is overwhelming. Such a fleet should strike at the enemy weaknesses in order to change the naval situation in the principal theater of war or operations to its advantage. The enemy must lose the initiative and thereby reduce his pressure on one's forces in the principal maritime theater.

Sometimes a stronger fleet may also conduct a strategic diversion with the aim of changing the enemy's fleet distribution, thereby facilitating other operations in the main theater. A stronger fleet has a major advantage over an inferior fleet in that, even after suffering high losses, it could in the end prevail because its margin of strength is so much greater.

Defensive Mining

Counter-blockade actions in a narrow sea also encompass the defensive use of mines and counter-fire by coastal anti-ship missile and gun batteries. These actions were conducted in all enclosed sea theaters in both world wars. Mine barriers can be very effective in controlling sea areas and usually have operational importance while minefields are usually tactical in their significance.[28] In general, defensive mine barriers are established along the probable routes of movement of enemy submarines and surface ships, on the flanks of one's own shipping routes, at choke points of merchant shipping, at the approaches of one's own naval bases and commercial ports, and in front of coastal defense sectors and probable sites for enemy amphibious landings. These barriers are normally larger than offensive mine barriers and consist of a simpler type of mine.

An anti-submarine mine barrier is intended to destroy or prevent the penetration of hostile submarines near the combat area of one's own surface ships or merchant vessels, or to protect naval bases, ports, and coastal shipping routes. For example, in World War I the British laid a defensive mine barrier across the Irish Channel. Not only did the barrier limit freedom of movement for German U-boats, but ten of them were sunk while trying to penetrate the defended area.[29] Likewise, the German–Finnish anti-submarine mine barriers in the Gulf of Finland in 1941–44 were very successful in preventing the exit of Soviet submarines from the Gulf of Finland.

An anti-submarine mine barrier can be established either close to one's controlled shores or in more distant areas – that is, beyond the range of coastal anti-ship missiles and gun batteries. The effectiveness of an antisubmarine mine barrier is greatly increased if it comprises a mix of diverse types of moored and bottom-influence mines. Mine barriers intended to protect one's maritime traffic are usually laid on the flanks of the sea routes and in the assembly areas of convoys. Because of their inferiority in naval strength in the Black Sea, the Germans in 1941 used defensive mining extensively to prevent Soviet surface forces and submarines from attacking the Axis coastal convoy routes. The mine barriers were laid down off the Romanian port of Constanta and the Danube estuary. A number of smaller mine barriers were established off the Bulgarian ports of Burgas and Varna, the estuary of the Don River, and at Bugaz. A deep-water anti-submarine mine barrier was laid at the entrance to the Bosporus. German–Romanian convoys sailed in relative safety between Burgas and the Turkish Straits protected behind the defensive wall of mine barriers. However, this shield had a number of holes, notably off the Danube Estuary, Constanta, Varna, and Burgas, which allowed Soviet submarines to attack the Axis transports from time to time. The Germans also laid a number of anti-submarine mine barriers off Sulina (Danube Estuary) and Yevpatoriya (Crimea). New mine barriers were laid off Yalta, Feodosiya and Novorossiysk in 1943, intended to prevent Soviet landings and attacks by Soviet light forces.

The last large German mining action in the Black Sea was renewed mining off the Danube estuary in the summer of 1944. The Axis forces in the Black Sea laid about 20,000 mines in 1941–44. The question remains, however, whether the mine barriers were necessary in such numbers and whether they were properly positioned. Many barriers were in fact established too close to the coast and the places free of mines were too narrow. Several Axis ships were lost on these barriers. Yet the German defensive mining did prevent the Soviets from carrying out attacks on Axis shipping with higher intensity than they did.[30]

Mines are extensively used to defend straits and narrows against hostile ships or submarines penetrating the defended area. The Allied effort to force the Dardanelles in 1915 failed in part because of the defensive minefield laid by the Turks. A small Turkish steamer avoided a British destroyer patrol and laid a new line of mines well outside the main minefield, dropping them parallel to the shore in Even Keni Bay. The new line was undiscovered and unsuspected while the Allied ships advanced past it to engage the forts. These mines were responsible for sinking three out of 18 Allied battleships, while three more battleships were badly damaged. Afterwards, the Allied commander, Admiral John de Robeck, ordered a general withdrawal from the Narrows and that ended the naval effort in the Dardanelles.[31]

Mines are key weapons in organizing beach defenses on the mainland coast and offshore islands. They can be laid at the approaches to coastal areas that are accessible to enemy landings or in front of some operationally important part of the coast. As part of beach defenses, mines are used to isolate the landing zone and adjacent coastal waters and to prevent the enemy's surface combatants from penetrating these zones.

The Soviets used defensive mining to dispute German control of the sea in the initial phase of the war in the Baltic in 1941. Soviet mine barriers were laid in the Bay of Riga. Some 2,150 mines were laid between Hangö and Odensholm. The Soviets began to lay mines by using destroyers and torpedo boats in the Irben Passage after the Germans had already seized the surrounding land area. They also laid some mines off Libau and the Dvina estuary in July. By the end of September the Soviets had mined the southern entrance to Moonsund. The Soviets laid about 8,000 mines in 1941, but most of their minefields were not properly defended. Nevertheless, a total of 18 German and three Finnish warships and ten merchant ships were sunk by Soviet mines.[32]

Normally, defensive mine barriers laid in one's coastal waters are defended with coastal anti-ship and gun batteries and fleet forces. Those established beyond the range of coastal anti-ship missile and gun batteries should be made difficult to clear by the enemy's MCM forces. They should be indirectly protected, however, by one's fleet forces, mostly light surface ships. In the Korean War the North Koreans extensively used mines and counter-fire from coastal gun batteries that were tunneled deeply into the rocky hills of the Korean coast, and were often able to defy the overwhelming naval strength operating close to the coast.[33] One of the most dramatic examples as to how a small navy can temporarily deny the command of the sea to a much stronger fleet was the US fiasco at Wönsan in October 1950. The North Koreans received from the Soviets large quantities of contact and bottom-influence mines, torpedoes and technical support as well as training in using them. Most of these deliveries arrived in July 1950 by rail. Some 3,000 of the Soviet-made mines were laid within three weeks at the approaches to the ports of Wönsan and Chinnanpo starting in August. These minefields were laid in excellent patterns and integrated with coastal defenses. The US planned an amphibious landing at Wŏnsan in October to speed up the advance of the First ROK Corps along the coast. The troops of the First Marine Division embarked on amphibious ships arriving off Wŏnsan on 19 October. The sweeping of the channel started on 10 October and it was expected to be completed within ten days but the US Navy encountered great difficulties in sweeping the mines and suffered high losses. The landing was delayed and it was then decided, on 18 October, to carry out an unopposed assault, because the South Korean troops had made more rapid progress on the coast than originally anticipated. However, the embarked troops were unable to land until 25 October because of unswept mines in the channel.[34]

In a relatively recent example, the Iraqis relied exclusively on mines to dispute the US-led Coalition's control of the northern part of the Gulf, laying a total of about 1,170 mines, mostly off the Kuwaiti coast.[35]

Defense of the Sea Coast

Defense of the coast is the joint responsibility of naval, ground, and air forces. Specifically, fleet forces and aviation cooperate with ground forces in anti-landing defense to ensure the safety of one's shipping, the deployment of forces to their staging area, and the safety of ships in training areas. They are responsible for the placement of defensive mine barriers and they must organize and secure the safety of ship basing areas and naval vessels and transports against weapons of mass destruction. Coastal defense forces composed of anti-ship missile and gun batteries and troops are the principal element in defense of one's coast in a narrow sea. These forces are tasked to defend one's naval bases, ports, and important parts of the sea coast, islands, straits and narrows from enemy attack across the sea. Normally, coastal defense forces are organized into independent batteries and squadrons of coastal artillery and are directly subordinate to the respective naval base commanders.

Defense of the sea coast encompasses a series of combat actions and measures aimed to prevent a hostile fleet from seizing coast or coastal facilities. The chief elements in defense of a coast are the anti-amphibious or beach defenses and the protection of one's important coastal installations from enemy attack. One's ground troops defend the sea coast in coordination with naval forces and land-based aircraft.

The most important parts of the coast that have beaches suitable for enemy naval landing should be prepared for the conduct of anti-amphibious defense. The objective is to prevent the debarkation of enemy amphibious forces and airborne assault landings. The primary responsibility of one's naval forces in the beach defense is to reconnoiter and engage the enemy amphibious landing forces before they reach their designated debarkation area. Naval forces bear the responsibility for organizing the continuous sea and air-surveillance system and for neutralizing enemy air and naval reconnaissance forces. It also establishes defensive mine barriers and beach obstacles at sea and ashore in the most threatened part of the coast. The fleet cooperates with other services of the armed forces in organizing a common firing system at the approaches to a defended coast.

The beach defense of the sea coast can be organized on active or mobile principles. The important parts of the coast or installations are normally held by only a part of available forces, while the main forces are held in reserve to carry out powerful counter attacks intended to annihilate debarking enemy forces in their assigned area. One of the most critical elements of beach defense is the system of anti-amphibious obstacles at the approaches to the beaches and on the coast.

The ability of a weaker fleet to challenge command of the sea by a stronger fleet is generally much higher in a typical narrow sea than in the open ocean. The smallness of the area, with corresponding short distances and, generally, the presence of a large number of offshore islands and shallow waters, allow a fleet composed of small surface combatants, and acting with determination, successfully to contest control by a stronger fleet. The threat posed by enemy land-based aircraft, conventional submarines, coastal anti-ship missile batteries, and mines can make operations of a stronger fleet extremely difficult in a typical narrow sea. The inferiority in one's naval strength can sometimes be compensated for by land-based aircraft, thus thereby enhancing the chances of success in disputing control of the sea even by a blue-water navy when operating in a typical narrow sea.

To sum up, it is possible for a weaker fleet operating in a typical narrow sea to contest successfully the control of a stronger fleet. The key prerequisites for success are a full integration of all available forces and assets and the closest cooperation among the services. Disputing control of the sea will be primarily conducted by land-based aircraft, submarines, fast attack craft, and coastal anti-ship missile and gun batteries. Defensive mining is perhaps one of the most effective and surely the most economical method of reducing the stronger fleet's freedom of action in a narrow and shallow sea.

NOTES

1 M. G. Cook, 'Naval Strategy', 2 March 1931, Air Corps Tactical School, Langley Field, VA, 1930–1931, Strategic Plans Division Records, Series, Box 003 (Washington, DC: Naval Operational Archives), p. 12.

2. Geoffrey Till, ed., *Maritime Strategy and the Nuclear Age* (New York: St Martin's Press, 2nd edn, 1984), p. 117.
3. Cook, 'Naval Strategy', p. 10.
4. Phillip H. Colomb, *Naval Warfare: Its Ruling Principles and Practice Historically Treated*, Vol. I (originally published W. H. Allen, London, 1891) (repr. Naval Institute Press, Annapolis, MD, 1991) p. 153; Till, *Maritime Strategy and the Nuclear Age*, p. 113.
5. Julian S. Corbett, Vol. I: *To the Battle of the Falklands 1914, Naval Operations, History of the Great War Based on Official Documents* (London: Longmans, Green, 1920), p. 226.
6. John Creswell, *Naval Warfare: An Introductory Study* (New York: Chemical Publishing, 2nd rev. edn, 1942), pp. 87–8.
7. Corbett, Vol. I, *Naval Operations*, p. 163.
8. Arthur J. Marder, Vol. II: *The War Years: To the Eve of Jutland, From the Dreadnought to Scapa Flow: The Royal Navy in the Fisher Era, 1904–1919* (London: Oxford University Press, 1965), p. 130.
9. Ibid., p. 131.
10. Till, *Maritime Strategy and the Nuclear Age*, p. 118.
11. Ibid., p. 119.
12. Ibid., p. 120.
13. Ibid., p. 119.
14. Ibid., p. 115.
15. Macintyre, *Sea Power in the Pacific*, pp. 141–2, 151.
16. Cresswell, *Naval Warfare*, p. 103.
17. Till, *Maritime Strategy and the Nuclear Age*, p. 118.
18. Friedrich Ruge, *The Soviets as Naval Opponents 1941–1945* (Annapolis, MD: Naval Institute Press, 1979), pp. 27–9, 34–5.
19. Michael Salewski, *Die deutsche Seekriegsleitung 1935–1945*, Vol. 2: *1942–1945* (Frankfurt, a.M: Bernard & Graefe, 1975), p. 453.
20. Hans Fuchs, 'Die Diversion als strategisches Mittel zur Erzielung eines Kraefteausgleiches, dargelegt an geschichtlichen Beispielen', *Marine Rundschau*, 4 (April 1938), pp. 238–9.
21. Ibid., p. 239.
22. Ibid., p. 246.
23. Ibid., p. 245.
24. Ibid., pp. 245–6.
25. Elmer B. Potter and Chester W. Nimitz, eds, *Seemacht: Eine Seekriegsgeschichte von der Antike bis zur Gegenwart* (Herrsching: Manfred Pawlak, rev. edn, 1986), pp. 346–7.
26. Ibid., pp. 347–8.
27. Fuchs, 'Die Diversion als strategisches Mittel zur Erzielung eines Kraefteausgleiches, dargelegt an geschichtlichen Beispielen,' p. 239.
28. Thomas H. Robbins, *The Employment of Naval Mines* (Newport, RI: Naval War College), 20 March 1939, p. 10.
29. Robert C. Duncan, *America's Use of Mines* (White Oak, MD: US Naval Ordnance Laboratory, 1962), pp. 68–9.
30. Juerg Meister, *Der Seekrieg in den osteuropaeischen Gewaessern 1941–45* (Munich: J. F. Lehmans Verlag, 1958), pp. 295–6.
31. B. H. Liddell Hart, *The Real War* (Boston, MA: Little, Brown, 1st edn, 1930, repr. 1964), pp. 157–8; Duncan, *America's Use of Mines*, p. 67.
32. Meister, *Der Seekrieg in den osteuropaeischen Gewaessern 1941–45*, pp. 41–4.
33. Malcolm Cagle and Frank A. Mason, *The Sea War in Korea* (Annapolis, MD:

United States Naval Institute, 1957), p. 283.
34 Andrew Patterson, Jr, 'Mining: A Naval Strategy', *Naval War College Review* (May 1971), p. 56; Malcolm W. Cagle and Frank A. Mason, *The Sea War in Korea* (Annapolis, MD: United States Naval Institute, 1957), pp. 145, 148–50.
35 Department of Defense, *Conduct of the Persian Gulf War: Final Report to Congress, April 1992* (Washington, DC: Government Printing Office, 1992), p. 274.

12

Attack on Maritime Trade

Attacks on the enemy's and defense of one's own maritime trade, generically called 'trade warfare', are an integral part of the navy's mission to exercise and contest control of the sea. They are conducted by both stronger and weaker fleets throughout the entire conflict. The difference is that focus of an inferior fleet in contesting control will be on attacking the enemy's maritime trade, while the stronger navy will protect the country's and friendly maritime trade. These tasks are an integral part of a much broader task of weakening the enemy's or protecting one's own military-economic potential. Protection or attack on maritime trade cannot be accomplished quickly or cheaply. The greatest care should be taken, however, not to give too much emphasis to these tasks to the detriment of any other.

Attack on maritime trade in general has always been a considerable feature of any war at sea. It constitutes a way of bringing pressure to bear on a country dependent on trade with overseas for the necessities of life. Attack on maritime trade is usually highly effective against an island nation, but less so against a continental country with well-developed land communications with neighboring countries. A country faced with the prospect of having its overseas links seriously threatened or completely cut off may in fact choose to redress the loss of seaborne trade by increased trade on land. While an industrialized nation can absorb damage and destruction of a finite number of what may be termed economic and military-economic high-value targets, a developing country may receive a crippling blow by their destruction and put the clock back years or even decades.

Objectives and Prerequisites

The objectives of attack on enemy maritime trade are reduction of the traffic volume, usually expressed in percentages, in a given sea or ocean area for a certain period of time. In generic terms, this is accomplished by 'interfering' with or 'interdicting' enemy maritime trade. The degree to which enemy maritime trade is reduced ranges from hampering to curtailing, interrupting, and cutting-off the enemy maritime trade. The effectiveness of actions by one's forces in attacking the enemy maritime trade can also be expressed in numbers of ships or total tonnage sunk and damaged over a certain time and in a given sea or ocean area.

'Hampering' means that one's forces inflict such losses on the enemy shipping and shipping-related facilities ashore that they result in the enemy troops being unable to carry out their planned combat actions within the assigned timetables. In practice, this is accomplished if the enemy maritime transport volume is reduced by 25–30 per cent. The enemy maritime trade is 'curtailed' if reduced from 30 to between 50 and 60 per cent in terms of volume. 'Interruption' means a serious disruption of enemy maritime trade in a given sea or ocean area. This objective is accomplished by reducing the volume of enemy maritime traffic by 60 to 80 per cent. 'Cutting-off' maritime trade in practice means at least 80 per cent losses in bottoms and a corresponding reduction in the capabilities of loading and offloading shipping terminals. Under these circumstances the enemy forces and civilian economy function only by drawing on existing stockpiles of material. This objective can normally can be accomplished only by instituting a full naval and air blockade.

To be effective, attack on the enemy's maritime trade must be conducted methodically over a relatively large area, and against all the elements of maritime trade. Specifically, this includes attacks against enemy merchant ships at sea and in ports, on escort forces, on commercial ports/anchorages, on cargo loading and offloading facilities, on rail or road junctions in the coastal area, on shipyards and ship-repair facilities, and on shipping-related industries. In enclosed seas, attacks on the enemy and defense and protection of one's maritime trade are normally conducted with the full participation of not only naval forces but also land-based aircraft and, often, coastal defense forces and ground forces.

The success in the attack on enemy maritime trade depends considerably on sound theory and doctrine developed and practiced in peacetime, on numerically sufficient forces, on suitable platforms and weapons, on a high degree of training, on unity of effort, on timely, reliable, and relevant intelligence, on reliable air support, on suitable bases, and on reliable logistical support and sustainment. The theory of methods and procedures of attacking enemy maritime trade should be developed in peacetime, otherwise tactical and operational concepts for the employment of one's forces will be improvised and tested in wartime, often at very high cost in terms of loss of personnel and *matériel*. Also, a sound service and joint doctrine cannot be developed if there is lack of debate among naval theoreticians as to how the attack on enemy maritime trade should be planned, prepared, and conducted.

Sufficient forces should be used in conducting attack on enemy maritime trade. This is often difficult to do, because of competing demands of the other essential strategic and operational tasks of the fleet. A fleet should be composed of diverse platforms with adequate speed, combat range, endurance and combat potential and capability of operating in any part of a given narrow sea. One of the reasons for the Iranians' poor results in attacking maritime trade in their war with Iraq (1980–88) was that they lacked the aircraft to strike at Iraqi oil production facilities and pipelines. Iraq itself could not risk the air power losses necessary to sustain full-scale attacks on Iran's oil facilities and lacked the forces and assets necessary to interdict Iranian shipping through the Gulf. The Iraqis also failed to use their land-based *Super Étendard* aircraft with an effective range of 530 miles against some of the key elements of the Iranian maritime trade. Iraq's most forward secure air base was at Nāsiriya, 150 miles northwest of the Gulf and 300 miles from Kharg Island. They instead alternated between attacks on Iranian cities and attacks on shipping. The combination of the lack of suitable platforms and sustained effort resulted in failure to inflict major damage on Iran's economic centers and to sustain blockade of Iranian export facilities.

The effectiveness of actions against enemy merchant shipping depends not only on the use of suitable weapons, but also on a high degree of training in peacetime. The experience shows that training cannot be done on the cheap and improvised in wartime. The performance of the British Coastal Command in World War II was less than expected, because neither the aircraft nor their weapons were designed for these tasks, nor had attacks on shipping been a part of the bombers' training in the 1930s. The German success in attacking enemy shipping would have been much higher if the *Luftwaffe* had devoted more time to the practice of attacks against ships at sea. At the beginning of war, the *Luftwaffe* had only two bromber units (KG.26 and KG.30). This was later increased into a full air corps specialized in attacking ships at sea.

One's forces engaged in the effort against the enemy maritime trade must be provided with continuous and reliable operational reconnaissance and intelligence. The data on the movement of enemy shipping and the cargo carried must be detailed and above all provided timely to the operational commander. Ideally, data on all elements of enemy trade should be collected, analyzed, and evaluated a long time before the conflict erupts.

In wartime, one of the prerequisites for a success in attacking enemy maritime trade in narrow seas is to possess control of the air. If the British battle fleet had commanded the sea in 1941–42 it would have probably stopped Axis traffic to Libya and provided an effective cover to shipping traffic through the Mediterranean. For both sides the strength of the air threat and the strength of the air cover available to counter the threat was the most important factor.[1]

An example of how the lack of organic air support can doom the efforts of fleet forces in attacking enemy shipping at sea was the case of the Italian Navy during attacks on the Allied convoy from Gibraltar to Malta (Operation *Pedestal*) on 13 August 1942. The Italian Navy was then almost immobilized through lack of oil fuel, failed to obtain the necessary support of the Italian air force, and was thereby robbed of the chance to win a decisive victory.

The availability of suitable bases for aircraft often has a critical role in the ultimate success of the attack against shipping in a given narrow sea. Malta had such a role in the central Mediterranean in 1940–43, and it was able to survive due to combined employment of air and sea power. Only through Malta's survival were the Allies able to continue their attacks against the Italian supply routes to the *Afrika Korps*, which made Rommel's advance in Libya difficult, and ultimately stopped it.[2] Likewise, the ability to threaten British coastal trade in the North Sea and the Channel was much increased due to German control of a large number of naval bases and airfields stretching from the Jutland Peninsula to Brittany.

To be effective, attack against enemy shipping should be conducted by all available forces to generate the maximum pressure on enemy maritime trade. Any new weapon should be used in mass. Attacks on selected elements of enemy maritime trade should be conducted continuously and spread over relatively large sea or ocean areas so that the enemy is never given an opportunity to recover or react in timely fashion. The focus of attacks should be the most important type of ships and shipping-related facilities ashore. The Iran–Iraq War is an example of how an attack against enemy shipping should not be conducted. Both sides never tried to commit all their forces to stop their opponent's oil shipments. Iraqi attacks against the shipping bound to and from Iranian ports lacked consistency and persistence. The Iranians were allowed plenty of time to recover after each new round of air attacks. And each time, the Iranians used the cease-fire or new peace initiative to reduce the level of Iraqi attacks on its cities and shipping traffic.[3] Iran for its part failed to take the opportunity to attack the Iraqi pipeline across Turkey, nor did it exert diplomatic pressure on Turkey to have it closed.

Attacks on enemy maritime trade should be conducted by diverse naval combat arms and the combat arms of other services. These attacks should be conducted over a long period and be consistent. One's forces should not be frittered away by pursuing multiple objectives as the Germans did in their attacks against British maritime trade in the Channel and off England's east coast in 1940–42. They failed to achieve decisive results against British coastal shipping, because of their lack of attention to the selection and attack on a proper set of targets. The Germans often used their available aircraft for bombing land targets of doubtful importance and with little effect. By the end of 1942 they no longer possessed the strength to make a sustained effort. The Germans showed a clear tendency not to adhere to a single purpose and single objective long enough to produce decisive results.[4]

To be effective, attacks against the elements of maritime trade must be properly sequenced and synchronized. Otherwise, no great successes can be expected. The lack of coordinated attacks against enemy shipping was evident by the British in the Eastern Mediterranean in the fall of 1943. Then only submarines conducted the attacks on the Axis coastwise traffic in the Aegean Sea because neither surface forces nor aircraft were available.

The focus should be on those elements of enemy maritime trade that if destroyed or neutralized can cause a *ripple effect* on the trade as whole. Hence, the bottlenecks of enemy trade should be identified early on, preferably in peacetime. One of the most effective ways to attack enemy maritime trade is to destroy or heavily damage enemy ship-repair facilities. Enemy ships that cannot be repaired are out of service and the effect is the same as if they had been sunk.

Sometimes too much stress on attack on shipping may divert strength from attacking troops on the ground. This happened to the Axis in the central part of the Mediterranean in the spring of 1942. The successes achieved at sea against Allied shipping were purchased at the price of easing the pressure on the British Eighth Army at a critical time during its retreat to El Alamein.[5]

Methods

The principal methods of combat force employment in attacking enemy maritime trade in a typical narrow sea are tactical actions and major naval operations. Any extensive attack on maritime trade is inherently attritional. The reason for this is that maritime trade is conducted almost continuously, and each convoy or independently sailing ship represents only a fraction of the traffic volume in a theater during a given time. Correspondingly, most naval actions in attacking the enemy's or protecting one's own maritime trade will be tactical in character. Major naval operations planned to interrupt or cut off enemy maritime trade should be conducted only occasionally and only in certain sea or ocean areas and when there is a need to mount a major effort to attack or protect a large convoy that would result in the accomplishment of an operational objective.

In contrast to such major operations conducted in the open ocean, those conducted in narrow seas will be conducted with the participation of combat arms of several services. Any such operation requires the commitment of considerable forces and assets that might be required elsewhere. It also requires all-encompassing support by other fleet forces. Hence, in practice, tactical actions, and strikes and attacks in particular, are a predominant form of combat employment of naval forces and aircraft against the enemy maritime trade in a typical narrow sea.

A blue-water navy can employ aircraft carriers, large surface combatants, and nuclear-powered attack submarines (SSNs) and maritime patrol aircraft (MPA) in carrying out attack on the elements of enemy maritime trade. In a typical narrow sea the most suitable forces for attacking enemy maritime trade are land-based aircraft, diesel-electric attack submarines, destroyers or frigates, light forces (missile-armed corvettes and combat craft), special forces, and coastal defense forces.

Because of the small size of the area and short distances, land-based aircraft are perhaps the most effective platforms for the attack on enemy shipping and shipping-related facilities on the coast in an enclosed sea theater. A high degree of readiness and maneuverability enable aircraft to concentrate their strikes against transports, warships or aircraft covering maritime convoys. In contrast to war in the open ocean, attack helicopters armed with anti-ship missiles are probably one of the most suitable platforms for the attack on enemy merchant ships at sea.

The influence of land-based aircraft on operations against merchant shipping proved critical in many narrow seas in World War II. This was especially the case in the Mediterranean. After the fall of France in June 1940 the situation in the Mediterranean changed dramatically. The sea route from the Sicilian Narrows as far as the sea area between Crete and Cyrenaica was within easy bombing distance of the bases of the Italian Air Force (*Regia Aeronautica*) in Sardinia, Sicily and North Africa. Moreover, French naval aircraft and bases in the western Mediterranean were denied to Britain. After Crete passed into German hands in May 1941, the convoy route to Malta was flanked from the east on both sides by the *Luftwaffe*. With the *Luftwaffe* based in Crete to the north of 'Bomb Alley', and Cyrenaica to the south, it was no longer practicable for Allies to send convoys from Alexandria to Malta.[6]

Submarines were used to great effect in both world conflicts for attacking enemy maritime trade in narrow seas. The German U-boats were highly successful in the English Channel and southwestern approaches to the British Isles and the Mediterranean in World War I. Specifically, after the German occupation of the Flanders coast in 1914, small U-boats were based there. They operated against traffic off England's east coast, north of the Humber River estuary and south of the Foreland, in the English Channel, in the Southwest Approaches and the Irish Sea, and in the Bay of Biscay. The U-boats subordinate to the High Seas Fleet operated from their bases in the North Sea and the Baltic. Their main operating areas were the northwest and southwest approaches of the British Isles. To reach these areas they used the route from the Heligoland Bight and Skagerrak, then rounding the tip of Scotland.[7] The Germans also deployed several U-boats by sea from the Austro-Hungarian naval base of Pola to the Bulgarian port of Varna via the Turkish Straits in July 1915. This was a great surprise for the Russians who believed that U-boats were transported from Germany by railway to Constantinople from Pola.[8] These U-boats operated against Sevastopol' and the Romanian coast.[9] They were so successful that transport of Russian troops to the front in Caucasus was almost completely paralyzed. The Russian destroyers were used almost continuously as escorts. Despite all their efforts the Russians lost about 15 per cent of their transports by the beginning of July 1916. By then the traffic to Odessa had almost stopped, while the actions of the U-boats contributed to the lack of food and ammunition for the Russian army in the Caucasus.[10]

The German 250-ton U-boats operated with some success in the Black Sea again in 1942–44. At no time were more than six of these boats operating from the Romanian base Constanta against Soviet traffic along the Caucasian coast between the ports of Tuapse and Batum (Batumi). After May 1944 the U-boats shifted their operating area to between the Crimea and the Romanian coast and after August 1944 the operating area was reduced to the waters off the Danube Estuary.[11]

One's surface forces are also effective in attacking elements of enemy maritime trade. However, to be successful, large surface ships must be employed systematically and provided with reliable air cover. Small surface combatants provide an almost ideal platform to attack enemy shipping in straits or narrows and in the archipelago type of coast. The best results can be expected if small surface ships or craft are employed jointly with land-based aircraft in attacking elements of the enemy maritime trade. They must be provided with strong air cover, however, because of the almost continuous threat from the air.

Sometimes, even a small number of combat craft can cause significant difficulties for the enemy in protection of its coastal shipping. In 1940–43 the British had their chief trouble in the Channel with the German S-boats.[12] The large size of British convoys and the frequency of traffic made the waters around Dover a target-rich area. At any time Germans could choose between one or perhaps half-a-dozen large convoys moving slowly along the English Channel coast. The British escorts were not so numerous as to provide full protection. The Germans never employed more than three dozen of these excellent boats in the Channel. Because of their high speed and maneuverability the S-boats could be switched rapidly from one convoy route to another. They selected an area for attack of their own choosing and achieved some remarkable successes.[13]

In World War II the British used their Coastal Forces to attack the German convoys off the Norwegian, Dutch and Belgian coasts and in the channel. The British used their torpedo boats and gunboats against German convoys from Norway by making raids into the Leads from the Shetlands.[14] In July–August 1943 there were many clashes between British light forces and enemy convoys, patrols, minesweepers off the Dutch coast and in the Channel. The fights off Ijmuiden, the Texel, and Terschelling continued throughout the summer of 1943.[15] The Germans had very little traffic by day, while by night their ships were heavily escorted and the British aircraft and coastal forces were only rarely successful in stopping that traffic.[16]

In a typical narrow sea, even the smallest craft armed with small arms can be used in attacks against enemy merchant ships and fishing vessels. The Iranians in their war with Iraq used not only mines, but after August 1987 intensified their attacks by using high-speed, small attack boats. These craft were heavily armed including bow-mounted machine guns and Soviet-made rocket grenade launchers.

In a typical narrow sea, coastal missile and gun batteries can be used to attack enemy port terminals and shipping-related installations within their effective range. Sometimes, these batteries, especially if deployed in a strait or passage, can be effectively used to attack transiting enemy merchant vessels. In the past the use of coastal gun batteries was not often very successful in inflicting damage on enemy shipping even in relatively small straits such as the Strait of Dover. Both the British and Germans used their long-range guns against each other's shipping, but with apparently little effect.[17] Yet the use of coastal gun batteries even when not very effective, invariably had a psychological effect on the enemy ship's crews, an example being the German use of long-range guns mounted on Cap Gris-Nez firing on the British shipping transiting the Strait of Dover in the summer of 1941.[18]

Today, the range and accuracy of coastal guns is so much greater that it was in World War II that their effectiveness against enemy shipping, especially in straits or narrows, will undoubtedly be much higher. Coastal anti-ship missile batteries pose potentially the most serious threat to the safety of one's merchant shipping especially in the straits. In some cases, these batteries, especially if deployed in a strait or passage, can be effectively used to attack transiting enemy merchant vessels. In the Iraq–Iran war, the Iranians started a blockade of the Shatt el-Arab and Basra on the first day of the war in September 1980. They used artillery fire to shut down the Iraqi oil-loading facility at Al Fāw. A total of 62 ships and tankers was trapped or sunk in the Shatt and the waterway was extensively mined. Six more ships were trapped at Khor Abdullah and several of these ships were sunk by Iranian aircraft on 6 October 1980.

Combined Arms

The most effective way to attack enemy merchant shipping is to use diverse naval combat arms and weapons against all elements of the enemy's maritime communications. This requires smooth and effective operational and tactical cooperation among all the participating combat arms. Past experience shows conclusively that success in attacking maritime trade can be achieved only using diverse combat arms and weapons. No single combat arm can accomplish that task. Each arm should complement the other. The British experiences in World War II show how important are interdependence of combat arms in conducting coastal warfare. The *Strike Wing*, the Nore and Dover MTB and MGB flotillas, the daylight and night bombing of enemy ports, the reconnaissance flights by Coastal Command and the fighter cover overhead were all used in combination to achieve the single aim of stopping the German shipping traffic.[19]

One example of highly successful employment of several naval combat arms was in the Sicilian Narrows in 1941–42. The Italians then exercised almost complete control of the 90-mile-wide Narrows by using submarines, light forces, and mines. The nearby island of Pantelleria was strongly fortified. Between Pantelleria and Sicily's coast, the waters were easily mined. South of Pantelleria the greater depths made mining more difficult, because this passage is only 30 miles wide and can easily be watched by light forces and submarines. Throughout the war the Italians laid about 115,000 mines between Sicily and Tunisia. The hazards of passage for Allied convoys were increased because of the presence of Italian aircraft on Sicily and Sardinia.

By the spring of 1943, Allies had control of the Sicilian Narrows due to their superiority in the air. They used it in combination their submarines, and surface ships to cut off the Axis' traffic to Tunisia. Allied aircraft became more effective as Allied troops advanced along the Tunisian coast, thereby allowing the use of airfields nearer the Sicilian Narrows. This led to a considerable increase in the losses of the Axis ships from both aircraft and surface ships. While about 25 per cent of the total supplies to Tunisia were sunk in January–February 1943, this increased to about 75 per cent in the following three months. Also, Allied aircraft inflicted about 48 per cent of all Axis losses. The aerial bombing of their ports destroyed more than half of the Axis' ships.[20]

The lack of coordinated attacks against enemy shipping was evident in the British naval offensive in the Eastern Mediterranean in the fall of

1943. British submarines attacking the Axis coastwise traffic in the Aegean were unable to interrupt it for any prolonged period of time. To do this, it would have been necessary to use also surface forces and aircraft. The prerequisite to employ them more effectively was to have a base much nearer than Beirut or Haifa, let alone Alexandria.[21]

Offensive Mining

Mines are one of the most effective weapons against merchant shipping. They can be used to maintain a continuous threat in the sea areas where bombs and torpedoes might pose a threat when ships or aircraft are present. They are also the only naval weapons capable to some extent of altering geographical circumstances by making certain areas impassable to enemy ships. A mine-dangerous area is normally treated with great respect by an adversary and avoided, as if it were a piece of land.[22]

The lack of awareness and preparation for offensive use of mines against merchant ships adversely affects one's entire effort aimed at weakening the enemy military-economic potential. Except for a small group of mine specialists, the US Navy was grossly unprepared to conduct a mining offensive after the outbreak of war with Japan in December 1941. This was the main reason that large-scale mining of Japanese controlled ports in the Pacific did not start in earnest until well into 1943. In a more recent example, the Iraqis failed to use mines on the shipping lanes used by tankers carrying Iran's oil exports down the Gulf. They instead used aircraft armed with anti-ship missiles. Although these strikes had a devastating effect, it would have probably been more effective and far cheaper to lay mines on the well-defined tanker routes off the Iranian coast, not used by neutral shipping.[23]

In a typical narrow sea, mines can be laid not only by aircraft and submarines, but also by surface ships including combat craft and even fishing trawlers. The Germans in World War I used both surface ships and U-boats to mine British coastal waters in the North Sea. They repeatedly mined the entrances off British ports on the east coast. Several thousand mines were laid each year and no part of the British coast was free of German mines.[24] Likewise, the Russians conducted extensive mining of the eastern entrances to the Bosporus in the fall of 1914 and July 1915 carried out by destroyers and a submarine. The Russians repeatedly mined the eastern entrance to the Bosporus in July 1916, January and May–June 1917.[25] They also mined the entrance to the port of Varna making it more difficult for Bulgarian merchant ships to use that port.[26] In 1914 the Russians laid some 1,600 mines at the choke points of the German shipping in the southern part of the Baltic and at the focal point of the shipping routes off Danzig. This almost paralyzed the movement of German forces and commercial shipping.[27]

In World War II, the British coastal defense flotillas in the Nore and Dover commands laid large numbers of mines in the German-swept channels at the entrance to German-controlled ports in the Channel and along the Belgian and Dutch coasts in late 1943.[28] One result of the British mining offensive against the Dutch coast in the spring 1943 was a significant reduction of the German traffic in and out of Rotterdam. This port was the best entry for the Swedish iron ore and also the best outlet for German coal and coke, which the Swedes demanded in return. The Germans were forced to divert much of the traffic to Emden. This put an additional strain on the German inland transportation system. Though this was uneconomical, the Germans accepted it because they badly needed Swedish iron ore.[29]

Mining of large sea areas greatly complicates the enemy's ability to counter the mine threat and sometimes it can lead to the serious strain or even breakdown of the entire naval defense effort in a given area. The British mining action in the Baltic forced the Germans to stop using the Gulf of Danzig entirely for the training of their submarines. Another important effect was felt by the Swedes. By August 1944 they had lost five ships totaling 100,000 tons from various causes and on 18 August they withdrew insurance on ships sailing to German ports, while on 27 September 1944 they closed all their Baltic ports to German shipping. A final and complete embargo was not placed, however, until 1 January 1945.[30]

For the last four months of 1944 the British Bomber Command conducted minelaying in the Baltic, the Kattegat and Oslo fiord. The sinkings rose to 52 German ships totaling nearly 50,000 tons and it was at this time that shortage of minesweepers first began to cause serious difficulties for the Germans. In fact, it was this shortage and not the ingenuity of British mines which caused most of the difficulties for the Germans. By January 1945 the most important result of this effort was that the Germans were forced to abandon the U-boat training area in the Gulf of Danzig.[31]

The German naval command was mistaken in restricting S-boats in the North Sea to the use of mines because of the fear of restricting the effectiveness of German midget submarines. However, the German North Sea Command protested and that restriction was finally waived in early 1945. The S-boats were hampered in their activity because of Allied air strikes against their bases. But they were active near Antwerp and the east coast of England almost every night. Mines accounted for more losses to Allied shipping than torpedoes and U-boats.[32]

Generally, mines can be used to blockade straits or narrows, on coastal shipping routes, and at the approaches to enemy ports before the arrival or departure of convoys. Once laid, mine barriers must be continuously renewed to maintain the same degree of threat to enemy shipping. Mines not only create an extended threat to enemy shipping, but also restrict the maneuverability of ships and their escorts on shipping routes.

The purpose of minelaying can also be to force an enemy's maritime traffic to abandon the use of more protected routes and come into the open where his ships are more vulnerable to attack by submarines or aircraft. This was the objective of the British mining offensive in Norwegian waters prior to the German invasion of that country in April 1940. The configuration of the Norwegian coastline was advantageous for defense of the German ship traffic to and from Narvik. The 500-mile route between Narvik and Stavanger passes through the Inner Leads, a series of channels running inside an almost continuous chain of offshore islands. Through the Inner Leads passed the vital iron ore traffic to Germany. This route emerged into the open sea at three points only. Thus, the British plan was to mine the route so that German ships would be forced to move further seaward and thereby be exposed to search and capture outside Norwegian territorial waters. In the end, the British decided to lay mines off Stadtlandet and Bud, and in the Vestfiord (Vestfjorden).[33]

To be effective, mining should be spread over as large an area as possible. Mines should be laid in mass, methodically, and over a long period of time. The value of persistent minelaying over a large sea area was evidenced in all maritime theaters in World War II. The British extensively mined the principal maritime routes and shipping terminals in the occupied part of Western Europe. By mid-April 1940 their aircraft had laid mines in the Great Belt, the Little Belt, the Sound, and off the Jade River estuary. Two months later, the British aircraft laid mines from the Scheldt in the west to Luebeck in the east. In addition, British aircraft laid mines off Oslo, in Kham Sound (between Karmö Island and the Norwegian coast), and in the Kiel Canal. At the same time, the British declared a large area of the Baltic to be a mine-dangerous area. They took care, however, to respect the limits of Sweden's territorial waters. The mining was rapidly extended to encompass the entire Danish coastline, the Skagerrak, the Kattegat, the German North Sea estuaries such as the Ems and the Weser, and all the Norwegian waters. British destroyers laid mines off the Dutch coast, while submarines laid mines in Norwegian waters.[34]

The best results are achieved when offensive mining is coordinated with other types of attack, specifically with attacks against merchant shipping at sea and ports and ship-related facilities ashore. The most effective method is the use of several types of mines to exert the greatest possible strain on enemy minesweeping forces. An example of what happens if the mining effort is not adequately diversified was the US mining of the Shimonoseki Strait in the spring of 1945. Because of operational reasons, the two ends of the strait were mined alternately on successive missions. However, the Japanese were quick to detect this and each night concentrated all of their minesweepers at the end of the Strait at which the mining attack had been anticipated. They were able to clear the channels more quickly than would have been possible had their forces been divided into two groups working at either end of the Straits.[35]

Much of the value of any new weapon lies in its surprise introduction and large-scale employment before an enemy can develop counter measures. Because of unfavorable circumstances, US mining of the Outer Zone in 1945 (in the war against Japan) violated that principle, because mines were used in small numbers.[36]

The Iranians contested the presence of the US Navy in the Arabian Gulf in 1987–88 by covert mining of shipping lanes in international waters. The initial US reaction to the mining of a frigate in mid-Gulf on 18 April 1987 was a strike carried out by five destroyers and frigates and an amphibious ship (with one marine helicopter-borne detachment) with distant cover by a carrier that resulted in the destruction of two Iranian platforms serving both commercial and military purposes.[37] The US ships also launched five missiles to sink an Iranian fast patrol craft. During the same day Iranian ships and aircraft attacked US and British merchant ships, and an American-operated oil rig. The US responded with a strike by carrier-based A-6 Intruder attack aircraft that destroyed one Iranian frigate, damaged another, sank one high-speed craft and damaged another. On 29 April the US announced that its warships would come to the aid of friendly and neutral ships when attacked by Iran outside its self-declared war zone.[38]

In a regional conflict, a weaker side can possibly use a friendly country to covertly lay mines on the shipping lanes outside the confines of a narrow sea where the main naval action is taking place. It is widely believed that Libya acted as a surrogate for Iran in covertly laying mines by a commercial ship at the southern approaches to the Suez Canal and the northern entrance to the Bab el-Mandeb Strait in early August 1984. About a dozen merchant ships of different flags ran into these mines, but fortunately none was sunk.[39]

Commercial Blockade

The most effective method of weakening the enemy's economic potential is by declaring and enforcing a commercial blockade. It is almost invariably an integral part of a naval blockade. Its immediate objective is to stop both the outbound and inbound flow of enemy seaborne trade, whether carried in his own or neutral bottoms. Commercial blockade can aim to stop some goods only, or all goods coming from or into the blockaded country. To be effective, the targeted country must have significant external interests and be vitally dependent on trade.[40] The first step is to put economic pressure on the targeted country, followed by intercepting its seaborne trade, cutting off its trade with neutrals and its allies overseas.[41]

A commercial blockade was used to great effect by the northern states against the South in the American Civil War (1861–65). President Abraham Lincoln formally declared a blockade (the 'Anaconda Policy' as characterized by General Winfield Scott) of the Confederate states on 19 April 1861. The South did not have a navy when the war started, and the northern states were determined that it should stay that way. The Confederate States also lacked the heavy industries essential to building a navy. The task of the Union Navy was to cut off the South's access to the industrial resources of Europe.[42] This meant in practice closing to any ship traffic the some 3,550 miles of the coast from the Chesapeake Bay to the Mexican border, including some 190 river estuaries. Another factor for the success of the blockade was control of the Mississippi River and its main tributaries. These actions isolated the trans-Mississippi Confederate states, extended the blockade along a third side of the military frontier, and seriously disrupted the internal communications of the Confederacy. The blockade of the Confederate seaboard was unquestionably the Navy's greatest contribution to the North's victory.[43]

After the turn of this century, with the distant blockade came the concept of combined voyage, that is, that the ultimate rather than the initial distinction of contraband goods was the criteria of determining whether such goods would be seized by a blockading navy. In World War I German commerce carried in German-flag ships practically ceased in the first days after the outbreak of hostilities. However, supplies continued to reach Germany through neutral shipping and neighboring neutral countries. This meant in practice that the German economy could only suffer a serious or even a fatal blow if Britain were able to control neutral shipping. This in turn required the Royal Navy to intercept neutral shipping en route to Dutch and Scandinavian ports, which reached the German ports either round the northern part of Scotland or through the Strait of Dover. The British closed the southern route by laying a minefield in the Dover Strait early in the war. This forced all merchant vessels to pass through the Downs and the Goodwin Sands, following the coast of Kent. Mining of the northern exit from the North Sea was not possible because some 450 miles separated Scotland from Norway, while Greenland lay approximately 160 miles from Iceland. Once neutral-flag ships passed the patrol lines, they were able to make for Norway's coast and thence proceed within territorial waters to their destination port in Germany. This was the reason that the British established the Northern Patrol composed of surface ships to watch the 610-mile-long blockade line.[44] The British blockade actually started on 29 July 1914 when the Grand Fleet sailed from Portland to its war station in Scapa Flow. From then on and until the end of the hostilities, the British pressure on Germany (and Austria-Hungary) never lessened. After the declaration of war a major part of the German merchant marine was interned or fell into the Entente's hands. Only 2.16 million tons remained under German control and was restricted to sailing the Baltic and the German coastal waters in the North Sea.[45]

Great Britain also declared a blockade of Germany shortly after the outbreak of World War II on 3 September 1939. The Royal Navy drove all German merchant ships into British homeports or into neutral harbors, where they were interned. The British quickly instituted an extensive contraband list of goods subject to seizure. By 27 November 1939 the British extended the blockade by prohibiting the importation of German goods into neutral harbors. As in World War I, the British blockade was incomplete because the Germans were able to use the Baltic Sea as a transit route for iron ore imports from Sweden for the entire duration of the conflict.

The US blockade of Japan in World War II was highly effective, although it did not stop the Japanese from using their maritime routes to Korea and Manchuria. The US started a submarine blockade of Japan in 1942, and by January 1945, Japan's links with its newly acquired possessions in the south were cut off by the capture of the island of Luzon. By then the Japanese were suffering heavy losses and very serious delays to shipping on the main convoy route between Singapore and their homeland, on which sailed the tankers loaded with oil from Borneo and the Netherlands East Indies. US submarines on patrols in the South China Sea were the principal factor that forced Japanese convoys to hug the coast of the Asian mainland and to try to slip undetected from one port to another.

By 1944 Japan depended on sea traffic to bring almost all of the key raw materials and food to the country.[46] Thus, a reduction in the amount of food supplies would have meant in effect starvation for a large part of the Japanese population. Inside the islands an estimated 75 per cent of transportation was waterborne. This traffic distributed over half of the coal used in the great Japanese industrial regions that initially caused delays and then completely stopped the flow of imports so urgently needed by Japan's war industries. The Allied blockade was instrumental in stopping the outward movement of supplies to the Japanese armies in Burma, Malaya and elsewhere in the Pacific.[47] By 1944, because of the US blockade, the traffic in Japan's Inner Zone was sharply reduced. Japanese imports fell from about 48 million tons in 1941 to only 17 million tons in 1944. The shortages of fuel were severely felt both in Japan's homeland and in its theaters of operations.[48]

The Allied close blockade of Japan did not start until the island of Iwo Jima had been captured in March 1945. Its effectiveness increased steadily after April when Okinawa was invaded. To block all the routes and close all the ports in the Japanese Inner Zone, Admiral Chester W. Nimitz decided to send submarines into the Sea of Japan. Simultaneously, the US XXI Bomber Command was directed to start mining the entrances to northeastern Honshū. US aircraft laid mines at both ends of the Shimonoseki Strait, in the approaches to the ports of Sasebo, Kure, and Hiroshima, at the entrances to all major ports in the Sea of Japan from Niigata to Yuya Bay, in northern Kyūshū, and in the Inland Sea.

By the beginning of July 1945, Allied aircraft ranged almost continuously over the Japanese mainland and its sea approaches, and the blockade was virtually complete. Thus, the ability of the Japanese to move reinforcements to the Ryūkyūs (Nansei Shotō) and also bolster homeland defense was effectively thwarted. After the beginning of spring 1945, Japanese shipping was confined to the East China Sea and the Sea of Japan. The Japanese used only those shipping routes to Korea and northern China through the Tsushima Strait. This situation in turn enhanced the importance of ports on the northwestern coast of Honshū and those on the Inland Sea and in the Shimonoseki Strait. At the same time, the importance of Japanese ports on the Pacific coast, such as Tokyo, Yokohama, and Nagoya, as well as the use of exits from the Inland Sea on either side of Shikoku was sharply reduced.

The effects of the aerial mining offensive on Japanese maritime traffic were almost immediate. The Japanese stopped the movement of all merchant ships in their harbors until minesweeping could be organized. While in March 1945 about 320,000 tons of shipping left Kōbe and Ōsaka, by July a mere 44,000 tons left these two ports. In another example, the Shimonoseki Strait (through which a very large percentage of traffic passed between the Sea of Japan and the Inland Sea and the coasts of Korea and China) was closed completely for 16 days between 1 July and 14 August. In the first 15 days of August, the traffic passing through the Shimonoseki Strait had fallen to 7 per cent of the March figures. The Japanese were forced to divert vitally needed supplies to Northern Honshū ports, where much of it remained waiting to be hauled over an already overloaded transportation system to its final destination.

In the mining offensive against Japan, code-named Operation *Starvation* some 12,000 mines (all influence type) were laid from 7 March through 15 August 1945.[49] The operation had been more successful than was realized at the time. Although US submarines were the first and main means of enforcing the blockade, it was in fact the air-laid mines that finally strangled Japan. The effect of mines on merchant shipping, the loss of use of ports and passages, the tying up of ships in areas where they were useless, the failure to support armies, and the final starvation of the Japanese economy, played a much larger part in losing the war than the sinking of the ships themselves.[50]

Another example of the successful weakening of an enemy's military-economic potential through the blockade was the UN blockade of the North Korean coast, formally declared on 4 July 1950.[51] However, both the Soviet Union and the People's Republic of China refused to acknowledge the blockade's existence or legality, although both countries observed it in practice.[52]

Blockade and bombardment of the North Korean sea coast lasted more than three years. The UN blockade was successful because it cut off three out of five main North Korean supply lines. The North Koreans were unable to use deep-water shipping along the east coast, shallow-water coastal shipping on the west coast, or deep-water shipping routes to the ports in Manchuria and mainland China. In addition, they were denied use of the sea for transporting supplies and for fishing. Only a negligible portion of seaborne traffic escaped the blockading line (mostly at night) around the Korean Peninsula. However, the UN blockade of North Korea was a special case because there was no opposition by enemy surface ships, no hostile submarines were used, and there was almost no opposition in the air. UN naval forces had complete control of the entire 1,500 miles of the North Korean coastline. There is little doubt that if the Communists had had the effective means to contest UN command of the sea, siege of the ports of Wŏnsan, Songjin, and Hüngnam would have been more difficult, if perhaps not impossible, to carry out. Moreover, UN forces would have required much larger forces and the blockade would have been far more costly for the US Navy.[53]

The UN blockade of North Korea was neither easy nor simple because geography and hydrography were very unfavorable to the sustained employment of naval forces. The western coast consisted of numerous estuaries, large mudbanks, and hundreds of offshore islands. Many shoals and rocks were unsurveyed. These difficulties were compounded by 30-foot high tides. North Korea's eastern coast is high, precipitous, and mostly barren. Many of Korea's rivers empty into the Yellow Sea. Nowhere was the water more than 60 fathoms deep and within ten miles of the shore, the water's depth was less than 20 fathoms. That meant in practice that large ships could not operate as close inshore off the western coast as was often possible off the eastern coast.[54] The waters off both coasts were suitable for laying all types of mines, including drifting mines.

The Vietnam War provides among other things the best proof of what the consequences can be when a strong navy does not utilize its strength to institute a blockade of the coast to cripple the enemy's effort to supply its forces on the ground. The US failure to blockade North Vietnam's ports earlier in the conflict allowed the North Vietnamese to move arms either directly by sea to the Vietcong in South Vietnam or indirectly via the Cambodian port of Sihanoukville (Kâmpóng Saô) and thence overland to destinations in the lower part of South Vietnam. Before 1965, the

Vietcong had received an estimated 70 per cent of supplies through maritime infiltration routes. In addition, some inland shipments originated in Chinese ports. However, after the US Navy initiated operation *Market Time* the direct seaborne share of Vietcong supplies was reduced to less than ten per cent by the end of 1966. The Americans failed, however, to extend their naval patrols to encompass the Cambodian sea coast. This was a necessary step to make the blockade more effective, and thereby to reduce the combat activities of the Vietcong in South Vietnam. As it happened, the North Vietnamese were able to use Sihanoukville freely for several more years to support their armies in the field. The blockade of North Vietnam's coast might also have rendered unnecessary the substantial US naval effort in the Mekong Delta. Moreover, the absence of a US blockade off North Vietnam's sea coast allowed undisturbed use of North Vietnam's ports by Soviet/Eastern Bloc ships and other ships as well. About 85 per cent of North Vietnamese arms imports entered through Haiphong (Hâi Phòng).[55] The USA did not impose a full blockade of North Vietnam until May 1972, when the US aircraft of TF-77 conducted a massive mining of the port of Haiphong and other North Vietnamese ports.[56] The results of that action were immediate and highly beneficial to the US war effort in the south. No ships were able either to get in or get out of the mined ports for almost a year.

Potentially decisive influence of a blockade on a country highly dependent on seaborne trade was shown in the Yom Kippur (Ramadan) War of 1973. The Egyptian aims from the outset was to cut off Israel from all seaborne traffic. The Egyptians declared a naval blockade in the Red Sea, along the 23rd parallel, and mined both the Gulf of Aqaba and the waters off the Israeli-controlled port of Sharm el-Sheikh. Two Egyptian submarines were based in Port Sudan (Sudan). In the Gulf of Suez, the Egyptians blockaded the Abu Rueis to Eilat route used by Israeli tankers carrying oil from the fields in the Sinai to the port of Eilat. Egyptian destroyers and some auxiliary craft patrolled the Bab el-Mandeb Strait. These ships put an end to Israeli oil supplies from Iran. However, mainly due to the shortness of the war, the blockade in the Red Sea did not have significant influence on the war's outcome. Israel was unable to sustain itself with airborne supplies alone. Moreover the Israeli Navy was unable to translate its tactical successes in the Mediterranean into local mastery of the sea by cutting off Egypt's line of communications in the area.[57]

The effectiveness of a commercial blockade depends on the blockaded country's maritime position, dependence on seaborne trade, and the existence (or lack of) a well-developed land communications network. A continental country bordering a narrow sea is usually much less vulnerable to a commercial blockade than an island nation. And a peninsular country such as Italy or Korea is more dependent on maritime trade than a landlocked continental country. A country that is situated far from the sea's only exit is highly vulnerable to a blockade even by an inferior power that controls its only link with the outside world.

A maritime country rich in resources is less likely to rely heavily on maritime trade for its economic well being, and therefore is usually less vulnerable to commercial blockade. In general, an island nation poor in natural resources, such as Japan or Britain is much more vulnerable to any prolonged interruption of trade with overseas. In the past, a large continental country rich in resources was largely impervious to commercial blockade. However, this situation has drastically changed in the modern era, because even the largest countries depend on the import of some critical strategic materials. Also, the countries which rely almost exclusively on the export of a single commodity – as are all the oil-producing countries in the Arabian Gulf – are extremely vulnerable to stopping the flow of oil either by tankers or via pipelines to port terminals.

The strategic importance of the Arabian Gulf is largely due to its oil and gas resources, which are vital to the industrialized world.[58] They provide about 70 per cent of all oil imports by the West and Japan. Shortly after the outbreak of the Iran–Iraq War, Iraq declared on 7 October 1980 that waters along the Iranian coast north of 29°30'N were a prohibited war zone.[59] On 12 August 1982 the Iraqis declared a maritime exclusion zone some 35 nautical miles off Kharg. Its boundary began about 30 nautical miles east of Kuwait and then ran southwards along the center of the Gulf to a point south of Būshehr.[60] Any ship entering this zone was subject to attack. Initially the Iraqi attacks were not sustained and ships continued to sail to Iranian ports.[61]

The Iranians also declared a war zone stretching from the coastline and covered a line stretching up to 12 nautical miles south of each of Abū Mūsá Island, Sirrī Island, the Cable Light and southwest of Fārsī Island. The Iranians also claimed an exclusion zone that ran along the Gulf at a point roughly 40 miles from the shore. In practice, that zone was so large that it created points at which traffic to Kuwait was confined to a very narrow channel outside the Iranian zone. This was especially true of the waters south of Fārsī Island where the exclusion zone permitted a main navigation channel only two miles wide, although the actual depth of the Gulf allowed passage over a 60-mile-wide area. The Iranian exclusion zone forced tankers into a vulnerable area and made their routes predictable. Before passing Ras al-Kuh in the Straits of Hormuz, all ships had 48 hours to seek official permission from Bandar-e 'Abbās and also to declare their destination, route and speed.[62] Bandar-e 'Abbās was used as the main base to intercept shipping at sea, carry out inspections, and if necessary bring in ships. By April 1987 Iranians boarded about 1,200 ships and seized 30 cargoes. However, they lacked sufficient naval strength to impose a total blockade of the strait in defiance of neutrals.[63]

In general, a commercial blockade of an oil-producing country should stop the flow of oil both through the country's shipping terminals and pipelines overland. Iraq is virtually a landlocked country with only a narrow coastline. In the Iran–Iraq War, the Iranians effectively cut off Iraq from the Gulf by seizing and subsequently closing the Shatt el-Arab waterway. Iraq's only port, Basra, was blockaded from the first days of the conflict. Its oil installations were made unusable because of the threat of Iranian attack. Important Iraqi oil installations at Al Fāw and on the Fāw peninsula were put out of action by Iranian air raids soon after the start of hostilities in September 1980. In addition, Iran's ally, Syria, closed the Iraqi pipeline to the Mediterranean over its territory. All this seriously undermined the Iraqi economy. It was only with huge financial help from the Gulf States, particularly Saudi Arabia and Kuwait, that Iraq was able to maintain its war effort as long as it did. Only in 1985 with new or improved oil pipelines through Saudi Arabia and Turkey was Iraq able to replace its lost outlets. The Iraqis tried to lessen the impact of the Iranian blockade by intensively using the improved 700-mile highway from Aqaba (Al 'Aqabah), Jordan, to Baghdad. Convoys of trucks carried tons of needed supplies to Iraq. Although each truck could carry 36-metric tons of food, supplying Iraq alone required about 9,000 shipments a month.[64] The practical impossibility of doing this was evident.

The vulnerability of Iraq to blockade was illustrated again in the Gulf Crisis of 1990–91. By then more than 90 per cent of Iraq's revenue and almost half of its national income derived from oil. Iraq imported about 60 per cent of food in good harvest years, and 70 per cent in bad. By 6 August 1990 the UN Security Council had ordered a sweeping trade and financial boycott of Iraq and occupied Kuwait. Three days later US-led Coalition naval forces, deployed in the Red Sea, Indian Ocean, the Mediterranean, and the Arabian Gulf, began to enforce the blockade. The US Navy began to enforce its naval blockade of Iraq on 14 August 1990. Initially the United States described the effort as 'interdiction' rather than a blockade. The order covered vessels sailing to and from Iraq and Kuwait as well as vessels from other ports carrying goods and supplies bound for these countries. Vessels headed to or from other destinations were stopped only if the US Navy had reason to suspect that the ships were carrying cargo for Iraq. However, Jordan was a loophole in the naval blockade of Iraq and Jordan's main trading partner was Iraq. Among other things, Iraq provided Jordan with 90 per cent of its oil needs before the blockade started.[65]

Blockade of maritime trade is usually very complicated if the targeted country has access to overseas markets through neutrals or if neutral countries are not cooperative. The blockade of Germany in World War I was made difficult for the Allies to enforce because the Scandinavian states, the Netherlands and Switzerland, all continued to trade with Germany. It was only by means of diplomatic pressure, including the adoption of the 'rationing' system against these neutrals, that the economic blockade of Germany was made effective.[66] The British announced that the area between the Thames estuary and the Belgian coast was mined. To get direction a neutral ship had to stop at the British ports and clear contraband control. The British also announced that the ocean area from Ireland to Norway was a war zone and one safe passage was declared through the area. The Royal Navy's 10th Squadron was used on contraband control.[67]

A *choke point control* also aims to cut off the enemy from its links with overseas. This is especially favorable for a weaker force guarding one or both shores of an enclosed sea's only exit. Geographic circumstances make commercial blockade much easier to enforce against a regional power in a typical narrow sea than against a country bordering a semi-enclosed sea or the open ocean. The closure of a sea's only exit by one power has the immediate effect of cutting off completely all seaborne trade to the riparian states.

Control of enemy shipping by closing a sea's only exit to a large narrow sea or the open ocean has been practiced with great success on many occasions in the past. The British blockaded the Strait of Hormuz in World War I to ensure the safety of the invaluable oil supply and to prevent the Arabs and Persians from joining the 'Holy War' (*Jihad*).[68] In the Iraq–Iran War (1980–88), the Iranians were unwilling to close the Strait of Hormuz, because such an action was bound to provoke an extreme response on the part of major out-of-the-region powers, and the US in particular. The Strait of Hormuz is a choke point of crucial importance to the Western industrialized nations.[69] It is even more crucial to the Gulf States because the oil revenues constitute the backbone of their very existence.

The Iranians on 1 October 1980 declared the closure of the Shatt el-Arab estuary where some 70 merchant ships were berthed. Additionally, eight merchant ships were at the Iraqi main naval base of Umm Qasr. The access to these vessels was impossible until the end of hostilities in 1988 and many of them became the targets of attack by both sides in the conflict.[70] In October 1980 the Iranians carried out a series of air attacks that resulted in damage to the Iraqi oilfields. Their naval commando raids struck the oil outlets of Mina al-Bakr, Khor al-Amaya as well as port facilities at Al Fāw, Basra and Umm Qasr.[71]

The Arabs in their war with Israel in October 1973 achieved great success by closing the Red Sea's only exit, the Strait of Bab el-Mandeb, to Israeli shipping. Thus, in a single stroke, the Israeli oil imports from Iran were cut off and Israel became entirely dependent on oilfields in the southwestern part of Sinai. Moreover, the Israeli Navy lacked the range and combat power to break the blockade of the straits. Likewise, the Israeli aircraft based in Sinai lacked sufficient range to attack targets in the Bab el-Mandeb area.[72]

A distant blockade of a semi-enclosed sea or larger narrow sea for enemy traffic can hardly ever be complete, even if all its exits are closed. The Germans made free use of the Baltic and eastern part of the North Sea during World War I, despite British general command of the sea. In 1940 the British deployed one ship squadron at Gibraltar and another in Alexandria, but that action did stop the Axis from moving troops and *matériel* to North Africa. The extreme narrowness of the Mediterranean at several points, and the fact that a battle fleet cannot spread itself out and still remain superior to the adversary added to the difficulties. Moreover, because of the German superiority in the air in the central Mediterranean, Malta became unusable as a base for the British surface forces. Consequently, the Italian communications were only partially interrupted.

Among other things, a commercial blockade is more effective if the blockaded country has only a limited number of major ports. Blockading forces must have an adequate number of bases favorably situated in the proximity of the blockaded sea area. They must be adequately supplied and sustained. The long coastline, numerous harbors, long estuaries, and extensive inland communications of a blockaded country require correspondingly larger forces to make a blockade effective. The capture of the enemy's key ports and the prevention of sorties of blockade runners from and to the enemy ports is another key factor in the ultimate success of any commercial blockade. The ability to locate and utilize neutral ports and prize courts enhances the blockade. Cooperation among the combat arms of the several services is the key for success. The stronger fleet must be able in timely fashion and effectively to counteract the enemy's innovative approach in using technology.

Commercial Counter-Blockade

A blockaded country often resorts to *commercial counter-blockade* to inflict losses as large as possible on the maritime trade of a stronger sea power. The British blockade of the coasts of France from Brest to the mouth of the Elbe in 1806 was answered by Napoleon's counter-blockade. This, in turn, forced Napoleon to expand French power to Spain and Portugal, to occupy part of Germany, and finally to invade Russia. The French counter-blockade, however, failed in its stated purpose because it frittered away France's resources and ultimately doomed Napoleon's attempt to establish his domination on the Continent.

A weaker fleet has usually not been very effective at striking a crippling blow to an enemy's maritime trade by conducting a counter-blockade. Nevertheless, the blockaded fleet is always a potential menace whenever the superior force allows it to be uncovered and free to move. For example, if the Grand Fleet had allowed the High Seas Fleet to move freely in the

North Sea, Allied traffic would have been stopped in a matters of weeks or even days.[73]

In World War I Germany attempted a counter-blockade with its U-boats. Yet, despite many successes, the Germans ultimately failed. They were unable to provide adequate protection for their U-boats and thereby prevent the actions of numerous small craft, destroyers, sloops, not to mention drifters, trawlers and aircraft used to escort the convoys. Similarly, the Germans responded to the British blockade in September 1939 with a counter-blockade. The instructions to the German Navy issued in May 1939 by Hitler emphasized that it was hopeless to try to keep open Germany's North Sea communications. Consequently, German naval forces would be used to attack enemy merchant ships, and this could be successfully done only on the oceans. While the German Navy could not expect to use the English Channel, it was capable of interrupting British efforts to blockade the northern exits from the North Sea. The *Kriegsmarine* was also expected to conduct commerce raiding in the Atlantic by occasionally sending surface ships through the blockade and creating diversions. Small-scale warfare, surprise attacks on inferior forces and constant harassment actions were the order of the day. Close cooperation with the *Luftwaffe* was considered essential.[74]

The Germans and Italians tried to blockade the straits and narrows in the Mediterranean to interrupt Allied shipping traffic in World War II. The use of air bases on Sicily and Crete allowed the Axis to establish control over the Sicilian Narrows and the eastern part of the Mediterranean. This in turn forced the Allies to give up use of the Suez Canal from 1940–43. All primary troop shipments in Northern Africa went around the Cape. This caused the transshipment time to be increased by a factor of four, equivalent to a reduction of almost nine million tons of cargo. In fact, the operational capabilities of the British shipping were reduced more by the Axis blockade of the maritime routes in the Mediterranean than as a result of actual loss of shipping.[75]

In another example, the Israeli counter-blockade in the Yom Kippur (Ramadan) War of 1973 was based on Sharm el-Sheikh and the Sinai coast of Suez. This affected the Egyptian economy to a greater degree than previously thought. The Morgan oil fields in the Gulf of Suez and those on the Egyptian coast produced about eight million tons of oil per year out of a total of ten million tons produced by Egypt. Part of this production was exported and accounted for 20 per cent of Egyptian hard-currency earnings, the remainder being shipped to the pipeline at Ras Sadaat and thence to Cairo. To make up for the oil lost due to the war, Egypt depended on Saudi Arabia and Libya. However, these supplies were not always forthcoming and were in any case debited against Egypt's credit. A further economic threat was the fact that the Gulf of Suez ports were Egypt's outlets to East Africa and to Asia, no less than was the port of Eilat for

Israel. This threat was increased because the Israelis were able to operate freely in the Gulf of Suez as a result of their position along the Sinai coast, and particularly at Sharm el-Sheikh.[76]

In the Iran–Iraq War the Iranians declared on 22 September 1980 that all waterways near their coasts were a war zone, and that they would not allow any merchant ship to carry cargo to Iraqi ports. Subsequently, the Iranians were largely successful in preventing the resupply of Iraq through the Strait of Hormuz. However, the Iraqis were able to use the Jordanian port of Aqaba and Saudi Arabia's ports in the Red Sea and thereby minimize the effects of the Iranian blockade in the Arabian Gulf. Iraq was also successful in sending some oil to the world market via pipelines through Syria and Turkey. Meanwhile, Iran succeeded in exporting some oil from ports on the lower reaches of the Gulf, it being apparently beyond the capabilities of the Iraqi navy to take any type of preventive action.[77]

Commerce Raiding

In contrast to attacks against merchant shipping, *commerce raiding* (*guerre de course*) is directed purely against enemy merchant shipping. This is a logical course of action for a weaker fleet. Britain's enemies on the Continent frustrated in their attempts to obtain command of the sea or to invade the British Isles, usually turned to commerce raiding.[78] A navy conducting commerce raiding essentially prolongs the conflict. This happens even if the weaker navy is eventually defeated, as the German Navy was in both world wars.

The main objective of commerce raiding is not to cause the economic ruin of the stronger opponent at sea, but to inflict as much damage as possible to his maritime trade. Commerce raiding succeeds even when conducted on a very small scale by pushing up marine insurance rates and freight costs. It also may force the enemy to abandon some of his economic activity. This, in turn, allows non-belligerents to pick up trade, thereby injuring the enemy's long-term trade prospects. Commerce raiding also reduces the enemy's revenues and the credit available to wage a war. Hence, in some cases, commerce raiding can possibly have strategic significance.

Commerce raiding sometimes can have great value for a weaker side at sea because the stronger fleet must divert its forces to provide protection for its merchant shipping. For example, in World War II the German offensive against Allied shipping in coastal waters forced the British to keep a large number of fighter aircraft and escort vessels permanently in home waters at a time when they were desperately needed in the Mediterranean, the Far East, and later in the Atlantic. This indirectly had repercussions on the British ability to defend Malta, whose ordeal might have been shortened if reinforcements had been sent in time.[79]

ATTACK ON MARITIME TRADE

One of the fundamental principles of commerce warfare is to direct concerted action against the warships that protect enemy merchant shipping. This is nothing but a reaffirmation of an old rule of naval warfare, enemy warships are the primary objectives in war. Commerce raiding when conducted methodically and on a large scale, as well as with persistence, has been very successful.

In a typical narrow sea, commerce raiding can possibly be used against another riparian state. Sometimes, this method of contesting sea control may be the only way of causing sufficient displacement of forces for a weaker navy ever to hope to defeat a stronger fleet, though this is only in the case of conflict between two blue-water navies. Hence, the utility of commerce raiding in a typical narrow sea is generally limited.

Attack on enemy maritime trade will be conducted throughout a war. To be successful, suitable platforms and weapons should be available. The importance of having a favorable strategic position for the employment of one's forces cannot be underestimated. All the elements of enemy maritime trade should be attacked simultaneously and sequentially. Hence, not only the navy, but other sister services, and the air force in particular, should be involved in the attack on enemy maritime trade. This operational task is inherently 'attritional' and hence most of the actions by naval forces and aircraft are tactical in their character. Major naval operations in a typical narrow sea intended to interrupt or cut off enemy maritime trade are relatively rare. Control of the air is the key prerequisite for success in attacking enemy maritime trade in enclosed or semi-enclosed seas. Large and diverse forces and assets are required for these tasks and they must be provided with strong and continuous air cover. Operational reconnaissance and intelligence should be all-encompassing, reliable, timely, and, above all, relevant.

NOTES

1. Arthur Hezlet, *Aircraft and Sea Power* (New York: Stein & Day, 1970), pp. 275–6.
2. S. W. C. Pack, *Sea Power in the Mediterranean* (London: Arthur Barker, 1971), p. 193.
3. Anthony H. Cordesman, *The Iran–Iraq War and Western Security 1984–87: Strategic Implications and Policy Options* (London: Jane's Publishing, 1987), p. 68.
4. S. W. Roskill, *The War At Sea 1939–1945*, Vol. II: *The Period of Balance* (London: Her Majesty's Stationery Office, 1956), p. 258.
5. Ibid., p. 72.
6. Pack, *Sea Power in the Mediterranean*, p. 92.
7. Eric J. Grove, ed., *The Defeat of the Enemy Attack on Shipping 1939–1945*, revised edn of the Naval Staff History, Volumes 1A (Text and Appendices) and 1B (Plans and Tables), (Aldershot, Ashgate, The Navy Records Society, 1997), p. 148.
8. Lorey, 'Der Minenkrieg im Schwarzen Meer: Eine kritische Betrachtung', *Marine*

Rundschau, 5 (May 1933), p. 212.
9   Ibid., p. 217.
10 Ibid., p. 214.
11 These U-boats were responsible for sinking a dozen Soviet merchant ships with about 29,300 tons in 1943 (no sinkings were reported in 1942) and three merchant vessels with less than 5,000 tons in 1944; Juerg Meister, *Der Seekrieg in den osteuropaeischen Gewaessern 1941–45* (Munich: J. F. Lehmans Verlag, 1958), pp. 286–7.
12 German fast boats (*Schnellboote*) called by Allies E-[Enemy] boats.
13 S. W. Roskill, *The War At Sea 1939–1945*, Vol. III, *The Offensive* pt. 1, 1st June 1943–31st May 1944 (London: Her Majesty's Stationery Office, 1960), p. 284.
14 Roskill, *The War At Sea 1939–1945*, Vol. II, p. 257.
15 S. W. Roskill, *The War At Sea 1939–1945*, Vol. III, *The Offensive* pt. 1, 1st June 1943–31st May 1944 (London: Her Majesty's Stationery Office, 1960), pp. 98–9.
16 Roskill, *The War At Sea 1939–1945*, Vol. II, pt. 1, p. 256.
17 Ibid., p. 99.
18 S. W. Roskill, *The War At Sea 1939–1945*, Vol. I: *The Defensive* (London: Her Majesty's Stationery Office, 1954), p, 325.
19 Roskill, *The War At Sea 1939–1945*, Vol. II, p. 391.
20 Hezlet, *Aircraft and Sea Power*, p. 274.
21 Roskill, *The War At Sea 1939–1945*, Vol. III, pt. 1, p. 189.
22 Friedrich Ruge, *Sea Warfare 1939–1945: A German Viewpoint*; trans. M. G Saunders (London: Cassell, 1957), p. 13.
23 James Bruce, 'Gulf War – A Dangerous Legacy', *Jane's Defence Weekly*, 12 November 1988, p. 1200.
24 Thomas H. Robbins, *The Employment of Naval Mines* (Newport, RI: Naval War College), 20 March 1939, p. 14.
25 In the last ten days of July 1916, Russian destroyers laid about 1,000 mines off the Asian and European coasts of the Straits. The mining continued in January 1917 when about 475 mines were laid off the Bosporus. In May and June 1917 about 1,200 mines were laid by the Russians; Lorey, 'Der Minenkrieg im Schwarzen Meer: Eine kritische Betrachtung', pp. 215–16, 218.
26 One mine barrier by a torpedo-boat flotilla (in fact, 1,200-ton destroyers) was laid in October 1914. Four destroyers laid a total of 240 mines on the night of 4–5 November 1914. Two mine barriers totaling 857 mines were laid on the night of 21–22 December 1914. The Russian submarine *Krab* laid some 60 mines off the Bosporus on 10 July 1915 aimed to block the route Bosporus–Sevastopol' and Bosporus–Zonguldak. However, these mines proved to be ineffective because they were laid too far from the entrance. The German battle cruiser *Goeben* ran on the Russian mines on 26 December 1914 and the cruiser *Breslau* was damaged on a mine on 18 August 1915. Lorey, 'Der Minenkrieg im Schwarzen Meer: Eine kritische Betrachtung', pp. 212–13.
27 N. B. Pavlovich, *The Fleet in the First World War*, Vol. I: *Operations of the Russian Fleet* (Moscow, Voyenizdat, 1964; trans. from Russian, published for the Smithsonian Institution and the National Science Foundation, Washington, DC by Amerind Publishing, New Delhi, 1979), p. 120.
28 Roskill, *The War At Sea 1939–1945*, Vol. III: pt. 1, p. 96.
29 Ibid., *The War At Sea 1939–1945*, Vol. II, p. 390.
30 In January 1945 the British aerial minelaying resulted in the loss of 18 German ships totaling 42,673 tons; ibid., *The War At Sea 1939–1945*, Vol. III, pt. 2, pp. 140–1.
31 Ibid., pp. 269, 271.

32 In the first two months of 1945 the mines laid by the S-boats sank 15 ships with about 35,900 tons while torpedoes from S-boats and *Seehund* midget submarines accounted for only seven ships sunk with 13,000 tons.; ibid., pp. 271–2; The mines laid by the S-boats from January through May 1945 accounted for 32 Allied ships sunk with 76,490 tons, while nine ships with 41,924 tons were damaged. Ibid., p. 279.
33 J. S. Cowie, *Mines, Minelayers and Minelaying* (London/New York: Oxford University Press, 1949), p. 140.
34 Ibid., pp. 145–6.
35 US Strategic Bombing Survey (Pacific), *The Offensive Mine Laying Campaign against Japan* (Washington, DC: Government Printing Office, 1947), pp. 21–2.
36 Ibid., p. 21.
37 Frank Uhlig, Jr, *How Navies Fight: The US Navy and Its Allies* (Annapolis, MD: Naval Institute Press, 1994), p. 382.
38 Ibid., p. 383.
39 To search for these mines a multinational force of 28 minesweepers/minehunters and 11 support ships were deployed; Martin S. Navias and E. R. Hooton, *Tanker Wars: The Assault on Merchant Shipping during the Iran–Iraq conflict, 1980–1988* (London/New York: I. B. Tauris Publishers, 1996), pp. 90–1; the Soviet-designed KMD-1000 mines were reportedly used by the Libyans; ibid., p. 92.
40 David T. Cunningham, 'The Naval Blockade: A Study of Factors Necessary for Effective Utilization' (Fort Leavenworth, KS: US Army Command and General Staff College, unpubl. MA Thesis, June 1987), p. 25.
41 M. G. Cook, 'Naval Strategy' 2 March 1931, Air Corps Tactical School, Langley Field, VA, 1930–1931, Strategic Plans Division Records, Series, Box 003 (Washington, DC: Naval Operational Archives), p. 12.
42 Allan Westcott, ed., *American Sea Power since 1775* (Chicago: J. B. Lippincott, 1947), p. 109.
43 Harold M. Sprout, and Margaret T. Sprout, *The Rise of American Naval Power 1776–1918* (Princeton, NJ: Princeton University Press, 1939), p. 155. During the war the Union Navy seized or destroyed about 1,504 Confederate ships (1,149 vessels were captured including 210 steamers, while 355 ships were burned or sunk of which 85 were steamers; William N. Still, Jr, 'A Naval Sieve: The Union Blockade in the Civil War', *Naval War College Review* (May–June 1983), p. 39.
44 Arthur J. Marder, *From the Dreadnought to Scapa Flow: The Royal Navy in the Fisher Era*, 1914–1919, Vol. II: *The War Years: To the Eve of Jutland* (London: Oxford University Press, 1965), pp. 372–3.
45 In 1914, Germany and Austria-Hungary controlled 6,200,000 tons of shipping or about 14.7 per cent of the world's total; after the war started 2,875,000 tons took refuge in foreign ports and 1,165,000 tons were seized by the Entente; A. De Weerd, 'Blockade, Ultimate Weapon of Sea Power', *Proceedings*, 8 (August 1933), p. 1142.
46 Specifically, maritime trade accounted for about 80 per cent of Japan's oil supplies, 88 per cent of iron ore and ingots, 24 per cent of all coal (90 per cent of her coking coal), and 20 per cent of food; Robert C. Duncan, *America's Use of Mines* (White Oak, MD: US Naval Ordnance Laboratory, 1962), p. 159.
47 Roskill, *The War At Sea 1939–1945*, Vol. III, *The Offensive*, pt. 2: 1st June 1944–14th August 1945, p. 231.
48 Ibid., p. 233.
49 Ibid., pp. 370–1.
50 Duncan, *America's Use of Mines*, p. 159. When operation *Starvation* began, Japan's merchant marine consisted of 1.8 million tons (ships larger than 1,000 tons), of

which all but 150,000 plowed the waters of the inner zone. This fleet transported between 1.0 and 1.5 million tons of food and raw materials to Japan from China and Korea. Out of this total, a very large percentage of cargo transited the Shimonoseki Straits on the way to the industrial centers of Kōbe and Ōsaka. Between March and September 1945, a total 670 ships (including 65 warships) of 591,700 tons were sunk or heavily damaged by US mines laid in the Inner Zone. Specifically, 294 ships were sunk, 137 damaged beyond repair, and 239 damaged but repaired. During the 4.5 month-long offensive two air wings of 80 to 100 aircraft were used for minelaying, but only 16 B-29s were lost from all causes; ibid., pp. 145–6, 156–7; Roskill, *The War At Sea 1939–1945*, Vol. III, pt. 2, pp. 370–2; S. Woodburn Kirby, *The War Against Japan*, Vol. V: *The Surrender of Japan* (London: Her Majesty's Stationery Office, 1969) p. 164; The United States Strategic Bombing Survey, *The Offensive Mine Laying Campaign Against Japan* (originally published by Naval Analysis Division, November 1946; reprinted by Department of the Navy, Headquarters, Naval Material Command, Washington, DC, 1969), p. 14.

51 The blockade encompassed the area from latitude 39°35'N on the west coast and 41°51'N on the east coast of the Korean Peninsula.

52 Malcolm W. Cagle and Frank A. Mason, *The Sea War in Korea* (Annapolis, MD: United States Naval Institute, 1957), pp. 281–2.

53 Ibid., pp. 370–1.

54 Ibid., pp. 283, 295.

55 Bruce Palmer, *The 25-Year War: American Military Role in Vietnam* (Lexington, KY: 1984), pp. 98–9; Tran Dinh Tho, *The Cambodian Incursion* (Washington, DC: US Army Center for Military History, 1980), p. 22; William C. Westmoreland, *A Soldier Reports* (Garden City, NY: Doubleday, 1976) p. 184. Edward J. Marolda and J. Wesley Price III, *A Short History of the United States Navy and the Southeast Asian Conflict 1950–1975* (Washington, DC: Naval Historical Center, 1984), p. 88.

56 The mining was carried out on 8 May; port of Haiphong with its 12-mile-long and 200–250-feet-wide channel was mined in three places; a total of 75 Mk 52 mines were dropped; outside the channel the US carrier-based A-6 *Intruder* aircraft laid some 700 *Destructor* mines (in fact, a 500 or 1,000-pound bomb fitted with a sensor to act as a mine); the mines were to be activated after three days to give enough time for foreign ships to leave the port; five ships left the harbor within that time, but 27 remained until the US blockade was lifted some nine months later; Ulrik Luckow, 'Victory Over Ignorance And Fear: The US Minelaying Attack on North Vietnam', *Naval War College Review* (January–February 1982), pp. 23–5.

57 Frank Aker, *October 1973: The Arab–Israeli War* (Hamden, CT: Arcon Books, 1985) p. 62; Chaim Herzog, *The War of Atonement, October 1973* (Boston, MA: Little, Brown, 1975) p. 267. However, the Israelis contend that they maintained local control of the northern part of the Red Sea, and that the flow of oil to Israel continued. An alternative route by way of the Strait of Tiran was prepared as a contingency long before the war began; Almog, Ze'ev, 'Israel's Navy Beat the Odds', *Proceedings*, 3 (March 1997), p. 108.

58 In 1996, the proven crude oil reserves of the Arabian Gulf states were estimated at 648 billions of barrels or 62.3 per cent of the world's total (1,043 billions of barrels); Maritime Assessments Division, Civil Maritime Analysis Department, US Office of Naval Intelligence, *The Strait of Hormuz: Global Shipping and Trade Implications in the Event of Closure*, August 1997, p. ii.

59 Martin S. Navias and E. R. Hooton, *Tanker Wars*, p. 28.

60 Ibid., p. 52.
61 William J. Olson, 'The Iran–Iraq War and the Future of the Persian Gulf', *Military Review*, 3 (March 1984), p. 25.
62 Navias and Hooton, *Tanker Wars*, p. 31.
63 Nicholas Tracy, *Attack On Maritime Trade* (Toronto: University of Toronto Press, 1991), p. 225.
64 *The New York Times*, 12 August 1990, p. A13.
65 Ibid., 15 August 1990 p. A19.
66 M. G. Cook, 'Naval Strategy', 2 March 1931, p. 13.
67 David T. Cunningham, 'The Naval Blockade', pp. 72–3.
68 Julian S. Corbett, et al., *Naval Operations: History of the Great War based on Official Documents* Vol. III: *The Dardanelles Campaign* (London: Longmans, Green, 1923), p. 181.
69 Approximately one-eighth of the world's merchant ships over 1,000 gross tons and nearly one-third of the world's shipping capacity passed through the Strait of Hormuz in 1994. Then, about 91 per cent of the world's Cape-size tanker fleet (ships over 160,000 DWT) and 12 per cent of the tankers smaller than 160,000 DWT passed through the Strait. About half the 28 ships which enter the Arabian Gulf each day are tankers. The shipping transiting the Strait of Hormuz represents some 17 per cent of world seaborne trade by volume and 3 per cent in terms of value. In 1994, about 32 mbpd of internationally traded petroleum transited the Strait. Some 24 per cent of Europe's needs in oil transited the Strait, 15 per cent of North America's, and 55 per cent of Asia's (70 per cent for Japan); US Office of Naval Intelligence, *The Strait of Hormuz: Global Shipping and Trade Implications in the Event of Closure*, pp. 5, 11, 15–16.
70 Navias and Hooton, *Tanker Wars*, pp. 32, 44.
71 Ibid., p. 28.
72 Walter Jablonsky, 'Die Seekriegfuehrung im vierten Nahostkrieg', *Marine Rundschau*, 11 (November 1974), p. 656.
73 Herbert W. Richmond, *Naval Warfare* (London: Ernest Benn, 1930), p. 67.
74 J. R. M. Butler, *Grand Strategy*, Vol. II: *September 1939–June 1941* (London: Her Majesty's Stationery Office, 1957), p. 81.
75 G. Morozov and B. Krivinskiy, Captains 1st Rank, 'The Role of Straits in Naval Warfare', *Morskoy Sbornik* (Moscow), 8 (August 1982), p. 20.
76 Herzog, *The War of Atonement, October 1973*, p. 269.
77 William L. Dowdy, 'Naval Warfare in the Gulf: Iraq versus Iran', *Proceedings*, 6 (June 1981), pp. 115–16.
78 Stephen W. Roskill, *The Strategy of Sea Power: Its Development and Application* (London: Collins, 1962), p. 41.
79 Roskill, *The War At Sea 1939–1945*, Vol. II, p. 261.

13

Defense and Protection of Maritime Trade

One of the main purposes of command of the sea in the past was, and remains, ensuring the safety of one's own seaborne trade. A country unable or negligent in the provision of protection to its seaborne trade will suffer not only great damage but may also find its entire war effort doomed.

All maritime countries depend in varying degree for their economic wellbeing on overseas trade. Generally, the importance of maritime trade depends on a country's geographic location and the level of its economic self-sufficiency. Obviously, an island nation depends much more on uninterrupted seaborne trade than a continental state bordering a narrow sea. Likewise, countries whose main export commodity is oil depend considerably on the uninterrupted functioning of their oil-exporting industries and shipping.

Riparian states in a given narrow sea depend to a large extent on the continuous flow of trade over coastal routes for the everyday functioning of their economy. This is especially the case for a country with undeveloped land communications in the littoral area, but sometimes the value of routes in a narrow sea has not been adequately appreciated by a country's high command in wartime.

The maintenance of uninterrupted seaborne trade in a narrow sea can often be of strategic importance for a continental great power. In both world wars, control of the Baltic was critical to Germany's wartime effort. In World War II out of approximately 11 million tons of iron ore imported annually by Germany, about six and a half million tons or approximately 60 per cent came from Sweden via the Baltic routes. The rest of the iron ore imports were transported from Narvik and Kirkenes through the Inner Leads and the Skagerrak. The Baltic and Norwegian routes accounted for approximately 80 per cent of Germany's total iron ore needs.[1]

A failure to protect maritime traffic in one sea area often has adverse strategic consequences in the adjacent sea or ocean area. If the Germans had been unable to use the Baltic to import iron ore from Sweden, then more ore would have had to be transported from Narvik via the Inner

Leads and the Skagerrak, which would have led the Allies to intensify their attacks against Norway, thereby forcing the Germans to reinforce their defenses in Norwegian waters.[2]

Actual hostilities at sea have often led to quite different demands for the protection of shipping from those envisaged in prewar plans. New routes might come into use which were more difficult to defend and which involved considerably longer distances to be transited than anticipated. Axis control of the air in the central part of the Mediterranean in 1941–42 forced the Allies to use the route around the Cape, instead of the much shorter route from Gibraltar to Alexandria. The Allies did not foresee that they would be forced to use the northern route to bring ship supplies to their besieged Russian ally. After Japan entered the war, the sailing of British and US ships to Vladivostok was no longer possible and, although the Soviets were not at war with the Japanese, only a few Soviet ships were available in the Far East. The only practical way to help the Soviets was to use the Arctic route and the one through the Persian Gulf. However, each of these routes had some serious drawbacks. The most limiting factor for the northern route was its high vulnerability to attack by German land-based aircraft, U-boats and surface ships. Ironically, the Soviets consistently failed to admit or to appreciate the difficulties and dangers attending the course of the Arctic convoys.

Coastal Seaborne Trade

Coastal shipping forms an integral part of a country's communications system. Its importance depends on the length of the country's coastline, the size of population in the coastal area, the level of development of the land communications network, and the degree to which the offshore waters and seabed have been endowed with natural resources (oil, gas, and minerals).

Coastal seaborne trade belongs generally to two main categories, inter- and intra-regional trade. *Inter-regional trade* mostly bulk cargo moves between production and distribution centers. *Intra-regional* trade distributes the commodities, provides communications between outlying settlements, and maintains the offshore-resources industries.

Defense and protection of merchant shipping depends chiefly on the number and mutual distances between ports of loading and offloading, the number of ships to be defended, the number and positions of one's naval bases and airfields, and the number and location of the enemy's naval bases and airfields. A prerequisite to ensure the safety of one's maritime shipping is possession of command of a particular sea area. This was realized even in ancient times. Command of the sea allowed the Athenians to secure the transport of food from Pontus (what is today southern Russia and Ukraine) to the port of Piraeus (Pireás). Enemy trade passing up to the Saronic (Saronikós) Gulf or along the Gulf of Corinth (Korinthia) was greatly hindered so long as there was an Athenian naval station near its entrance. It also allowed the Athenians to transport their armies to any part of their large domain and carry raiding forces to any point on or off the coast of the Peloponnesus.

Prerequisites

One's resources are by definition finite. Consequently, too much emphasis on the need to secure control of the sea may easily lead to lack of attention to the difficult business of exercising it. Moreover, one's objectives should never be too ambitious. Temporary local control of the sea is usually adequate to ensure the safety of one's shipping, and sometimes it is sufficient merely to obtain control of a focal area of maritime trade, as the example of Allied protection of shipping in the Mediterranean in 1940–43 shows.

One of the main prerequisites for providing reliable defense and protection of one's shipping is prior possession in peacetime, or acquiring in wartime, of a string of bases that flanks the enemy lines of operations. Optimally, one's naval forces should be based within striking distance of the sea's exit(s). The presence of a British fleet decisively superior to the German would cover effectively all British sea lines of communications from the Atlantic with the possible exception of the Baltic countries. At the same time, the British fleet in the North Sea would be able to interrupt German overseas communications except those with Baltic and cover the British Isles against an invasion in force.[3] The Italian Navy enjoyed a much more favorable position than the Royal Navy with respect to the defense of maritime trade in 1940–43. The few British convoys that passed from time to time to Malta or along the entire length of the sea were normally provided with close cover either by the Gibraltar-based or the Mediterranean Fleet. In those restricted waters, the covering force did not necessarily need to provide direct support during the entire voyage, but it had to be deployed somewhere between a convoy and the Italian fleet.

Destruction or neutralization of hostile forces that pose a threat to one's maritime trade in a typical narrow sea can best be achieved through a combination of offensive and defensive actions. Optimally, the enemy surface ships and aircraft should be destroyed at their bases or during the transit to their respective operating areas. These objectives can best be accomplished by mounting a series of offensive major naval operations at the very outset of hostilities. Subsequently, offensive tactical actions and occasionally major naval operations will be conducted when the operational situation in the theater is favorable. One of the best ways to protect one's maritime trade is to go on to the offensive, as the British did in the Channel and in the North Sea in the summer of 1944. The result was that the Germans did not have either the time or the forces to carry out attacks on British shipping in the Channel and off the east coast as they did in the summer of 1941.[4]

To ensure the safety of maritime shipping, one's naval forces must be able to destroy, or at least neutralize, enemy forces posing a threat to the security of one's maritime routes. In the past, a naval blockade or decisive victory made it much easier to protect one's own shipping. In World War I, the British Grand Fleet neutralized the threat posed by the German surface forces, but not by the U-boats which escaped the blockading line in the North Sea. While the threat of the U-boats proved to be much greater than anticipated, the task of protecting the Entente's shipping in the Atlantic and the contiguous seas would have been almost impossible to resolve if the Germans had been able to operate surface ships in large numbers.

Defense and protection of one's maritime trade in a typical narrow sea is conducted almost continuously once hostilities begin. This operational task cannot be accomplished by a single or even two services; the efforts of all services of the armed forces are needed. In many cases, active support and participation of allied and/or coalition navies is also required. In a typical narrow sea, not only large surface combatants but also all types of fast attack craft, including commercial vessels and fishing trawlers, could be used for the protection of merchant shipping at sea and in ports. Also, all types of fixed and rotary wing aircraft can successfully be used for these tasks.

Perhaps one of the key prerequisites for success in wartime is to have a sound and tested doctrine for the naval control of shipping. Defense of one's maritime trade should also encompass defense of all other elements of trade, specifically port terminals, storage depots, shipbuilding and ship repair facilities, and railway and road communications in the coastal area. This means that not only the navy, but also other services and sectors of the economy must be fully integrated into the overall concept. Methods and procedures for the defense of shipping should be fully practiced in peacetime through war games and exercises. Defense of maritime trade is not a secondary but one of the most important tasks of the armed forces. Furthermore, this task is not 'defensive'. It is in many ways also an offensive task for navies to accomplish in wartime.

Naval Control of Shipping

Normally, control of shipping in wartime should be in the Navy's hands. Any other solution is bound to cause considerable difficulties in ensuring unity of effort due to overlapping responsibilities and poorly delineated authority. The British learned that lesson in World War I and the Royal Navy took control of shipping at the outset of World War II.[5] On 26 August 1939 the Naval Control of Shipping was authorized and Naval Control Service officers were activated in the United States. The Reserve Fleet was mobilized upon the outbreak of war and the Baltic and the Mediterranean were closed to British shipping. The Home Fleet was at Scapa Flow and Coastal Command began daily air reconnaissance of the North Sea on 24 August. Also, the plans to mine the Strait of Dover were ready. A small Channel Force in the Western Approaches was tasked to protect merchant shipping against German surface raiders.[6]

A country that faces both narrow seas and an open ocean must organize protection and defense of both oceanic and coastal shipping as the British did in 1939–45. Up to the entry of the United States into the war in December 1941, the British organized their shipping into several convoy systems: the Ocean Convoy System, the East Coast Convoy System and the Scandinavian Convoy System. The latter was dissolved after the German invasion of Norway in April 1940. After the fall of France in June the South Coast and the Scottish Coastal Convoy Systems were introduced.[7] The Coastal Convoy System linked up to some extent with the Ocean Convoy System and ended at the west coast ports of the Clyde and at Liverpool. The homebound convoys for the east coast were routed around the northern tip of Scotland to their ports of destination, while those for the west coast were brought in through the North Channel.[8]

One of the most important measures in organizing defense of merchant shipping is to classify available shipping according to their purpose, importance, and availability. The ships most suitable for overseas trade should not be used on coastal routes, and vice versa. Defense of each ship category should be organized in a different way. Therefore, freighters and military transports should not be grouped together, nor should slow ships sail together with very fast ships.

The British measures in the Channel in 1942 provide a good example of a successful method of protecting one's merchant shipping. All the available shipping was classified into five categories ranging from the exceptional or irreplaceable ships, through such groups as the main fleet or troop convoys down to vessels of minor importance. Protective measures were also divided into five categories. A fighter escort was provided for irreplaceable ships, while sweeps or patrols were made over routes taken by important vessels. Distant protection or cover, which meant that aircraft were only available if attacks developed, was given to the least valuable ships. The system was later controlled from shore headquarters when coastal radar posts became capable of detecting low-flying aircraft. The fighters normally withdrew at dusk, when ships had to rely on their guns.[9]

Safety of one's maritime trade may be considerably improved by diversifying the routes on which the critical raw materials or troops are moved. The Germans used two main routes to import iron ore from Norway and Sweden prior to and during World War II. In a study prepared by the German Naval Staff in February 1940 for the Supreme Command of the *Wehrmacht*, it was stated that Sweden had agreed to supply Germany with ten million tons of iron ore, while one or two million tons of inferior iron ore were due from Norway, mainly from Kirkenes. Some two or three million tons of iron ore from Sweden would come through Narvik, though this amount might be reduced to one million if arrangements could be made for storing the mined ore at the Baltic ports. Admiral Raeder told Hitler that any cutoff of this traffic would mean a loss of between two and two and a half million tons of iron ore annually.[10]

Methods

The main methods of protecting one's shipping in a narrow sea are convoying, short-range independent sailing, and evasive routing. The principal advantages of convoys are that they provide a concentration of forces, as well as an economy of force, and often bring the enemy to a decisive action. Convoying enhances the morale of merchant seamen. On the open ocean, sailing in convoy offers individual ships the greatest mathematical chance of avoiding detection and attack. Individual ships can escape after the first ships were attacked.

The main shortcoming of convoying is that the speed of a convoy is limited by that of the slowest ship. Moreover, fully loaded vessels must wait idly for the convoy to form, and sailing in a convoy increases the risk of collision. Convoying also reduces the usefulness of available shipping because merchant vessels are delayed in departure and the arrival and departure of many vessels simultaneously presents the most inefficient use of port facilities. The arrival of ships in large numbers causes alternate congestion and slackness in ports. For example, the British calculated the loss of cargo capacity in 1941 due to movement in convoys to be about 25 per cent.[11]

In a typical narrow sea, convoying is more difficult than on the open ocean because the enemy can relatively easily obtain timely information regarding the ports of departure, assembly areas, and the composition and route of a convoy. In World War II, the most suitable size for a coastal convoy was from a few up to ten ships. Today, because of the much greater threat from the air, the optimum size for such a convoy is perhaps smaller than that. The small expanse of the area, dependence on relatively fixed routes and the proximity of the coast in enclosed seas adversely affect one's forces' maneuverability and consequently their ability to avoid enemy strikes. Regardless of size, any convoy presents a large area target to enemy aircraft. Therefore, a small number of ships in any one convoy but a larger number of convoys should be the rule in narrow seas. Fast and relatively small ships can sail independently in narrow seas under the protection of night or bad weather. This is an especially suitable method on a coastal route guarded by a large number of offshore islands and dotted by many harbors situated at relatively short distances from each other.

Coastal convoys should not be too large otherwise they may unduly complicate the management of the entire system. Regardless of their size, convoys represent a relatively large area target to enemy aircraft. Hence, the solution is to reduce the number of ships in a convoy and increase the number of convoys and to prohibit independent sailing during daylight along the most dangerous stretches of the coast, as the British Admiralty did in the early 1940s.[12]

One advantage of a convoy in a typical narrow sea is that the distance to the next port can be transited in most cases during a single night. The Germans moved their coastal traffic in the Channel in 1942 almost entirely during the night and in short stages from one port to the next. They took full advantage of bad weather. For especially important convoys, very powerful and numerous escorts were provided.[13]

In enclosed seas, the small expanse of the area, dependence on relatively fixed routes and the proximity of the coast adversely affect one's own forces' maneuverability and consequently their ability to avoid enemy strikes.

A regularly running *interlocking convoy system* is an ingenious and highly efficient method of running maritime traffic in a typical narrow sea. This is especially the case when coastal convoys serve as the feeder to the transoceanic trade. This system was first instituted by the British in October 1942. Thereafter, the British reorganized their convoy system in the Channel and off their east coast. Convoys of 20 ships each sailed every two days between Plymouth and the Bristol Channel. Small convoys of about seven ships ran at the same interval between Portsmouth and Plymouth. Between the Thames and Portsmouth convoys ran every six days; and between the Thames and the Forth convoys of 36 ships ran on six out of every seven days. This system was made possible by the British control in the Channel.[14] The US Navy adopted a similar system in the summer of 1942. It involved scheduling the movements of thousands of ships and the use of scores of maritime routes in the Caribbean Sea and off the East Coast. Northbound convoys were timed to arrive at New York just before a transatlantic convoy sailed to the United Kingdom. The two main convoys, into which all others were tied, were the Key West–New

York convoy, and the Guantanamo–New York convoy. These two convoy routes could be compared to express trains. Other local trains feeding freight into the two big southern termini, Key West and Guantanamo, on carefully worked out schedules kept lost shipping days to a minimum. If a coastal convoy arrived one day late, ships destined for an express convoy would lie idle until the next one came through. Local convoys ran at ten-day intervals, and usually met every other express convoy, since these were run at intervals of four or five days.[15]

One way of reducing the losses of large ships is to use smaller ships on that part of the run within the range of hostile aircraft. The Iranians used this method during their war with Iraq (1980–88). After the Iraqis began to intensify their attacks on the oil terminal at Kharg Island, putting the large oil tankers under increased risk, the Iranians organized a tanker shuttle between Kharg and Sirrī Island in the spring of 1985. Sirrī is situated about 430 nautical miles from Kharg. The smaller tankers sailed close to the Iranian coast and were protected by Iranian land-based fighters.[16]

A country's main ports and connecting waterways are the most vulnerable points in maintaining the flow of essential shipping. Hence, protection of the infrastructure, which supports the shipping industry, is just as important as protecting the actual ships. The experience of the Iran–Iraq War showed that overland oil transportation via pipelines was more resilient to attacks from the air, than port terminals and vessels at sea.

For an oil-exporting country bordering a typical narrow sea, an effective way to counter enemy attacks against its oil trade is to shift oil exports over to land pipelines. Following the closure of the pipeline across Syria, Iraq in its war with Iran moved swiftly to increase the capacity of the line across Turkey, and to acquire a new outlet by building a pipeline across Saudi Arabia. A pipeline across Jordan to Aqaba was also considered but was abandoned because of the proximity to Israel.

Defense of maritime trade is difficult, if not impossible, when the enemy possesses control of the air. However, in a typical narrow sea, land-based air can be very successful in defending one's merchant shipping. The arrival of the German X Air Corps to the airfields of Reggio Calabria and the Sicilian fields at Catania, Comiso, Trapani, and Palermo was aimed among other things at protecting the Axis sea and air routes to Northern Africa. The Axis fighters provided convoy protection during daytime only, retiring before twilight and returning only after dawn.[17] However, after the German X Air Corps was redeployed from Sicily to Russia, the Italian convoys sailing to Libya in July–December 1941 had very little air protection during daylight hours. They were practically helpless against attacks by British aircraft based on Malta, and the British were able to increase their air strength on Malta. The British aircraft covered most of the central Mediterranean. After October 1941, the British surface ships and submarines based on Malta also intensified their attacks against Italian convoys to Libya.[18]

The example of Malta in World War II also shows how important it is both for offensive and defensive purpose to possess naval and air bases that dominate most of the enemy shipping routes. The activities of the British aircraft based on Malta forced the Italians to frequently vary their routes from the mainland of Italy to Libya. This task became ever more complicated as the war progressed because of the steady increase in the range of the British aircraft based on Malta. In 1940, the effective range of aircraft from Malta was 130 miles. By the end of 1941 their range was 160 miles and reached the Tunisian coast. This in turn led the Italians to abandon routes west of Malta. By late 1942, the range of aircraft from Malta was 400 miles, and thus they were able to cover the entire Central Mediterranean. This situation forced the Italians to make wide detours for their convoys. Moreover, they also split up their convoys into smaller groups. This in turn put greater demands on the Italian navy to provide more escorts. An additional problem was that because of the perceived British superiority in night fighting, the Italians had to use their battleships and heavy cruisers for distant protection of convoys. These ships required large quantities of fuel, which was in very short supply for the Italian navy throughout the war.[19]

By mid-1942 the situation in the central Mediterranean had changed drastically because Malta was under virtual siege by the Axis. This allowed the Italians to land a large number of troops and *matériel* in Libya. Between June and July 1942 the Italians transported 442,000 tons of supplies and 10,900 troops. The losses from Allied attacks amounted to only six per cent of all cargo transported. This success allowed Field Marshal Rommel to launch his offensive in May 1942. However, after Rommel's offensive in North Africa stopped at El Alamein in July 1942, the situation began slowly to change for the better for the Allies. Despite its very precarious situation, Malta did not fall. By the fall of 1942, the Italian losses of ships bound to Libya began to rise again. In November 1942, the Italian Navy brought about 67,000 tons of supplies to North Africa, but the losses from Allied aircraft, submarines, and surface forces amounted to about 26 per cent.[20]

In a typical narrow sea, especially when control of the air is in dispute, defense of one's maritime trade becomes a very complicated task. The attacks by enemy land-based aircraft against port terminals can force the defender to relocate part or even his entire maritime trade to other ports outside the effective range of enemy aircraft. This in turn causes more congestion of ship traffic in these ports and creates easier situations for enemy submarines or light forces to find and attack their targets. The British were on the defensive strategically between June 1940 and March 1941 and the struggle in the narrow seas was also defensive. For the British it was an absolute necessity to maintain the flow of shipping up and down the east coast and, and to lesser extent, along the English Channel. Because of the attacks by the *Luftwaffe* the British were forced to reroute much of their shipping traffic as far as possible to the west coast, and the ports of the Clyde, the Mersey and in the Channel became severely congested. Because of the closure of the Channel ports to ocean-going traffic, and the diversion of all traffic round the north of Scotland, it became imperative to keep the port of London and the North Sea ports open and convoys had to be protected and kept sailing regularly.[21]

The strategic situation in a narrow sea on land considerably affects the protection of one's shipping in a narrow sea. The withdrawal or retreat of one's troops can result in the loss of valuable airfields from which air cover has been provided. The last Allied convoy to Malta from Egypt sailed in June 1942. No further attempts were made to run a convoy to Malta from Egypt until the British Eighth Army had driven the Axis forces out of Libya. The principal reason for this decision was the unfavorable strategic situation on land, which in itself made any fleet operation in the central part of the Mediterranean very risky. While the convoy was actually at sea, the Eighth Army had to make a further withdrawal involving loss of the vital desert airfields. For all the brilliance of Admiral Philip Vian's actions, it was the lack of cover by heavy ships that delayed the arrival of merchantmen at Malta and gave the *Luftwaffe* the chance to destroy the convoy. In fact, the success or failure of a convoy operation was to a very large extent decided by the fighting in the air.[22]

Defense of one's convoys is very difficult if the enemy possesses a large number of airfields flanking the route as the situation in the central Mediterranean in the summer of 1942 amply illustrates. The presence of the *Luftwaffe* on airfields at Sicily, Sardinia, and Libya greatly facilitated its attacks against Allied convoys. This by itself also made any fleet operation in the central Mediterranean basin too hazardous. The principal reason for the failure of the Axis to resupply and eventually evacuate their forces from the Tunisian bridgehead in the spring of 1943 was the lack of control of the air. Without it, the entire Axis system of sea communications literally collapsed and the Allies captured some 240,000 prisoners.[23]

Sometimes it is possible to move some elements of the seaborne trade out of the range of enemy aircraft. In the Iraq–Iran War, the Iranians countered Iraqi raids against oil terminals by using the terminal at Sirrī as a trans-shipment point for foreign tankers unwilling to venture as far as Kharg. They believed that Sirrī, about 400 miles from Kharg and 150 miles from the Strait of Hormuz, was beyond the range of Iraqi aircraft.

After Iraqi reconnaissance aircraft appeared over Sirrī in June 1986, the Iranians moved most of their trans-shipment activities to a makeshift installation at Lārak Island, some 100 miles beyond Sirrī. As it turned out, the Iraqis extended the range of their *Mirage* fighter-bombers (whose normal combat range was about 500 miles) by using converted ex-Soviet An-12 transport aircraft as aerial tankers. Thus, they were able to attack the Iranian-bound tankers at Sirrī.[24] In 1985 the Iranians designated three new terminals at Jāsk, Qeshm, and Lengeh, as well as new loading points such as Kangah/Asalūyeh. They established a shuttle of chartered tankers to move oil to transloading points near Sirrī and Lārak in the eastern part of the Gulf. To avoid the loss of its oil exports the Iranians also announced plans in 1984 to open a new port at Ganāveh, northeast of Kharg, to build pipelines to Lāvān island or Jāsk near the Strait of Hormuz, and to speed up efforts to build a new pipeline from the pumping station at Gūreh to Būshehr and Asalūyeh. However, these plans proved to be more ambitious and far more costly than Iran had initially estimated.[25]

Effective defense of shipping in a typical narrow sea also requires extensive use of mines to protect the approaches to ports, assembly areas of convoys and the seaward flank of coastal shipping routes. The British laid extensive minefields in 1917 and 1918 off the Thames River estuary and east of a line Orfordness–North Foreland. These minefields accounted for some six U-boats destroyed.[26] The British also planned to establish an east coast barrage in 1917 to limit the activities of the U-boats in the Flamborough Head area which saw considerable traffic of independently sailing ships from Scandinavia on their way to the Humber and the Thames. In November 1917 the British laid an extensive minefield between 15 and 30 miles off the coast of Durham and Yorkshire, but there was no evidence that any U-boats were destroyed.[27]

Defensive mine barriers off the coast can often considerably reduce the room for maneuver for convoys and independently sailing ships in coastal waters. The narrowness of the channels off the coast makes it difficult to properly deploy convoys. The merchant vessels must sail in a single column and this in turn extends the length of the entire convoy, with more ships becoming vulnerable to enemy attacks. The British were forced in early 1940, because of the narrowness of swept channels inside the mine barrier, to arrange that the convoys steam in double or even single line, thus stringing themselves out over a long distance.[28]

Defense of one's maritime trade encompasses protection and defense of all elements of maritime trade not just shipping at sea. This task is conducted by exercising or contesting sea control. In a narrow sea, defense of shipping is accomplished by conducting tactical actions with the close cooperation of the diverse combat arms of the several services. It is an integral part of obtaining sea control or an integral part of other missions in support of the army on the coast. To resolve this task successfully it is necessary to conduct offensive actions by naval forces and combat arms of other services aimed at the timely destruction or blockading of those enemy surface forces which pose a potential threat to one's own shipping, at searching out and destroying enemy submarines, and at attacking the enemy's airfields and naval bases.

NOTES

1. B. B. Schoefield, *The Russian Convoys* (Philadelphia, PA: Dufour Editions, 1964), p. 18.
2. Michael Salewski, *Die deutsche Seekriegsleitung 1935–1945*, Vol. I: *1935–1941* (Frankfurt a.M: Bernard & Graefe, 1970), p. 367.
3. Alfred T. Mahan, *Naval Strategy Compared and Contrasted with the Principles and Practice of Military Operations on Land* (Boston, MA: Little, Brown, 1919), p. 129.
4. S. W. Roskill, *The War At Sea 1939–1945*, Vol. I: *The Defensive* (London: Her Majesty's Stationery Office, 1954), p. 330.
5. S. W. Roskill, *The War At Sea 1939–1945*, Vol. II: *The Period of Balance* (London: Her Majesty's Stationery Office, 1956), p. 258.
6. Eric J. Grove, ed., *The Defeat of the Enemy Attack on Shipping 1939–1945*, revised edn of the Naval Staff History Volumes 1A (Text and Appendices) and 1B (Plans and Tables) (Aldershot: Ashgate, The Navy Records Society, 1997), p. 29.
7. Ibid., p. 29.
8. Ibid., pp. 30–1.
9. Roskill, *The War At Sea 1939–1945*, Vol. II, p. 149.
10. J. R. M. Butler, Vol. II: *Sept. 1939–June 1941*, in John P. W. Ehrman, ed., *Grand Strategy* (London: Her Majesty's Stationery Office, 6 Vols, 1956–76), p. 92.
11. Bernard Brodie, *A Layman's Guide to Naval Strategy* (Princeton, NJ: Princeton University Press, 4th edn, 1958), p. 145.
12. Roskill, *The War At Sea 1939–1945*, Vol. I, p. 142.
13. Ibid., *The War At Sea 1939–1945*, Vol. II, p. 163.
14. Ibid., *The War At Sea 1939–1945*, Vol. II, pp. 255–6.
15. Samuel E. Morison, *The Battle of the Atlantic, September 1939–May 1943*, Vol. I in *History of United States Naval Operations in World War II* (Boston, MA: Little, Brown, repr. 1984), p. 260.
16. Anthony H. Cordesman, *The Iran–Iraq War and Western Security 1984–87: Strategic Implications and Policy Options* (London: Jane's, 1987), p. 209.
17. Raymond de Belot, *The Struggle for the Mediterranean 1939–1945*, trans. James A. Field Jr (Princeton, NJ: Princeton University Press, 1951), pp. 95, 155.
18. M. A. Bragadin, 'Mediterranean Convoys in World War II', *Proceedings*, 2 (February 1950), p. 147.
19. De Belot, *The Struggle for the Mediterranean 1939–1945*, p. 154.
20. Bragadin, 'Mediterranean Convoys in World War II', pp. 151, 153; up to 1 November 1942 the Italians made 1,785 trips to Libya in 883 convoys and transported 2,105,815 tons of *matériel* and 54,282 vehicles to Libya; the losses were 15.7 per cent of the tonnage shipped; De Belot, The *Struggle for the Mediterranean 1939–1945*, p. 155.
21. Roskill, *The War At Sea 1939–1945*, Vol. I, pp. 321–2.

22 Roskill, *The War At Sea 1939–1945*, Vol. II, pp. 71–2.
23 De Belot, *The Struggle for the Mediterranean 1939–1945*, p. 205.
24 'If Your Enemy Squash You, Strangle Him First', *The Economist*, 23 August 1986, p. 33.
25 Cordesman, *The Iran–Iraq War and Western Security 1984–87*, pp. 51–84.
26 Grove, *The Defeat of the Enemy Attack on Shipping 1939–1945*, p. 149.
27 Ibid., p. 149.
28 Roskill, *The War At Sea 1939–1945*, Vol. I, p. 139.

14

Support of the Army Flank

One of the main characteristics of a war in a typical narrow sea is close cooperation between naval forces and the ground troops operating on the coast. Very often, support of one's naval forces has been vital for the success of ground troops operating in the littoral area. Because the navy will carry out other missions at the same time, support of the army flank will require thoughtful distribution of forces and efforts. Command and control must be flexible so that one's naval forces can be released and redeployed to carry out other missions at short notice.

The task of supporting the army flank has been carried out by almost all fleets operating within the confines of an enclosed or semi-enclosed sea or on large lakes and river estuaries since ancient times. Whenever the cooperation was well organized and executed, the results were most beneficial for the progress of the war as a whole. A fleet can facilitate the advance of one's own and friendly troops on the coast by obtaining command, even temporarily, of a body of water through which a critical supply line passes. During the American Civil War, cooperation between the Union Army and Navy in the Chesapeake Bay and Virginia was vital for the success of the war on the ground. The Navy's support played a large role in General George B. McClellan's Peninsula Campaign of 1862, and again in General Ulysses S. Grant's victorious offensive against Richmond in 1864–65. The Union naval forces covered the army's movement across otherwise impassable terrain and guarded the vital line of water communications to that Army's ultimate source of supplies.[1]

The British and French navies provided extensive support to the troops operating along the French coast of the English Channel and the Belgian coast during World War I. Their role was especially valuable in protecting the transport of British troops across the Channel in the early phase of the war. The task of naval forces was to protect the army troops from German raids in the area between Dunkirk and Nieuport.[2] Likewise, one of the principal tasks of the Russian navy in the Baltic and the Black Sea in World War I was to provide support to the troops fighting ashore.

During World War II, all the navies that operated in narrow seas were extensively used to carry out diverse operational and tactical tasks in support of ground troops. This was especially the case in the Allied campaign in North Africa in 1940–43 and in Italy in 1943–45. The campaign in North Africa was characterized by highly successful cooperation between the British Army and the fleet in both offensive and defensive operations.[3] Similarly, Allied naval forces provided significant support to the troops on the coast during the New Guinea campaign in 1942–44. In 1944–45, Soviet naval forces and aviation took part in protecting the army flank in a series of major offensive operations along the Baltic and the Black Sea coast.

Failure or inability to use naval forces to support one's troops on the coast has had an adverse effect on actions on the ground, and in some cases doomed the entire war effort. In World War I both the Germans and the British failed to employ their naval forces effectively to support army troops on the Western front.[4] This lack in appreciating the value of cooperation between the army and the navy was also obvious in the German plan for the war against the Soviet Russia in 1941. The Supreme Command of the *Wehrmacht* believed that the main task of the *Kriegsmarine* should be limited to coastal defense, while support of naval forces for the flank of Army Group North in the Baltic and Army Group South in the Black Sea was not envisaged. The *Kriegsmarine* expected that the army and *Luftwaffe* would capture Soviet bases on land, and that its role would be to restrict the actions of the Soviet Baltic Fleet by laying mine barriers off the Soviet main naval bases in the Gulf of Finland and the Gulf of Riga. The *Kriegsmarine* required the support of the Army in the Arctic because the capture of the naval bases of Polyarnoye and Murmansk could not be carried out by its own resources.[5] The Germans seized the Baltic Islands in the fall of 1941 because the Soviet Baltic Fleet was unable to provide support to the ground troops. The Germans were surprised at how fast the Soviet resistance collapsed under their assault.[6]

The importance of naval support of troops in their operations along the coast has been shown in many regional conflicts fought since 1945. In the early stages of the Korean War, the US Navy's support of troops on the coast included amphibious landing at Pohang, attacks by naval/marine aircraft, the effective and timely operations of the First Provisional Marine Brigade, and the bombardment of targets on the east coast of the Korean Peninsula. The firepower of the US Navy's blockading forces was used for the twin tasks of supporting ground forces with naval gunfire, and destroying North Korean forces during their movements to the front. Eventually, three main roles for UN naval forces emerged during the war, providing fire support to the fixed positions at the front, securing both flanks of the UN battleline, and interdicting rail and road lines along Korea's northeast coast.[7]

SUPPORT OF THE ARMY FLANK

Experiences of local conflicts since 1945 have shown that the army needs the support of naval forces even in situations when there is no threat of hostile naval forces. In the Korean War, the North Korean Navy represented a minor threat to UN naval forces conducting the naval blockade of the Korean Peninsula. The US Seventh Fleet used 300 combatants and 200 transports in support of ground troops. In the Vietnam War (1964–75) a similar situation prevailed but, the US Navy committed large forces, including aircraft carriers, cruisers, and destroyers in support of the flank to ground troops on the coast.

In all the Arab–Israeli conflicts, naval forces played a not insignificant role in support of forces on the ground. In the invasion of the Lebanon in June 1982 the main tasks of the Israeli Navy were to obtain freedom of action at sea, carry out landings along the Lebanese coast, provide fire support to ground forces, and isolate the combat area on the coastal flank.[8] The US Navy and its coalition partners supported the coastal flank of their ground troops in the northern part of the Arabian Gulf in the Gulf War of 1990–91.

Naval forces can cooperate with ground forces operating along the coast both offensively and defensively. The type, variety, and duration of tasks then differ considerably. In general, one's naval forces can support army troops on the coast by providing, cover, support, and supply. Cover is provided by preventing enemy air, missile, or gunnery strikes, and by preventing enemy amphibious landings on the flank of one's ground troops and in their rear areas. These tasks are carried out by destroying enemy forces that threaten one's own ground troops from the sea, and destroying enemy amphibious forces at their embarkation area, during sea transit and in the landing area. Support by one's naval forces encompasses a range of tasks, from destroying important targets on the coast and in the enemy's defense in depth, to attacking enemy maritime traffic in coastal waters flanking his troops on the coast. Supply tasks include the transport of reinforcements and supplies, securing crossings of large water areas (estuaries, rivers, straits, etc.), and evacuation of troops from beachheads, etc.

One's naval forces can provide support of the army flank by conducting major naval operations as a part of a land or maritime campaign or by conducting a series of tactical actions in support of a major operation by the ground forces in the coastal area. Major operations in *support of ground forces on the coast* are conducted in both marginal seas of oceans and in semi-enclosed or enclosed seas. Participation of one's naval forces in support of the army flank is most effective within the framework of a single major operation. Success in operations by ground troops along the coast is affected not only by naval forces supporting the ground troops but also by naval forces conducting major independent naval operations aimed at destroying enemy naval forces at sea or in their bases and thereby obtaining sea control. Tactical actions are the predominant method for naval forces and aviation to support ground forces on the coast. A major naval operation may be planned to destroy an enemy's fleet threatening one's troops operating along the coast, to seize, by an amphibious landing, a major island or strait or a part of the enemy-held coast, to defend one's own coast against an enemy amphibious landing, to attack the enemy or protect one's maritime trade, and to conduct a large-scale evacuation of troops or civilians.

Prerequisites

Offensive operations in a typical narrow sea are usually conducted after one's fleet, land-based air or the army on land, acting independently or in combination, have obtained at least local command of the sea. Naval support of a major defensive operation in the coastal area is conducted when the adversary holds command of the sea. Success is achieved by coordinating the actions of all forces in terms of the objective to be accomplished or the task to be carried out, and its place, time, and method of combat employment.

Command of the sea has often had only one purpose, that of preventing an enemy army from being reinforced in troops and *matériel* from across the sea or, in the modern era, also by the air.

How lack of command of the sea can limit the ability of one's army and force it to take an unfavorable course of action was shown in the Russo-Turkish War of 1876–77. The Russians did not have a Black Sea Fleet and could not attain command of the sea. Therefore, the supplies for the Russian army had to come overland via a single line leading from Russian territory down the Danube to Galati. At the same time, Turkish command of the sea meant that the Russian attempt to seize Constantinople and to advance in the Caucasus had to be made overland, for there was no possibility to land troops by sea. This led to fierce and costly fighting at Plevna in July–November 1877.[9] The Axis offensive to the Nile Estuary in 1942 depended chiefly on the naval situation in the central Mediterranean. The offensive could be carried out only with successful resolution of the supply problem for the *Afrika Korps*. For their part, the British realized that a catastrophe in Egypt could be averted only if the German supply routes to North Africa were cut off, and for that, Malta had to be held at all costs.[10]

One of the key prerequisites for successful employment of one's naval forces in support of the army flank is to have a theoretical concept and an operational and tactical doctrine developed and tested in peacetime. The higher staffs of all services operating in the littoral area should establish sound procedures to ensure the highest degree of cooperation and interoperability. Support of the army flank requires a thorough understanding of the capabilities of each service at every command echelon. Normally, the ground force commander should be in overall command when one's fleet forces are envisaged for the provision of support to troops on the ground. Command and control should be simple, and above all flexible, allowing speedy reaction to any change in the operational or tactical situation. Joint training should be conducted and tested in peacetime through exercises, maneuvers, planning games and war games. Otherwise, the Navy may find itself confronted suddenly with tasks for which neither the forces nor concept exist. The German *Kriegsmarine* was called upon to support the flank of Army Group North in 1944. This task often included shelling land targets, for which the German ship crews had not been trained. The tactical employment of cruisers, destroyers, and torpedo craft in the restricted waters of the Baltic was rather difficult. In the end, the *Kriegsmarine* developed an entirely new operational-tactical concept of fleet employment in support of the troops on the coast, which was subsequently applied until the last days of the war. The employment of German naval forces in the central and eastern part of the Baltic, despite many difficulties and sometimes painful losses, justified the continued existence of German surface forces.[11]

Ground forces should operate within the effective range of naval forces, otherwise operations on the coast can be adversely affected, as happened in Allied invasion of Sicily in July–August 1943. After the initial landings, the British Eighth Army made a swing out from the coast after it had been held up before Catania. This action slowed down its advance to Messina considerably. It was not easy for the troops on the ground to disengage and re-embark with their equipment in a short time. To add to the difficulties, the coast north of the port of Catania was precipitous and its few beaches had poor exits, making it very difficult to land or deploy a substantial assault force. Also, the strength of Axis defenses in the Strait of Messina made it dangerous to pass large troop and supply convoys through these narrow waters. The Royal Navy expected to land comparatively small forces to harass and disorganize traffic on the coastal road and railway. The one landing which actually took place, on the night of 15–16 August, was too late to prevent the retreat of Axis forces toward Messina.[12]

Offensive Major Operations

The principal tasks of naval forces in supporting ground forces on the offensive include attacks on troop concentrations and installations/facilities deep behind the enemy's defenses, conducting amphibious landings, destruction of enemy naval forces posing a threat to the advance of one's troops along the coast, carrying out tactical amphibious landings and raids behind enemy lines, preventing the arrival of enemy reinforcements by sea or land, the transport of one's troops and *matériel*, capture of enemy naval bases and ports, provision of support in crossing straits, bays, or river estuaries, and destroying enemy naval forces that pose a threat to one's troops on the coast.

The primary focus of the above-listed tasks is to facilitate the advance of one's ground troops along the coast. Sometimes, one's ground troops may be called upon to seize some objectives on or off the enemy-held coast that are of little or no significance for the army, but are critical in allowing naval forces to operate. Also, the actions of ground troops should be directed against objectives, which in the further course of a major operation are used by naval forces, and therefore must be seized undamaged and in the shortest period of time possible.

Naval forces can blockade a large stretch of an enemy-held coast to prevent reinforcement of the enemy's army with troops and *matériel*. In the Crimean War (1853–56), the first effective allied naval intervention was blockading the estuary of the Danube River, thus cutting off the rear of the Russian troops advancing southward against the Turks. This blockade relieved pressure on the Turks and the Franco-British expeditionary force (sent first to Gallipoli on the Dardanelles and then to Varna on the Black Sea).[13] Similarly, the Russian naval blockade of the Anatolian coast in 1915 prevented the deployment of Turkish troops and supplies to the Caucasus front by sea at a time when the situation there threatened the Russians. The only coal available to Constantinople and therefore to the Turkish–German fleet and war effort came from the Anatolian port of Zonguldak. There was not a single railway link, and hence the coal had to be brought to Constantinople by ships. The Russians interfered, pretty successfully, with that trade.[14]

The most effective way to interfere with enemy reinforcements by sea is a blockade of straits or narrows. In late January 1943 the British recaptured Tobruk and the westward advance continued rapidly. The Royal Navy poured supplies through Benghazi. As soon as the attack on land started, the British destroyers and MTBs from Malta intensified their nightly sweeps in the Sicilian Narrows to frustrate enemy attempts to run reinforcements across the sea. The Allied stranglehold on the Sicilian Narrows tightened in the spring of 1943 when almost every night the destroyers and MTBs from Malta or Bône were out seeking targets. Despite Italian opposition, Hitler insisted that sea reinforcements to Tunisia should be continued to the end despite mounting losses in ships and personnel. By the beginning of May 1943, the Allied blockade of the Sicilian Narrows was almost complete and only a small fraction of cargoes carried by the Axis ships reached Tunisia. The Axis tried to compensate for its loss of command of the surface by transporting troops and *matériel* by air, but with little success.[15]

Another method of preventing the arrival of enemy reinforcements to frontline troops is to interdict enemy rail and road communications by strikes conducted by carrier and land-based attack aircraft and heavy surface vessels against each element of the enemy's land communications network. Interdiction is generally more successful against an army depending on a few rail or road lines for bringing up supplies to the front, and if the terrain is flat rather than mountainous. The UN interdiction effort in Korea was only partially successful because the Communists used the features of the terrain to their advantage and even increased the flow of supplies to their frontline troops despite heavy punishment by UN aircraft. They also used their literally unlimited human resources to keep the roads and rail lines in operation. Another reason for the partial success of the UN interdiction effort was a self-imposed limitation not to attack the actual sources of the enemy supplies. In addition, UN aircraft were unable to find and destroy at night and in adverse weather the small individual targets not destroyed in daylight raids.[16]

In the past, straits have often served as the main highway for large-scale invasions. The Japanese used the Strait of Korea to transport troops in their invasion of Korea in 1588. The war had been caused by the refusal of the King of Korea to assist the Japanese warlord Toyotomi Hideyoshi in attacking Korea's suzerain, the Ming Emperor of China. The Japanese dispatched 240,000 men in 4,000 small vessels to invade Korea. The Koreans were taken by surprise and the Japanese landed unopposed at Pusan. Afterwards, they marched overland, seized the capital of Seoul, and quickly overran the entire peninsula. However, the Japanese ultimately were forced to leave Korea after their navy suffered a crushing defeat in the Tsushima Strait. A new Japanese army landed unopposed in 1597. After initial setbacks, the Koreans were able to ambush the Japanese fleet and destroy it and thereby forced the Japanese in 1598 to evacuate their armies from the peninsula.[17]

In World War II a large number of troops were transported across straits and narrows. For example, during the Normandy invasion about 5 million Allied troops crossed the English Channel. In the same conflict, the Germans transported about 1.5 million troops across the Strait of Tunis and 800,000 across the Danish Straits. During the Allied operation *Baytown* on 3 September 1943 two divisions of XIII British Corps of the Eighth Army carried on 22 LSTs and 270 smaller landing craft landed across the Strait of Messina on the Calabrian coast near Reggio and Villa San Giovanni.[18] Also, large contingents of troops were redeployed from one theater to another through the Strait of Gibraltar, the Suez Canal, the Strait of Malacca, and the Formosa Strait.

Defensive Major Operations

The main tasks of one's naval forces and aviation in providing support of an army flank in a major defensive operation include blocking or slowing down an enemy advance along the coast of a sea or a large lake, providing fire support, participating in anti-amphibious defense of the coast, defense of naval bases and ports, conducting commando raids in the enemy's rear, and evacuating one's own troops and civilians.

Defense of naval bases is one of the prerequisites to employment of one's fleet to contest the sea control of a stronger navy. A prolonged defense slows down an enemy advance along the coast and often ties down significant enemy forces. It also results in a significant gain of valuable time to stabilize the front. In some cases time is gained for preparing defense of more important naval bases or ports. The loss of naval bases due to defeat on the ground, on the other hand, has always been costly and forced one's naval forces to be deployed from less-well equipped and more distant rear bases.

One of the most important reasons for the generally poor performance of the Soviet naval forces and aviation in the Baltic and Black Seas in 1941–45 was the loss of almost all their naval bases in the first few weeks of the war. In the Baltic, Soviets lost all major naval bases and ports, except the Leningrad–Kronstadt basing area, within three months of the start of the German invasion in June 1941. Likewise, because of the withdrawal of the Soviet ground troops, the Black Sea Fleet lost all its major bases within 12 months of the beginning of the hostilities. In retrospect, if the Germans had focused their efforts on seizing the few remaining Soviet naval bases in the eastern part of the Black Sea, as part of their summer offensive of 1942, they would probably have forced the Soviet Black Sea Fleet to self-destruct or suffer internment in Turkey. However, these ideas were apparently lost on the German Supreme Command of the *Wehrmacht*, although the Soviets were well aware of the threat.[19]

Successful defense of a naval base requires ideally that rudiments of a multi-echeloned defense system should have been established in peacetime. Defense systems should be adequate to defend a base from all types of threats, but especially from the air. Combined area and point defense should be used in defending naval bases and major commercial ports. Unity of command is perhaps most critical for success. Normally, a ground commander should be in command of all forces assigned to defend a naval base. Command organization should not undergo drastic changes in the course of defending a base from the attack of enemy forces. There should always be the closest possible cooperation among ground, sea, air, and special force elements taking part in the base's defense. The main tasks of fleet forces in defense of naval bases and major ports include among other things, creating a favorable operational situation at sea, providing fire support to troops on the ground, taking part in defense against all types of attack from the air, conducting landings or raids on the coastal flanks of the enemy's ground forces, keeping access open to a naval base or port besieged from the land side, protecting sea approaches to a base or port, providing transport of reinforcements in troops and *matériel*, evacuating sick and wounded, and possibly the final evacuation of the troops and civilians.

Support of naval forces in defense of naval bases and large ports from the seaside has proved invaluable on many occasions in both world conflicts. The loss of naval bases due to defeat on the ground, on the other hand, was always costly and forced fleet forces to be deployed from less-well equipped and more distant rear bases. In June 1942 the British Naval Staff warned the First Sea Lord to be ready for the worst, the loss of the main naval base of Alexandria. Preparations were made to move some parts of the fleet to Haifa, and some south of the Suez Canal. Although the British faced a similar situation in April 1942, because of their defeat on the ground, this time the threat was far more serious. On the 24 June, Sollum was evacuated and the Army of the Nile fell back to Mersa Matrûh. The Axis capture of the airfields close to the Egyptian border threatened the British naval base at Alexandria, which could be attacked by fighter-escorted bombers.[20]

The Soviets stubbornly defended their naval bases and major ports during the speedy German advance along the Baltic and the Black Sea coasts. This not only slowed down the German advance, but also gained invaluable time for the defenders. In almost all cases, the Soviet army and navy cooperated very closely and successfully in the execution of this often complicated operational task. The newly acquired base of Reval was defended by the Soviets between 22 July and 15 August 1941. The Soviets defended a small naval base at Hangö on the southeastern coast of Finland for almost four months in the fall and winter of 1941 (September–December). This fortress, manned by a force of 25,000 troops, was loosely encircled by a much weaker Finnish force. Hangö was supplied by sea, partly by Soviet submarines and partly by smaller ships and craft.

In the Black Sea, the Soviets defended the port of Odessa from mid-August through to mid-September 1941. Five Soviet divisions had held their ground against German–Romanian forces of 160,000 men with five army corps of 21 divisions. The port of Odessa had not been fortified before the war, and thus its defense was only possible with the help of the Black Sea Fleet. Romanian troops were fired on by Soviet cruisers and destroyers as they approached Odessa, considerably delaying the Romanian advance. Support of the base's defenders by Soviet naval forces prevented the Romanians from gaining favorable positions for bringing the port and its facilities under fire. Simultaneously, the Soviets brought at least two of their divisions and much material across the sea into the city, whereby the Germans were prevented from using an important port for their supply traffic.[21]

The large Soviet naval base of Sevastopol' in the Crimea was successfully defended from 29 October 1941 to 1 July 1942 or for 290 days. Soviet surface ships and submarines kept the garrison supplied through this period until, with complete command of the air, the Germans began their final attack on the fortress on 7 June 1942. Seven German and two Romanian divisions took part in the operation supported by German VIII Air Corps, about 208 batteries of all calibers being brought to bear on the Soviet fortifications. Yet the Soviet Black Sea Fleet continued to carry reinforcements and supplies to the fortress, provide its defenders with gunfire, and evacuate the wounded.[22]

Another prolonged defense of a naval base in which fleet forces took a prominent part was the German defense of the Crimea and the naval fortress of Sevastopol' in the spring of 1944. By the beginning of November 1943 German troops on the Crimea had been cut off on the land side because of Soviet troop advances to the Dnieper River. For political and military reasons, Hitler ordered defense of the Crimea by the forces of the German 17th Army (five German infantry divisions and two assault gun brigades), three Romanian mountain divisions, and two infantry and two cavalry divisions. In addition, the *Luftwaffe* deployed one fighter and one fighter-bomber group, one short- and one long-range reconnaissance squadron, and one AA division, while the *Kriegsmarine* manned 23 coastal gun batteries.[23]

In a more recent example, the Israelis used their naval forces and aircraft in successfully defending their naval base at Sharm el-Sheikh against the Egyptian attack in October 1973. About 1,200 Egyptian naval commandos landed by helicopters along the Gulf of Suez. The Egyptian 3rd Army was supposed to link with these forces and occupy the rest of southern Sinai. However, the Israeli air force reacted quickly and shot down seven Egyptian MiG fighters and eight helicopters, while the Israeli patrol craft held off Egyptian commandos and eventually chased them away from Sharm el-Sheikh.[24]

During the retreat of German troops from the Kuban' Peninsula to Novorossiysk in the summer and fall of 1942, their flank was protected by German and Italian S-boats and submarines, while the *Luftwaffe* carried out strikes against Soviet ships in the naval bases at Tuapse, P'ot'i, and Batumi.[25] In 1944–45, the *Kriegsmarine* successfully protected the German flank troops during the withdrawal of Army Group North in the Baltic in 1944–45.

One's naval forces should be prepared to carry out evacuation of one's own and friendly troops and civilians when their lines of retreat are cut off on land or they are isolated in a pocket and threatened with destruction. The problem in this type of support is that the timing of all the tasks must

SUPPORT OF THE ARMY FLANK

be carefully planned, even though naval forces may have meanwhile been assigned to carry out some other important tasks.

One of the most successful evacuations of troops in World War I took place in Albania from December 1915 to February 1916. The Serbian Army reached the Albanian coast after long and arduous withdrawal across the mountains of Albania. The Allied ships, primarily Italians, in 322 trips evacuated about 136,000 men and 36,350 horses of the Serbian Army from the ports of Durazzo and San Giovanni di Medua.[26] This action offered great opportunities for the Austro-Hungarian fleet, but, except for occasional raids by light forces, their battle fleet did not appear on the scene.[27]

In the Mediterranean, the British successfully evacuated some 43,000 Allied troops (out of a total of 57,000 men) from Greece (Operation *Demon*) between 24 and 29 April 1941. About 27,000 men landed on Crete partly to reinforce it and partly to economize shipping by using the shortest passage until the evacuation was complete. Another 16,000 men were taken directly to Egypt under cover of the British fleet. The evacuation of Allied troops from Greece was an exceedingly hazardous undertaking because the *Luftwaffe* rendered the main embarkation port of Piraeus almost unusable by devastating raids. The Germans enjoyed superiority in the air, which by the time of evacuation had become air supremacy. The entire Allied effort was carried out in five nights. Because Piraeus was put out of action, the embarkation took place from the beaches, three in Attica and three in the Peloponnesus. There was no air cover for Allied troops except for a handful of long-range fighters operating from Crete. British losses were four transports and two destroyers sunk.[28]

One of the largest and most successful evacuations of troops across a small stretch of water took place in the English Channel in 1940. Despite almost constant attacks by the *Luftwaffe* and the advancing German troops, the British successfully extricated from Dunkirk (Operation *Dynamo*) about 338,000 troops (including 139,000 French troops). About 240,000 troops were evacuated from the port of Dunkirk, and the rest from the neighboring beaches. The evacuation took place between 27 May and 4 June 1940.[29]

Soviet evacuation of their fortress of Hangö in December 1941 was an example of successful cooperation between ground troops and fleet forces. Hangö posed a serious threat to German–Finnish coastal shipping because it controlled one shore of the entrance to the Gulf of Finland. However, after the loss of the Baltic Islands and the destruction of or damage inflicted on most of their surface ships, the Soviets decided that the fortress had to be abandoned before ice froze the Gulf of Finland. Despite enormous difficulties, they made great efforts first to supply and then to evacuate their personnel. The Soviet garrison in Hangö numbered about 20,000 men at the time of evacuation, and most of these troops reached Leningrad by sea. The difficulties of getting through the minefields and ice from December through April forced the Soviets to evacuate Hangö in four large and several small convoys between the end of October and the first days of December. They also decided to evacuate smaller garrisons on the islands of Odensholm, Suursaari, and several smaller islands in the Gulf of Finland. Soviet losses in ships and men were heavy.[30]

In the Black Sea, Soviet naval vessels evacuated about 86,000 soldiers and 150,000 civilians from Odessa between 1 and 16 October 1941. While the best divisions were brought to Sevastopol', enough Soviet troops remained to continue a stubborn defense of the city. The final evacuation took place on the night of 15–16 October 1941 when the Soviets extricated some 35,000 fully armed men from Odessa. The Romanians discovered this so long after that the Soviet transports were already near Sevastopol' by the time they were attacked from the air. Only 6,000 Soviet troops were taken prisoner.[31]

The Germans conducted several large-scale evacuations of troops and civilians in the Black Sea and the Mediterranean in 1943–44. By early September of 1943 there was a very critical situation for the Germans in the northern part of the Sea of Azov. A decision was made on 4 September to organize withdrawal of Army Group A. Three days later, the Germans began evacuation of their 17th Army and of the Romanian divisions. By 9 October a total of 200,000 troops (plus 16,000 wounded and 27,000 civilians) with most of their weapons and equipment were successfully evacuated from the Kuban' bridgehead. This was done despite intensive Soviet attacks and the presence of mines.[32] After the retreat of the 17th German Army to the Crimea, the Soviets deployed their gun batteries on the Kerch Strait. This forced the Germans to discontinue sending supplies through the strait, and to organize evacuation of their remaining ships and craft (Operation *Wiking*). Approximately 240 ships and craft were transferred to Sevastopol' in three large convoys on 8–10 October 1943. The last ships to leave the area laid mines at the southern exit of the Kerch Strait.[33]

One of the most successful evacuations of troops forced to retreat took place across the Strait of Messina in August 1943. In a brilliant feat of organization, the Germans safely evacuated 39,660 troops with several thousands of vehicles and a large quantity of ammunition. In addition, about 62,000 Italians along with some guns and vehicles also succeeded in escaping over the strait. The Axis lost only 15 craft, while an additional six were damaged. Before the two converging Allied columns entered the city of Messina on 17 August, the last ferry had left.[34]

The Axis success in its evacuation of Sicily was achieved by a skillfully conducted retreat supported by an excellent naval organization. The shortness of these routes contributed considerably to Allied difficulties in stopping Axis traffic across the strait. Another factor in the Allied failure was the extraordinary strength of Axis coastal defenses on both sides of the straits. The German gun concentration in the straits prevented the Allies from carrying out low-altitude air attacks and protracted raids by surface ships. German and Italian AA fire was described as being heavier than in the Ruhr. The Allied air force was mistakenly used against targets on land and only medium and light bombers, together with some fighters and fighter-bombers, were released to help the Tactical Air Force in its round the clock attacks. To use air strikes to neutralize coastal gun batteries was also exceedingly difficult for they offered small targets and Allied aircraft encountered a tremendous volume of fire if they flew too low.[35] Allied intelligence was late in drawing the correct conclusions about the Axis preparations to evacuate their troops from Sicily. But even if that had been clear earlier, the Allied effort suffered from a lack of inter-service coordination.[36]

The Germans conducted another successful evacuation of their troops from Sardinia to Corsica. In late August 1943 about 25,000 troops, 2,300 vehicles, and 5,000 tons of supplies were carried across the Strait of Bonifacio to Corsica with virtually no interference from Allied navies or aircraft. Hitler ordered evacuation of Corsica on 12 September and by 3 October, as the French troops approached Bastia, the German evacuation from Corsica had ended. By then, about 6,250 troops, 1,200 prisoners of war, over 3,000 vehicles, and nearly 5,000 tons of stores were evacuated by sea. At the peak of the evacuation, the Germans used 15 steamers and 120 barges, landing craft, and miscellaneous craft, losing 18 ships and craft of all types totaling 17,000 tons.[37]

The evacuations of the Axis troops from Sicily, Sardinia, and Corsica in the summer and fall of 1943 were superbly organized and executed. One cannot but admire the German organization and flexibility that allowed them to shift their naval and air assets so rapidly from one task to another, all this being at very short notice and at a time of almost complete Allied superiority at sea and in the air.

The Germans also conducted several large-scale evacuations of troops and civilians in the Black Sea in 1943–44. Despite huge Soviet superiority on the ground, at sea, and in the air, the Germans were highly successful in extricating their troops and large quantities of weapons and equipment. As in the Mediterranean, the Germans improvised evacuations from major ports at short notice and used all available shipping, including barges and tugs. The Germans would have evacuated a larger number of troops if it had not been for Hitler's delay in ordering the evacuation. After the Soviet armies had resumed their advance toward Odessa, the Germans evacuated their troops from the city during the second half of March 1944. Surprisingly, Soviet naval forces did not seize their chance after advancing Soviet troops had cut all the land communications to Odessa. This omission enabled German naval forces to transport most of the defenders and floating *matériel* to Romanian ports. The seaborne evacuation began on 1 April and ended ten days later. All available shipping space and operational naval units were used and all other supply traffic to the Crimea was suspended. The weather was favorable. In the end, the Germans evacuated 24,300 troops (including 9,300 wounded) and civilian refugees as well as taking 54,000 tons of material.[38]

The Germans conducted yet another large-scale evacuation of troops in the Black Sea as Soviet troops threatened to capture the Crimea in the spring of 1944. Preparations to evacuate the Crimea started on 27 October 1943.[39] The Soviet offensive to retake the Crimea began on 8 April 1944. At that time about 130,000 German troops and 70,000 Romanian troops were on the peninsula. Another 10,000 were cut off in the Yalta hills but were evacuated by naval barges. The Soviet attack on the Crimea started on 7 May 1944, and two days later Hitler finally gave the order to evacuate the Peninsula. By 11 May only about 20,000 German troops remained on the Crimea. Two days later the last German ship left the Chersonnes Peninsula.[40] In one month about 130,000 Romanian and German soldiers had been evacuated by sea. German losses on the Crimea were 78,000 men killed or captured. A total of 190 German warships and merchant vessels took part in the evacuation from the Crimea. The shipping volume was sufficient to evacuate 123,000 men one way, but relatively large distances were involved; about 200 nautical miles separated Sevastopol' from Constanta, the main debarkation port. The Axis losses were surprisingly low.[41] The German Navy performed extraordinarily well in carrying out the evacuation effort. Its success was achieved despite the fact that the Soviets controlled both the sea and the air around the Peninsula. Despite their numerical superiority on the ground, in the air, sea the Soviets were unable to hinder the embarkation of troops at Sevastopol'.[42]

One of the largest and most successful evacuations of all times was conducted by the Germans in the Baltic in the closing months of World War II. Because of Hitler's obstinacy, the Germans waited too long to effect an orderly withdrawal of their troops from the Baltic. The German Naval Staff began preparations for a large-scale evacuation of the German troops and civilians on the Baltic coast in the late summer of 1944.[43] By the end of February 1945 the German front in Pomerania had collapsed. However, the Germans were able to hold the beachheads around Danzig and Gotenhafen for four more weeks. A smaller beachhead was formed around the Pomeranian town of Kolberg, which enabled the Germans to carry out what probably was the largest evacuation across the sea in history. All available ships including large liners carried an estimated 2.2 million people westward. The Courland beachhead was supplied throughout, and there was never any lack of targets for Soviet submarines and aircraft.[44]

Yet the Soviets were surprisingly ineffective in their efots to prevent a German evacuation.

In World War II the Japanese conducted several relatively large evacuations. The final evacuation of Japanese troops from Guadalcanal started on the night of 1 February 1943 when a number of destroyers embarked the first elements of the Japanese garrison from the island. During five successive nights, a total of 12,000 Japanese troops were evacuated.[45] In the final phase of the struggle in the Solomons, the Japanese decided in mid-August 1943 to evacuate all the islands of the Solomons chain except Buka, Bougainville, and the Shortlands. The American capture of Vella Lavella greatly concerned the Japanese, because that island flanked the route by which the garrison of 10,000 men on Kolombangara was to be evacuated. Japanese fears proved to be well grounded, because the Americans at once blockaded Kolombangara. Nevertheless, roughly 9,000 Japanese troops were successfully evacuated from Kolombangara, mostly by motor barges moving at night.[46]

During the Korean War, the UN forces carried out one of the most successful evacuations of troops and civilians. After the failure of the UN offensive in the north, and in the face of the Communist Chinese advance toward Wŏnsan, General MacArthur ordered on 8 December 1950 the evacuation of the X Army Corps from Hŭngnam on the eastern coast of Korea.[47] The US 3rd Division task force and Marine shore party totaling about 3,800 troops, 1,146 vehicles, some 10,000 tons of cargo, and 7,000 refugees were evacuated from the port of Wŏnsan on 3 December.[48] Two days later, troops of the South Korean Third Division and 4,800 refugees were evacuated by US ships of Task Force 90 from the port of Songjin. The evacuation from the Hŭngnam area started on 11 December and officially ended on 26 December. The US X Army Corps and the South Korean First Corps, totaling 105,000 troops plus some 91,000 civilians, were evacuated in an orderly fashion and sealifted to Pusan and Ulsan (some 30 miles north of Pusan). In addition, 17,500 vehicles and 350,000 tons of cargo were transported by sea on board 192 ships.[49] All the serviceable equipment and fuel were taken away. The US Navy provided air support with seven aircraft carriers and gunfire support from 13 ships. Neither Chinese nor North Korean forces made any serious attempt to interfere with the evacuation of the UN troops.[50]

There have been many examples where naval forces failed to evacuate their own troops threatened with encirclement. The Soviet Black Sea Fleet was unable to extricate the defenders of the naval fortress at Sevastopol' in May 1942. German–Romanian troops in October 1941 succeeded in seizing the Perekop position and defeating the Soviet defenders within a few days. They finally captured the Crimea with the exception of the large naval base of Sevastopol'. The Germans launched their attack on the fortress aimed at destroying three Soviet armies (47th, 44th and 51st) on the Kerch Peninsula (Operation *Trappenjagd*). The Soviets were completely surprised. After a few days of fighting, eight Soviet divisions were encircled. By 16 May the Germans had captured the city of Kerch. Two days later the bulk of the Soviet troops stopped fighting. The German final attack on Sevastopol' began on 7 June 1942. Stalin in mid-June rejected the proposal that the fortress be evacuated. The rest of the Soviet troops withdrew to Kherson on 1 July, where they expected to be embarked on the Soviet warships but the expected help from the Soviet Black Sea Fleet never came.[51]

Offense–Defense

One's naval forces also conduct a range of tasks regardless of whether their own troops are on the offensive or on the defensive. The principal tasks of naval forces are to protect the flanks of ground troops by providing fire support and conducting defensive mining, to transport troops and *matériel*, taking part in the defense of the coast and conducting commando raids in the enemy's rear.

Fire support from naval guns and coastal gun batteries is used to defend one's naval bases, to support amphibious landings, to provide flank support to advancing troops, and to defend important sectors of one's coast. The high mobility of ships and their strong and accurate firepower play an important role in defending naval bases. Fire against coastal targets is considered the most complicated task for shipboard artillery.

In 1941–42, heavy ships of the British Mediterranean Fleet provided fire support to Allied troops during their offensive operations along the North African coast. Allied warships were also extensively used in attacking enemy positions on the coast prior to and during amphibious landings. During the Allied invasion of Sicily (Operation *Husky*) the 8-inch and 5-inch fire from US cruisers and destroyers was critical in preventing the advance of the German armored division *Hermann Goering* toward the US Seventh Army's beaches at Gela on the first day of the landing, 10 July 1943.[52]

During the crossing of the British Eighth Army from Sicily to Italy's mainland, two battleships, one cruiser, and nine destroyers of the Allied forces conducted a preliminary bombardment of the landing area on 31 August 1943. Two days later two battleships, two cruisers, two destroyers, three monitors, and two gunboats shelled targets on the coast. During the actual landing, three monitors, two cruisers, six destroyers, and two gunboats as well as some army gun batteries in the Strait of Messina provided fire support to Allied troops on the coast.[53]

During the withdrawal of Army Group North in the summer and fall of 1944, the Germans deployed their 'pocket' battleships and heavy cruisers with great success in support of their troops on the coast by bombarding Soviet forward tank columns and gun positions. The bombardments from the sea helped to stop Soviet armies advancing on Courland along the southern shore of the Gulf of Riga and along the Baltic coast from the south toward Libau in the fall of 1944. On several occasions one or more German pocket battleships and cruisers along with a few destroyers and S-boats shelled the large Soviet units preparing to attack, helping the Germans to avoid a potential disaster.[54]

By January 1945 most of the German eastern front had collapsed under massive attack by well-equipped Soviet armies. The Soviets reached the Oder River inland and Eastern Prussia was cut off. The German troops on the Baltic coast were compressed in the Samland–Koenigsberg and Elbing area. Even the land connection between Koenigsberg and Pillau was interrupted. However, fire from German S-boats and gunboats forced the Soviets to retreat temporarily.[55]

In the Korean War (1950–53), the UN ships were involved on many occasions in providing gunfire support to the troops on both the east and west coasts of the peninsula. The gunfire support by the US Navy's ships was critical in holding the perimeter around Pusan in the early phase of the war, between June and September 1950. After the Communist Chinese troops intervened in the conflict in November 1950, the UN Eighth Army on the west coast and the Tenth Corps on the east coast were in danger of being encircled. In that critical situation, surface ships of Task Force 95 were extensively employed in providing gunfire and air support to the Eighth Army during its evacuation from Chinnanpo to Inchon.[56]

During the Hüngnam evacuation, troops on the ground were supported by the Hüngnam gunfire support group, consisting of one battleship, two heavy cruisers, seven destroyers, and three LSMRs. The first request for naval gunfire support came on 15 December. These ships delivered 8-inch fires against targets up to ten miles inland from preselected positions within the maneuvering area, swept clear of enemy mines, ten miles to the north and ten miles to the south of Hüngnam.[57]

In the Yom Kippur (Ramadan) War of 1973, the Egyptian attack across the Suez Canal was supported by coastal artillery. The Egyptians used both stationary and mobile 130-mm guns and former torpedo craft rearmed with army MRLs. Also, the *Osa* and *Komar* class missile craft were used to attack targets on the coast.[58] Likewise, the Israeli missile boats took positions behind the Egyptian lines and attacked targets on the coast. The Israeli landing craft used mortars in attacking Egyptian troops at Mersa T'lemet anchorage.[59]

Modern land attack cruise missiles and guns allow one's naval forces to carry out a range of tasks, from direct fire support against an enemy held coast to the destruction of operationally important targets in the enemy country's interior. One's fleet forces should be able to strike the elements of the enemy defense system, to destroy troops and reserves of *matériel*, and to support the seizure of coastal installations and ports by one's troops. Another task includes preventing the evacuation of besieged enemy troops by sea.

The most important task of naval forces in either offense or defense is to protect and defend ships transporting troops and *matériel* along the coast. The use of the sea for the transport of troops and supplies is not only the most economical, but sometimes also the only practical way of moving supplies up to the frontline. However, this is sometimes not fully appreciated by the highest politico-military leadership, as shown in planning of the invasion of Soviet Russia by the Supreme Command of the *Wehrmacht*. The Germans did not fully exploit the potentialities offered by maritime transport to support the advance of their troops along the Baltic coast. Instead, they relied on supplies carried over poor and inadequately developed Soviet roads and railroads. These problems could have been avoided, had the Germans deployed stronger naval forces in the Baltic, and had some units of the *Luftwaffe* moved cargo rapidly to the Estonian ports on the Gulf of Finland, thereby providing more impetus to the attack by Army troops on Leningrad. The main reasons for this failure were the inherent contradiction between the Army's offensive concept and the Navy's defensive plan, which was never overcome. The *Kriegsmarine* was more focused on laying defensive mine barriers in the Gulf of Finland than in employing its small surface warships to escort convoys loaded with troops and supplies.[60]

In the Black Sea, the Germans were also faced with the daunting task of how to transport and supply their troops along the coast in the face of Soviet superiority at sea and in the later phase of the war also in the air. They and their allies had only a few large ships available for the task. The plan to arm about 21 German merchant steamers found in the Romanian and the Bulgarian ports after the war started in September 1939 was only partially completed. The Romanian Navy was too small to provide an adequate protection for the Axis shipping in the Black Sea.[61] This was the reason that they had to bring overland, then assemble and build locally a large number of shallow-draft *Siebel* (*Luftwaffe*-designed) landing craft, naval landing craft (MFPs), and tugs.[62] The problem of supplying troops on the coast became especially important to the Germans after the fall of Sevastopol' in July 1942, and the subsequent German advance toward the Caucasus. The first German convoy route was established between Odessa and Yevpatoriya to Sevastopol'. After July 1942 this route was extended further eastward to the Kerch Strait and a month later the Germans initiated several convoy practice runs to Taganrog in the Sea of Azov. By the beginning of July 1942 the entire Crimea was in German hands. There was a possibility of moving troops via the Kerch Strait to the Taman

Peninsula and thereby to attack from the rear Soviet divisions in the Kuban area and either destroy them or force them to retreat. Also, for the planned advance to the Caucasus, the shortened supply route over the Crimea made it unnecessary to use the longer roundabout route along the northern coast of the Sea of Azov.[63]

Beginning in September 1942 the Soviets evacuated the Taman Peninsula sinking their large ships and moving their smaller ships to the Kuban' and up the River Don. Soviet naval forces ceased to exist in the Sea of Azov. This in turn, allowed the Germans to ship a major part of their troops and *matériel* over the Kerch Strait by MFPs, *Siebel* landing craft, and lighters. Later on, the Germans also used a causeway with a capacity of 800 tons per day, and planned to build a large bridge at Genichevs'k which was never completed. During the evacuation of the northern Caucasus the Germans transported about 450,000 men to the Crimean Peninsula. They also organized about 150 convoys to Taganrog and 190 convoys to Anapa. Surprisingly, the Soviets did not even try to interfere with this traffic. By September 1943 the convoy route was extended southward of the Kerch Strait to Novorossiysk, and afterwards to Anapa. The total length of these sea routes was about 1,000 nautical miles.[64]

In the Mediterranean, the German Naval Staff considered its most important task was to ensure the supply of German troops in the Balkans. Because of the inadequate road and rail communications and the sabotage actions by the Communist-led resistance in the former Yugoslavia and Albania, these supplies had to come primarily by sea. However, because the distances to the German-controlled Dalmatian coast across the Adriatic averaged between 60 and 80 nautical miles, the Allies were given the opportunity to carry out raids from their bases in southern Italy. This in turn forced the Germans to increase urgently their sea and coastal defenses in the eastern part of the Adriatic. In contrast, the German problem of supplying their troops in Italy was much less difficult because of the good road communications along the peninsula's western coast.[65]

German efforts to supply their troops in Italy in 1943–44, although made at short notice, were highly successful. The retreating German troops were supplied by a coastal route along the peninsula's eastern coast. The entire system was highly improvised because of the lack of suitable port facilities. In the fall of 1943 the Germans transported a monthly average of some 8,000 to 12,000 tons of military cargoes along the west coast of Italy to ports near the front line. In addition, some 4,000 to 6,000 tons were carried down the Adriatic each month, mostly from Venice to Ancona.[66] The German troops deployed along the Dalmatian coast were in an even more precarious position than those in Italy because land communications were very bad. In addition, the Communist-led resistance movement was active in the interior. The Germans eventually deployed

18 divisions in Yugoslavia and Greece, and these troops had to be supplied mostly by sea. The problem facing the Germans was formidable. Initially, they had hardly any naval force in the Adriatic and the available merchant shipping was unsuitable for the task at hand. The large ships, which the Germans assembled in Venice, were better suited for the Aegean than the Adriatic. For use in the Adriatic, the Germans needed small merchant vessels, motor-driven barges, and landing craft. The Germans acted with their characteristic energy, transporting the required vessels in sections. The assembly of a considerable fleet of small craft in this manner was indeed a remarkable feat.[67]

The withdrawal of one's army along a coast will, among other things, result in shortening the lines of communications. If the retreat is extensive, this considerably reduces the need for sea transport. In June 1944 the western flank of the *Wehrmacht* in Italy retreated to a line stretching from Lake Trasimeno to Piombino, thereby shortening the shipping routes and the German reliance on the sea to supply their troops. However, an unwelcome result was that the Allies shifted the focus of their air attacks from coastal traffic to the German-controlled ports on Italy's western coast.[68]

One of the most remarkable efforts to supply troops on a coast was undertaken by the *Kriegsmarine* in the Baltic in the fall of 1944 and the winter of 1945. By October 1944 the retreating Germans had formed a large beachhead around the ports of Libau and Windau containing 26 divisions and 600,000 men. These troops held the bridgehead to the end of the war against all Soviet attacks. More remarkably, the German troops were kept successfully supplied all the time and four of the besieged divisions were eventually transported across the sea to Germany.[69]

When the voyage by sea becomes perilous, transport of troops or *matériel* may be conducted by air. During the final campaign in Tunisia in late March 1943, the Axis transport aircraft became ever more important as a means of ferrying men and supplies to Africa as the difficulties of getting them there by sea grew. It has been estimated that the daily average arrivals of Axis transport aircraft, Ju-52s, Ju-290s, and the giant Me-323s, in Tunisia had risen to 200. Most flights used Naples and one of the airfields in western Sicily as staging posts, although some went directly to Tunisia, rendezvousing over Trapani with fighter escorts.[70]

In the Allied campaign in New Guinea, the main problem was to provide adequate and suitable sea transport for the support and supply of troops on the ground. General Douglas MacArthur's naval forces were small and generally lacked the light craft so essential for joint operations in the littorals. To use transports and to escort them with cruisers and destroyers, even if these had been available, would have been too risky because the Allies then still did not have secure control of the air. The

SUPPORT OF THE ARMY FLANK

solution was found by using Dutch and Australian coasters and native craft whose crews were familiar with New Guinea's coastal waters. Their services were of great value in ferrying Allied troops along the coast. They were supported by light warships of the Royal Australian Navy, and together proved adequate as extemporized substitutes for specially designed and properly equipped landing craft.[71]

The value of the sea as a supply route, even for armies which could bring their necessary supplies overland, was shown in the Korean War. Despite the UN blockade of both coasts, the North Koreans continued to use small craft to help carry military supplies forward for their advancing army. Nevertheless, South Korean Navy ships, mainly small wooden minesweepers which had been transferred from the US, eventually destroyed most of these craft.[72]

Support of ground troops operating along the coast is one of the most important, but also one of the most complicated operational and tactical tasks for naval forces and air forces. The problem is compounded by the need for naval forces to carry out other essential tasks elsewhere. Thus, setting the right priorities and their sound sequencing are critical factors in success. With the greatly increased capabilities of naval forces in recent years, the traditional tasks have been expanded to include also the coastal area beyond the ground troops' flanks and the contiguous sea areas. Today, carrier-based aircraft and cruise missiles are capable of attacking targets many hundred miles away from the coast. In a large narrow sea such as the Mediterranean, naval forces can support the maritime flank of an army and/or provide broader support such as transporting or evacuating troops to and from the mainland. In contrast, naval forces operating in a typical narrow sea such as the Arabian Gulf or the Black Sea can carry out a wide range of operational and tactical tasks in support of ground forces operating in the coastal area.

Support of an army flank also requires detailed planing and preparation and energetic execution. There must be unity of command in order to ensure smooth inter-service cooperation at all levels of command. Operational and tactical concepts for the employment of naval forces and aviation in support of an army flank must be developed and practiced in peacetime. An adequate number of platforms and suitable platforms and weapon systems should be available to do the job.

NOTES

1 Harold M. Sprout and Margaret T. Sprout, *The Rise of American Sea Power 1776–1918* (Princeton, NJ: Princeton University Press, 1939), p. 155.
2 Henry Newbolt, Vol. 5: *From April 1917 to the End of the War, Naval Operations*, (London: Longmans, Green 1931), pp. 223–4.

3 Raymond De Belot, *The Struggle for the Mediterranean 1939–1945*, trans. James A. Field Jr (Princeton, NJ: Princeton University Press, 1951) p. 87.
4 Otto Groos, *Seekriegslehren im Lichte des Weltkrieges: Ein Buch fuer den Seemann, Soldat und Staatsmann* (Berlin: E. S. Mittler & Sohn, 1929), p. 31.
5 Michael Salewski, *Die deutsche Seekriegsleitung 1935–1945*, Vol. I: *1935–1941* (Frankfurt a.M: Bernard & Graefe, 1970), p. 367.
6 Juerg Meister, *Der Seekrieg in den osteuropaeischen Gewaessern 1941–45* (Munich: J. F. Lehmanns Verlag, 1958), p. 32.
7 Malcolm W. Cagle and Frank A. Mason, *The Sea War in Korea* (Annapolis, MD: United States Naval Institute, 1957), pp. 286, 371.
8 Abraham Zohar, 'Die israelische Marine im Libanonkrieg 1982', *Marine Rundschau*, 5 (May 1985), p. 98.
9 David Woodward, *The Russians at Sea: A History of the Russian Navy* (New York: Frederick A. Praeger, 1966), p. 112.
10 Salewski, *Die deutsche Seekriegsleitung 1935–1945*, Vol. 2: *1942–45* (Frankfurt a.M: Bernard & Graefe, 1975), pp. 69–70.
11 Ibid., p. 456.
12 Roskill, *The War At Sea 1939–1945*, Vol. III: *The Offensive* pt. 1, 1st June 1943–31st May 1944 (London: Her Majesty's Stationery Office, 1960), p. 151.
13 Woodward, *The Russians at Sea*, p. 102.
14 Ibid., p. 175–7.
15 In April 1943, out of 26 Axis ships which sailed on that route 15 were sunk and four were damaged, most by air attacks; nevertheless, some 27,000 tons of supplies and 2,500 troops reached Africa; in May eight Axis supply ships and 15 small craft carrying about 7,000 tons of supplies were sunk, while only about 2,160 tons of supplies reached North Africa; in April 1943 the Axis succeeded in flying 18,000 men and 5,000 tons of supplies by air, though 117 aircraft were lost; Roskill, *The War At Sea 1939–1945*, Vol. II: *The Period of Balance* (London: Her Majesty's Stationery Office, 1956), p. 440.
16 Malcolm W. Cagle and Frank A. Mason, *The Sea War in Korea*, pp. 270–1, 279–80.
17 Donald Macintyre, *Sea Power in the Pacific: A History from the 16th Century to the Present Day* (London: Military Book Society, 1972), pp. 53–4.
18 Juergen Rohwer and Gerhard Huemmelchen, *Chronology of the War at Sea 1939–1945*, 2 Vols (New York: Arco Publishing, 1974), Vol. 2 (1943–1945), p. 347.
19 Friedrich Ruge, *The Soviets as Naval Opponents 1941–1945* (Annapolis, MD: Naval Institute Press, 1979), p. 123.
20 Roskill, *The War At Sea 1939–1945*, Vol. II, *The Period of Balance* (London: Her Majesty's Stationery Office, 1956), p. 73.
21 Ruge, *The Soviets as Naval Opponents 1941–1945*, p. 67.
22 Woodward, *The Russians at Sea: A History of the Russian Navy*, pp. 216–17; the *Luftwaffe* sank three Soviet destroyers and a number of smaller warships and transports; Ruge, *The Soviets as Naval Opponents 1941–1945*, p. 79.
23 Meister, *Der Seekrieg in den osteuropaeischen Gewaessern 1941–45*, p. 275.
24 Ze'ev Almog, 'Israel's Navy Beat the Odds', *Proceedings*, 3 (March 1997), p. 106.
25 Meister, *Der Seekrieg in den osteuropaeischen Gewaessern 1941–45*, p. 265.
26 The Allies lost 11 destroyers and auxiliaries and eight steamers; Henry Newbolt, Vol. 4: June 1916 to April 1917, History of the Great War Based on Official Documents: *Naval Operations* (London: Longmans, Green, 1928), p. 121.
27 Anthony E. Sokol, 'Naval Strategy in the Adriatic Sea during the World War', *Proceedings*, 8 (August 1938), p. 1088.
28 I. S. O. Playfair *et al.*, *The Mediterranean and Middle East*, Vol. II, p. 105; Butler,

Grand Strategy, ed. J. P. W. Ehrman, *Sept. 1939–June 1941*, Vol. 2, p. 457; John Creswell, *Sea Warfare 1941–1945* (Berkeley, CA: University of California Press, rev. edn, 1967), p. 95; Louis L. Snyder, *The War: A Concise History 1939–1945* (New York: Julian Messner, 1960, p. 157.

29 About 28,000 troops were evacuated prior to 27 May; some 80,000 Allied troops went into captivity; Hans Umbreit, 'Der Kampf um die Vormachtstellung in Westeuropa', *Das Deutsche Reich und der Zweite Weltkrieg*, Vol. II: *Die Errichtung der Hegemonie auf dem europaeischen Kontinent* (Stuttgart: Deutsche Verlags-Anstalt, 1979), p. 296.

30 Ruge, *The Soviets as Naval Opponents 1941–1945*, p. 23; Meister, *Der Seekrieg in den osteuropaeischen Gewaessern 1941–45*, pp. 34, 36.

31 Ruge, *The Soviets as Naval Opponents 1941–1945*, p. 66.

32 Ibid., p. 112; including 74 tanks, 21,000 motor vehicles, 28,000 horse-drawn vehicles, 1,200–1,800 guns, 75,000 horses, 6,500 cattle, and 115,000 tons of military equipment; Dietrich von Conrady, 'Die Kriegsmarine und der letzte Kampf um Sewastopol im Mai 1944', *Wehrwissenschaftliche Rundschau*, 6 (June 1961), p. 314.

33 Meister, *Der Seekrieg in den osteuropaeischen Gewaessern 1941–45*, p. 265.

34 A total of 9,600 vehicles, 47 tanks and 94 guns together with 17,000 tons of ammunition were transported; Field Marshall Alfred Kesselring put the figure at 60,000 men; Gilbert A. Sheppard, *The Italian Campaign 1943–45* (New York, F. A. Praeger, 1968), p. 69.

35 Roskill, *The War At Sea 1939–1945*, Vol. III, pt. 1, pp. 145–6.

36 Ibid., pp. 149–50.

37 In the meantime, the German airlift from Corsica took out 21,000 men and 350 tons of stores. The Germans lost 55 aircraft, most of them due to Allied bombardment of the Italian airfields; I. S. O. Playfair *et al.*, *The Mediterranean and Middle East*, Vol. II, p. 105; J. R. M. Butler, *Grand Strategy*, Vol. 2 (September 1939–June 1941), p. 457.

38 Although the evacuation was improvised at the short notice, all the ships totaling some 81,000 tons, with the exception of one 80-ton barge, were taken away; 18 steamers undertook 26 trips, 24 tugs and similar craft made 60 trips, and 25 naval ferry barges made 76 trips; Ruge, *The Soviets as Naval Opponents 1941–1945*, p. 127.

39 The German Navy Commander Admiral Karl Doenitz stated that the 17th Army needed 50,000 tons of supplies per month and 30,000 tons of ships and 60 naval ferries. For evacuation of 200,000 men 40 days were necessary and twice that many in the case of bad weather; Conrady, 'Die Kriegsmarine und der letzte Kampf um Sewastopol im Mai 1944', p. 315.

40 Between 14 and 26 April about 80,000 troops and 2,500 tons of *matériel* were carried from the Crimea to Constanta; by 12 May about 26,000 soldiers and 6,100 wounded had been evacuated, while about 8,100 men had been lost; in addition, some 16,000 men including 1,000 wounded were transported by air to Romania;. Ruge, *The Soviets as Naval Opponents 1941–1945*, p. 130.

41 A total of 12 German warships and auxiliaries, one Romanian craft, seven steamers, seven tugs, and 21 miscellaneous craft were sunk, mostly by Soviet air attacks; however, the losses of personnel were relatively high – 6,000 to 9,000 men; Meister, *Der Seekrieg in den osteuropaeischen Gewaessern 1941–45*, pp. 280–2; Ruge, *The Soviets as Naval Opponents 1941–1945*, p. 132; Salewski, *Die deutsche Seekriegsleitung 1935–1945*, Vol. 2: *1942–1945*, pp. 398–9.

42 Conrady, 'Die Kriegsmarine und der letzte Kampf um Sewastopol im Mai 1944', pp. 320, 330.

43 By the beginning of September 1944, the German Naval Staff thought that by using all available ports, halving the country's imports, abandoning all equipment, it would be possible to evacuate 520,000 men within five weeks; the evacuation of all men with their equipment would require five months. However, it was unrealistic to anticipate that the most important ports such as Riga, Reval, and Libau would not be seized by the Soviets, or that the German convoys would not be attacked at sea. Salewski, *Die deutsche Seekriegsleitung 1935–1945*, Vol. 2: *1942–1945*, p. 491.

44 Ruge, *The Soviets as Naval Opponents 1941–1945*, p. 51.

45 Samuel E. Morison, *The Two-Ocean War: A Short History of the United States Navy in the Second World War* (Boston, MA/Toronto: Little, Brown, 1963), p. 214; Roskill, *The War At Sea 1939–1945*, Vol. II: *The Period of Balance*, p. 417.

46 Roskill, *The War At Sea 1939–1945*, Vol. III, *The Offensive*, pt. 1, p. 233.

47 Some sources categorize evacuation from Hüngnam as a 'redeployment'; Cagle and Mason, *The Sea War in Korea*, p. 164.

48 Ibid., p. 184; Billy C. Mossman, *Ebb and Flow: November 1950–July 1951*, United States Army in the Korean War series (Washington, DC: Center For Military History, 1990), p. 165.

49 These included 6 APAs, 6 AKAs, 12 TAPs, 76 time-charted ships, 81 LSTs, and 11 LSDs; Cagle and Mason, *The Sea War in Korea*, p. 191.

50 In addition, air force and marine aircraft airlifted some 2,600 men, 200 vehicles, and 1,300 tons of cargo and several hundreds of Korean refugees were airlifted from the nearby Yonpo airfield; Ibid., pp. 186, 188; Mossman, *Ebb and Flow. November 1950–July 1951*, pp. 172–5.

51 Meister, *Der Seekrieg in den osteuropaeischen Gewaessern 1941–45*, pp. 251, 253.

52 Albert N. Garland and Howard McGaw Smyth, *Sicily and the Surrender of Italy, United States Army in World War II: The Mediterranean Theater of Operations* (Washington, DC: Center of Military History, United States Army, reprint, 1986), p. 154.

53 Rohwer and Huemmelchen, *Chronology of the War at Sea 1939–1945*, Vol. 2, *1943–1945*, p. 347.

54 Ruge, *The Soviets as Naval Opponents 1941–1945*, p. 49.

55 Ibid., p. 51.

56 Cagle and Mason, *The Sea War in Korea*, p. 301.

57 Ibid., p. 187; between 7 and 24 December the ships of Task Force 90 fired 162 rounds of 16-inch, 2,932 rounds of 8-inch, 18,637 rounds of 5-inch; 71 rounds of 3-inch; 185 rounds of 40-mm, and 1,462 rockets; ibid., p. 188.

58 Walter Jablonski, 'Die Seekriegfuehrung im vierten Nahostkrieg aus aegyptischer Sicht', *Marine Rundschau*, 7 (July 1978), p. 663.

59 Almog, 'Israel's Navy Beat the Odds', p. 107.

60 Juerg Meister, *Der Seekrieg in den osteuropaeischen Gewaessern 1941–4*, p. 24.

61 At the outbreak of war, the Romanian Navy had in service only four destroyers, one minelayer, three torpedo boats and three gunboats, one S-boat, and one submarine; Meister, *Der Seekrieg in den osteuropaeischen Gewaessern 1941–45*, pp. 236–7.

62 In early 1941 the Germans brought from Antwerp to Constanta about 50 *Siebel* landing craft; ibid., p. 239.

63 Ibid., p. 264; 'The Soviet Navy in World War II pt. IV: Soviet Submarine Operations 1941–1945', *ONI Review*, 1 (January 1953), p. 16; Conrady, p. 116; Ruge, *The Soviets as Naval Opponents 1941–1945*, p. 99–100.

64 Meister, *Der Seekrieg in den osteuropaeischen Gewaessern 1941–45*, p. 257, 261.

65 Salewski, *Die deutsche Seekriegsleitung 1935–1945*, Vol. 2, p. 381.

66 Roskill, *The War At Sea 1939–1945*, Vol. III, pt. 1, p. 316.
67 Ibid., p. 205.
68 C. J. C. Molony, *The Mediterranean and Middle East*, Vol. VI: *Victory in the Mediterranean* pt. 1: *1st April to 4th June 1944* (London: Her Majesty's Stationery Office, 1984), p. 362.
69 Ruge, *The Soviets as Naval Opponents 1941–1945*, p. 49.
70 Playfair, *et al.*, Vol. IV, *The Destruction of the Axis Forces in Africa*, p. 415.
71 Roskill, *The War At Sea 1939–1945*, Vol. II, p. 235.
72 A. Field Jr, *History of United States Naval Operations in Korea* (Washington, DC, 1962) pp. 58–9, 127, 332.

Conclusion

Many blue-water navies have clearly not given efficient consideration to the problem of fighting a war in narrow seas. Continental states bordering one or several narrow seas have also often failed to take into account the possibilities offered by the use of naval power to accomplish objectives on the land front. The navies of such continental powers have often relied on the army and air force to accomplish essentially naval objectives.

The foundations of naval strategy in a typical narrow sea considerably differ from those in the open ocean because of the proximity of a continental landmass and the small size of the sea area. These factors both facilitate and complicate command and control as well as the employment of naval forces in such waters. A sound naval strategy should always fully take into account the limitations that the physical environment imposes on the employment and sustainment of one's naval forces and aircraft operating in narrow seas. In peacetime, a maritime state should do everything possible to enhance its naval position within or in the proximity of narrow seas, where its naval forces and aircraft will be employed in wartime or even in peacetime. This should be primarily accomplished by diplomatic means and by building naval coalitions with countries bordering narrow seas.

The elements of operational warfare in the open ocean and in a narrow sea are essentially the same, but there are some critical differences. A coastal navy in a conflict with a blue-water navy inherently operates from a central strategic position. The advantages of such a position are far smaller if a coastal navy is unable or unwilling to control a sea's only exit from a larger narrow sea or the open ocean. A blue-water navy can operate within a typical narrow sea without necessarily having access to naval bases and airfields in the area. However, it is critical that a coastal navy possesses a sufficient number of major and minor naval bases and airfields. Also, there is considerable difference between the lines of operations used by a force on the open ocean and those used by a force operating within a typical narrow sea. Lines of operations in a narrow sea are normally much shorter, less numerous, and more rigidly laid than those on the open ocean. In contrast to those in the open ocean, they can follow one's own or a friendly coastline. Also, they may be quite circuitous if there are many islands fronting the shore. Moreover, for a large coastal navy, lines of operations with the larger narrow sea or the open ocean run through the sea's exits. Hence, these lines can be cut off by a fleet controlling the sea exit.

The principal differences between shipping routes on the open ocean and those in narrow seas concern their number and length. The routes on the open ocean are more numerous, very long, and intersect each other. They are also more vulnerable than the routes in a typical narrow sea, and hence require relatively large forces for their protection. Routes in an enclosed sea in contrast are much shorter, less numerous, and more channelized, especially along a coast fronted with numerous islands and islets. Such routes resemble land communications because they are rigidly laid and there are few if any alternate routes. Shipping routes running between islands are usually well protected from enemy attack.

The strategic objectives in a typical narrow sea to be accomplished by a blue-water navy and a coastal navy differ considerably. Whether a war in a narrow sea is conducted offensively or defensively significantly affects the initial strength of the opposing forces on the ground and in the air, as well as the progress of the war on land. A stronger navy in a typical narrow sea can be on the defensive from the outset of hostilities because the army has been forced to undertake a hasty retreat and the adversary also possesses superiority in the air. Besides naval forces, land-based aircraft play an extraordinarily great role in obtaining control of the surface, the subsurface, and in the air. Sometimes, land-based air can predominantly be employed to accomplish the main objectives in a given maritime theater with a relatively small contribution from one's naval forces. In contrast to war in the open ocean, not only ground forces, but also coastal defense forces can be used to obtain absolute local, though often temporary, control of straits or river estuaries.

The principal methods of combat-force employment for accomplishing tactical objectives in narrow seas are naval battles, engagements, strikes, attacks, and raids. However, the ever-longer range, precision, and lethality of missiles, torpedoes, guns, and mines allow even small combatants to strike a variety of targets on the surface, the subsurface, and deep in the interior of enemy-held territory. Strikes have already replaced naval battles as the main method of employment of coastal navies. One of the main features of warfare in a typical narrow sea is high intensity and almost continuous employment of available forces. These actions when conducted in a certain sea area will over time accomplish operational objectives and thereby change the operational situation in a theater.

The operational or strategic objectives in a theater are accomplished through major naval operations and campaigns. The types of major naval operations are most directly related to their main purpose, the predominant sea/ocean area in which they are conducted, timing, and the degree of participation by other services. While one's naval forces will always play the most critical role in the accomplishment of the assigned objectives in a typical narrow sea, other services and their combat arms will increasingly participate, because the proximity of the coast will allow not only land-based aircraft, but also ground forces to take part in major naval operations. Therefore, not only amphibious landings, but also major naval operations against the enemy's maritime trade or the protection of one's maritime trade will increasingly become a joint or even a combined activity. The focus of theory should be on the type of major naval operations that are most likely to be conducted at the present or in the near future. However, other types of major naval operations should not be neglected or simply ignored. Because of the absence of major naval opponents at the present time, major naval operations to destroy the enemy fleet at sea are not likely to occur in the near future. Yet naval theoreticians and practitioners should pay great attention to this type of major naval operation. Likewise, a blue-water navy while focusing on conducting major naval operations in the narrow seas, should not neglect the possibility of fighting fleet-on-fleet encounters on the open ocean.

The operational commanders and their planners should know and understand the purpose and basic features of each method of combat-force employment. Naval tactical actions and major naval operations differ principally in terms of the objectives to be accomplished, size and mix of forces, and scope and duration of combat actions. Differences between major operations and campaigns are more than just cosmetic, particularly with respect to ways in which major operations and campaigns are planned, prepared and executed.

In a large narrow sea such as the Mediterranean, it has usually been very difficult to obtain permanent and absolute control of the sea in modern times. In a typical narrow sea, in contrast, full control of the sea surface and to a lesser degree of the subsurface can be attained relatively easily. Control of a typical narrow sea requires close cooperation among all the services of the country's armed forces, and here control of the sea is most directly affected by the situation on the land front and in the air.

Control of the sea on the open ocean is obtained primarily by destroying or at least neutralizing a major part of the enemy's naval forces.

In a narrow sea, in contrast, this objective can be accomplished also by seizing a large stretch of, or even the entire length of, the enemy coast along with naval and air bases and ports, choke points, and the key offshore islands. The accomplishment of that objective encompasses not only independent actions by one's naval forces, but also operations by ground troops along the coast and on the offshore islands.

Blockade of the exits from a larger narrow sea such as the Mediterranean is essentially a form of distant blockade. However, blockade of a small enclosed sea such as the Red Sea or the Adriatic is basically a form of close blockade. Blockade of straits, narrows, gulfs, or bays was extensively used in both world conflicts. The main prerequisite to secure blockade of a strait or narrows is to seize (preferably from the land side) all or most enemy naval bases in its proximity. Air power should be applied in mass and consistently over time so as not to allow the opponent to recover from the losses. In contrast to war on the open ocean, projecting power ashore in a narrow sea has sometimes brought more decisive results than actions against the enemy naval forces. The reason for this is that actions against the enemy shore are the ultimate objective of a war at sea, while the destruction of the enemy fleet, or at least a major part of it, creates only the prerequisite to resolve the war's ultimate objectives.

For a blue-water navy, projecting power ashore within a narrow sea is not the main task, but only one of the tasks in its overall power projection capabilities. Projecting power ashore essentially consists of striking or seizing a wide range of objectives on the enemy coast. This capability is the greatest asset a sea power can possess, and without it, the utility of sea power is considerably reduced.

In a narrow sea a stronger navy can sometimes find itself contesting command of the sea with a much weaker navy that occupies more advantageous geostrategic positions. A weaker navy can sometimes blockade the sea's only exit or the exit from the basing area. Methods of contesting the control of the sea on the vast expanses of an ocean and in a typical narrow sea differ considerably. This is principally due to the differences in the size and features of the physical environment. On the open ocean, and even in a large narrow sea such as the Mediterranean or the Caribbean, an inferior fleet can successfully contest command by the stronger fleet. This is a much more formidable task in a narrow sea. A stronger navy can relatively easily blockade the sea's only exit and exercise strategic control of a given enclosed sea. This task is more complicated in a semi-enclosed sea, such as the North Sea or the Yellow Sea, because of the relatively open passages to the waters of an ocean. To exercise operational or tactical control, a blue-water navy must operate within the confines of a narrow sea, and that in turn will provide even a small coastal force with the opportunity to challenge a blue-water navy. Land-based aircraft, small surface combatants, and mines are ideal means for contesting the control of the stronger opponent at sea. Also, a country guarding one or, even better, both shores of a strategically important waterway can use all the services of its armed forces to make it difficult, if not impossible, for a blue-water navy to obtain any degree of real control within a given narrow sea.

The tasks of a navy contesting control of the sea with a stronger navy range from a 'fleet-in-being', to breaking through a naval blockade, to commerce raiding or defensive mining. In addition, defense of a coast and adjacent waters, defense of a strait or narrows, and anti-amphibious defense are also component tasks in contesting control of the sea. This is one of the important distinctions between strategy on the open ocean and in a narrow sea. Also, both the weaker and stronger navy in a typical narrow sea can defend their coast regardless of the strategic situation in a given maritime theater.

Land-based air is today perhaps the most effective means of attacking elements of enemy maritime trade ashore. A high degree of readiness and maneuverability enable aircraft to concentrate their strikes against transports, warships, or aircraft covering maritime convoys. The small expanse of the area, dependence on relatively fixed routes, and the proximity of the coast in enclosed seas adversely affect a force's maneuverability and consequently its ability to avoid strikes by an enemy's aircraft or submarines.

The strategic situation on land in a typical narrow sea considerably affects the defense and protection of one's own and friendly shipping. The withdrawal or retreat of one's troops may result in the loss of airfields from which air cover can be provided. And the opposite is equally true: success on land enhances the protection of one's shipping.

In contrast to war on the open ocean, a navy operating in a narrow sea is called upon to carry out diverse and multiple tasks, such as supporting troops operating in the coastal area. Here, naval forces and land- or carrier-based aircraft can provide support to the troops ashore engaged in either offensive or defensive operations. Offensive operations are possible after one's fleet, land-based air, or ground troops, acting either independently or in combination, have obtained at least local control of the sea. Support by naval forces of defensive operations by one's ground forces is conducted when the opponent holds control of the sea.

Support of the maritime flank of one's army is perhaps one of the most difficult tasks for one's naval forces to carry out successfully. These tasks require a thorough understanding of the capabilities of each service at every command echelon. Success is achieved by coordinating the actions of all participating forces with respect to the assigned task, the place, the time, and the method of combat force employment. This seems a

CONCLUSION

simple thing to do, but has repeatedly proven difficult to accomplish in practice.

A fleet best suited to accomplish the tasks in narrow seas is not one designed to fight on the open ocean. A small riparian state in a typical narrow sea should possess sufficient naval strength to protect its economic interests and legal regime in peacetime, and to defend its coast from enemy invasion in wartime. However, in most cases small coastal navies must rely on much stronger out-of-the area fleets for their maritime protection.

Failure to build a balanced navy has in the past created many difficulties in accomplishing strategic objectives in a war at sea. A navy for a narrow sea should be composed optimally of coastal diesel-electric submarines, light forces, amphibious craft, mine warfare craft, and land-based fixed- and rotary-wing aircraft. Such a navy should be composed of a large number of both multi-purpose and specialized ships and craft. Full use should be made of craft with unconventional hulls and stealth technologies. Large surface combatants, specifically aircraft carriers, cruisers, destroyers, and large amphibious ships and nuclear-powered submarines should not be employed for tasks that can be carried out by smaller and lower-cost surface ships/craft and submarines. In other words, a blue-water navy should avoid using large, high-capability and high-cost ships for routine tasks that can be carried out more effectively and more cheaply by smaller, less capable but less costly ships.

The 500 to 1,000-ton conventional attack submarines are ideally suited for employment in narrow seas and shallow waters such as the Baltic or the Arabian Gulf. They can be employed to carry out a wide range of tasks, from surveillance/reconnaissance off the enemy coast, and attacks by cruise missiles and torpedoes against enemy warships and merchant vessels, to minelaying, and carrying special force teams into action. Modern conventional submarines fitted with state-of-the art sensors and weapons can outperform SSNs in some situations, because of the diesel's inherently quieter noise signature and smaller size. The new unconventional propulsion systems such as air-independent propulsion (AIP) allow submarines to operate submerged for three to four weeks.

Surface combatants are indispensable for control of a typical narrow sea. They are capable of much longer continuous operations than aircraft and are far more flexible than land-based weapon systems. A blue-water navy operating in a typical narrow sea should not use surface combatants larger than 2,000-tons. The optimal force mix for operations in narrow seas are groups composed of light frigates, multi-purpose corvettes, and combat craft. The 1,200–1,500-ton light frigates with combined diesel and gas (CODAG) propulsions can have ranges of up to 5,000 nautical miles and speeds of up to 30 knots. They can be armed with a variety of weapons and associated sensors such as anti-ship missiles and SAMs, ASW torpedoes, and light multi-purpose guns. The 500–1,000-ton corvettes with combined diesel or gas (CODOG) propulsion have smaller endurance than light frigates but their maximum speed is over 30 knots. Because of their size they can carry a larger number and a more diverse range of weapons and sensors. They can be armed with several launchers for anti-ship missiles, multipurpose guns, and ASW weapons and may be fitted with extensive electronic warfare suites. Corvettes have much better seakeeping capabilities and greater endurance than combat craft. Both frigates and corvettes can be fitted with a helicopter pad. They can be designed as multipurpose, anti-surface, ASW or air-defense platforms. Specifically, they can be used for patrolling, defense of shipping, carrying out strikes against enemy surface combatants and merchant vessels, and for providing fire support for the troops ashore.

The 100–400 ton combat craft form perhaps the optimal platforms for carrying out offensive and defensive tasks in a typical narrow sea. They can be designed primarily for strike warfare, anti-submarine warfare or special forces support. Their small size, high speed, excellent maneuverability and high combat power makes them ideal for employment in coastal waters and especially along the archipelago type of coast. Their range is generally small but more than adequate for operations in a typical narrow sea. The endurance of combat craft varies from one to five days, and their typical combat mission lasts from six to eight hours. They are especially well suited for attacks on enemy merchant shipping and the protection of one's own and friendly shipping in coastal waters.

A typical enclosed sea is an ideal place for the use of fixed-wing aircraft and helicopters. Their short flying time allows longer periods of time in which to attack a naval target and to attain tactical surprise by approaching the target area at low altitude. The range of modern ASMs allows aircraft to make a standoff attack without being observed by enemy surface ships. The side with weaker air forces will be forced to operate its surface ships at night. Aircraft are the only combat arm capable of challenging the enemy's control of straits. The striking force of land-based air in a typical narrow sea should be fighter-bombers armed with ASMs with a range exceeding that of the SAMs of enemy surface ships. Maritime reconnaissance aircraft should be able to detect small surface targets by day and night on the open seas, in bases, or in ports, and in straits in bad weather. The unmanned aerial vehicles (UAVs) are ideal platforms for providing surveillance, reconnaissance, target acquisition, battle damage assessment (BDA), and electronic warfare. They can be launched and retrieved from the coast or carried by light frigates or even corvettes.

Narrow seas and other restricted waters are ideal for the employment of multi-purpose combat craft. These can be used for a variety of duties ranging from fishery protection, anti-smuggling, EEZ patrol, to sea denial and possibly interdiction. They are capable of defending themselves against much larger and better armed counterparts. Forces for defensive tasks should be composed of ASW corvettes and light craft, MCM ships and light craft, offshore patrol vessels (OPVs), and patrol craft. The OPVs generally displace from 500–1,500 tons, with maximum speeds of 20–25 knots. The emphasis is on sea-keeping qualities and autonomy rather than speed. These vessels are relatively lightly armed, but provision is usually made for fitting stronger armament and electronics. A major requirement for any OPV is to have facilities for helicopter operations, even if no permanent hangar is fitted.

MCM ships and light craft are costly, and hence smaller navies with limited budgets cannot acquire them. Such navies should instead rely on a larger number of craft of opportunity (COOP) fitted with containerized MCM systems. However, MCM ships fitted with light armament can be used for low risk patrol duties such as those required by the EEZ. They are also ideal for carrying out survey work or even oceanographic research.

Support forces should be composed of medium and small amphibious ships, landing ships and Roll on–Roll off (Ro/Ro) vessels, small underway-replenishment ships, harbor tankers, rescue ships, repair ships, and various auxiliary and service craft. The logistic/amphibious ships are favored by many smaller navies, because such vessels provide the only means to land troops on hostile shores. Many small countries also have poor land communications, and logistic vessels provide a readily available means of communications between remote sites with only limited shore facilities.

Blue-water navies need to pay more attention to the conduct of naval strategy and operations in narrow seas. Obviously, there is a world of difference between operating in the open ocean and in the waters much closer to the continental landmass. Enclosed and semi-enclosed seas present an especially difficult environment for the employment of large surface combatants and nuclear-submarines, but these waters also present opportunities for blue-water navies as well. The history of naval warfare in both world wars and in many regional conflicts fought since 1945 shows that a navy capable of carrying out the entire spectrum of operational tasks must be a balanced force. A blue-water navy may be unable, owing to either budgetary constraints or domestic political reasons, to build in peacetime the force of surface ships/craft and submarines required to operate in narrow seas in case of a major regional conflict or general war. However, at least a small number of such ships should exist to develop and test tactical and operational concepts and provide necessary training for the crews. Without such a modest force of naval combatants best suited to operate in narrow seas, a blue-water navy would also have great difficulties in developing theory and sound doctrine to operate in such waters, a task that cannot be improvised at short notice when the need arises. Nor can a blue-water navy rely on smaller friendly navies to provide such capability in time of crisis or major conflict. Very often, a blue-water navy will be forced to act alone, at least in the initial phase of a conflict, or if its allies prove unwilling to provide the necessary support. Perhaps the most serious problem is that, without a force of small combatants in peacetime, a blue-water navy may neither be mentally prepared to conduct combat in narrow seas nor may integrated operational and tactical concepts have been devised.

Select Bibliography

Books

Aker, Frank, *October 1973: The Arab–Israeli War* (Hamden, CT: Arcon Books, 1985).
Aston, George, *Sea, Land and Air Strategy* (London: John Murray, 1914).
Bacon, Reginald, H. S. and Francis E. McMurtrie, *Modern Naval Strategy* (London: Frederick Muller, 1941).
Baum, Walter and Eberhard Weichold, *Der Krieg der 'Achsenmaechte' im Mittelmeer-Raum* (Goettingen: Musterschmidt Verlag, 1973).
Bernhardi, Friedrich von, *On War of To-Day*, trans. Karl Donat, 2 vols (London: Hugh Rees, 1913).
Bragadin, Marc' Antonio, *The Italian Navy in World War II* (Annapolis, MD: Naval Institute Press, 1957).
Brodie, Bernard, *A Layman's Guide To Naval Strategy* (Princeton, NJ: Princeton University Press, 4th edn, 1958).
——, *Sea Power in the Machine Age* (Princeton NJ: Princeton University Press, 2nd edn, 1943).
——, *A Guide to Naval Strategy* (Princeton, NJ: Princeton University Press, 4th edn, 1958).
Cagle, Malcolm, W. and Frank A. Mason, *The Sea War in Korea* (Annapolis, MD: United States Naval Institute, 1957).
Churchill, Randolph and Winston S. Churchill, *The Six Day War* (Boston, MA: Houghton Mifflin, 1967).
Colomb, Philip H., *Naval Warfare: Its Ruling Principles and Practice Historically Treated* (Annapolis, MD: Naval Institute Press, 1991).
Corbett, Julian S., *England in the Seven Year War: A Study in Combined Strategy* (London: Longmans, Green, 1918).
——, *Some Principles of Maritime Strategy* (Annapolis, MD: Naval Institute Press, 1988; originally published 1911, Longmans, Green, London).
Corbett, Julian S. and Henry Newbolt, *Naval Operations: History of the Great War Based on Official Documents* (London: Longmans, Green, 1920–31), 5 vols.

Vol. 1: *To the Battle of the Falklands, December 1914* (London: Longmans, Green, 1920).
Vol. 2: *From the Battle of the Falklands to the Entry of Italy into the War in May* 1915 (London: Longmans, Green, 1921).
Vol. 3: *The Dardanelles Campaign* (London: Longmans, Green, 1923).
Newbolt, Henry, Vol. 4: *June 1916 to April 1917* (London: Longmans, Green, 1928).
——, Vol. 5: *From April 1917 to the End of the War* (London: Longmans, Green, 1931).
Cordesman, Anthony H., *The Iran–Iraq War and Western Security 1984–87: Strategic Implications and Policy Options* (London: Jane's Publishing, 1987).
Cowie, J. S., *Mines, Minelayers and Minelaying* (London/New York: Oxford University Press, 1949).
Creswell, John, *Sea Warfare 1941–1945* (Berkeley, CA: University of California Press, rev. edn, 1967).
De Belot, Raymond, *The Struggle for the Mediterranean 1939–1945*, trans. James A. Field Jr (Princeton, NJ: Princeton University Press, 1951).
Department of Defense, *Conduct of the Persian Gulf War: Final Report to Congress* (Washington, DC: US Government Printing Office, April 1992).
Derry, T. K., *The Campaign in Norway* (London: Her Majesty's Stationery Office, 1952).
Dull, Paul S., *A Battle History of the Imperial Japanese Navy (1941–1945)* (Annapolis, MD: Naval Institute Press, 1978).
Duncan, Robert C., *America's Use of Mines* (White Oak, MD: US Naval Ordnance Laboratory, 1962).
Ehrman, John P. W., gen. ed., *Grand Strategy* (London: Her Majesty's Stationery Office, 6 vols, 1956–76).
Vol. 1: *Rearmament Policy*, ed. N. H. Gibbs.
Vol. 2: *Sept. 1939–June 1941*, ed. J. R. M. Butler.
Vol. 3: *June 1941–Aug. 1942*, ed. J. M. A. Gwyer and J. R. M. Butler.
Vol. 4: *Aug. 1941–Sept. 1943*, ed. M. Howard.
Vol. 5: *Aug. 1943–Sept. 1944*, ed. J. Ehrman.
Vol. 6: *6 Oct. 1944–Aug. 1945*, ed. J. Ehrman.
Field, James A., Jr, *History of US Naval Operations: Korea* (Washington, DC: Government Printing Office, 1962).
Giamberardino, Oscar di, *Seekriegskunst*, trans. E. Mohr (Berlin: Verlag 'Offene Worte', 1938).
Goltz, Colmar von der, *The Conduct of War: A Short Treatise on Its Most Important Branches and Guiding Rules*, trans. G. F. Leverson (London: Kegan Paul, Trench, Trubner & Co./New York, NY: E. P. Dutton & Co., 1917).
Groos, Otto, *Seekriegslehren im Lichte des Weltkrieges: Ein Buch fuer den Seemann, Soldaten und Staatsmann* (Berlin: E. S. Mittler & Sohn, 1929).

SELECT BIBLIOGRAPHY

Grove, Eric, J., *The Defeat of the Enemy Attack on Shipping 1939–1945* (A revised edition of the Naval Staff History Volumes 1A (Text and Appendices) and 1B (Plans and Tables) (Aldershot: Ashgate, the Navy Records Society, 1997).

Gundelach, Kurt, *Die Deutsche Luftwaffe im Mittelmeer 1940–45* (Frankfurt, a.M: Peter Lang, 1981).

Herzog, Chaim, *The Arab–Israeli War* (New York: Vintage Books, 1983).

——, *The War of Atonement, October, 1973* (Boston: Little, Brown, 1975).

Hezlet, Arthur, *Aircraft and Sea Power* (New York: Stein & Day, 1970).

Jomini de, Antoine, *The Art of War*, trans. G. H. Mendell and W. P. Craighill (Philadelphia, PA: J. B. Lippincott, 1862; repr. Westport, CT: Greenwood Press, 1971).

Kennedy, Paul M., *The Rise and Fall of British Naval Mastery* (London: Macmillan, 1983; repr. 1985).

Lambi, Ivo Nikolai, *The Navy and German Power Politics, 1862–1914* (Boston: Allen & Unwin, 1984).

Ledebur, Gerhard F. von, *Die Seemine* (Munich: J. F. Lehmans Verlag, 1977).

Liddell Hart, B. H. *The Real War* (Boston: Little, Brown, 1st edn 1930, repr. 1964).

Mackesy, Piers, *The War in the Mediterranean 1803–1810* (London/New York: Longmans, Green, 1957).

Macintyre, Donald, *Sea Power in the Pacific: A History from the 16th Century to the Present Day* (London: Military Book Society, 1972).

Mahan, Alfred T., *Naval Strategy Compared And Contrasted With The Principles And Practice Of Military Operations On Land* (Boston, MA: Little, Brown, 1919).

——, *The Influence of Sea Power upon History, 1660–1783* (Boston: Little, Brown, 12th edn, 1918).

——, *Naval Strategy* (Westport, CT: Greenwood Press, 1972).

Marcus, G. J., *A Naval History of England*, Vol. 1: *The Formative Years* (Boston, MA: Little, Brown, 1961); Vol. 2: *The Age of Nelson: The Royal Navy 1793–1815* (New York: Viking Press, 1971).

Marder, Arthur J., *From the Dreadnought to Scapa Flow: The Royal Navy in the Fisher Era, 1914–1919*, 5 vols (London: Her Majesty's Stationery Office [HMSO], 1961–70).

Vol. I: *The Road to War, 1904–1914* (London: HMSO, 1961).

Vol. II: *The War Years: To the Eve of Jutland, From the Dreadnought to Scapa Flow: The Royal Navy in the Fisher Era, 1904–1919* (London: Oxford University Press, 1965).

Vol. IV: *1917: Year of Crisis* (London: Oxford University Press, 1969).

Vol. V: *Victory and Aftermath (January 1918–June 1919)* (London: Oxford University Press, 1970).

Marolda, Edward J. and Wesley J. Price III, *A Short History of the United States Navy and the Southeast Asian Conflict 1950–1975* (Washington, DC: Naval Historical Center, 1984).

Maurice, Frederick B., *Principles of Strategy: A Study of the Application of the Principles of War* (New York: Richard R. Smith, 1930).

McCartney, Eugene S., *Warfare by Land and Sea* (Boston, MA: Marshall Jones, 1923).

Meister, Juerg, *Der Seekrieg in den osteuropaeischen Gewaessern 1941–45* (Munich: J. F. Lehmans Verlag, 1958).

Monkhouse, Francis J., *Principles of Physical Geography* (London: University of London, 2nd edn, 1965).

Morison, Samuel E., *History of United States Naval Operations in World War II*, 15 vols (Boston, MA: Little, Brown, repr. 1984).

——, *The Two-Ocean War: A Short History of the United States Navy in the Second World War* (Boston, MA/Toronto: Little, Brown, 1963).

Mossman, Billy C., *Ebb and Flow: November 1950–July 1951*, United States Army in the Korean War series (Washington, DC: Center For Military History, 1990).

Navias, Martin S. and E. R. Hooton, *Tanker Wars: The Assault on Merchant Shipping during the Iran–Iraq Conflict, 1980–1988* (London/New York: I. B. Tauris, 1996).

O'Ballance, Edgar O., *The Third Arab–Israeli War* (Hamden, CT: Archon Books, 1972).

Pavlovich, N. B., *The Fleet in the First World War*, Vol. 1: *Operations of the Russian Fleet* (Moscow, Voyenizdat, 1964), trans. from Russian; published for the Smithsonian Institution and the National Science Foundation, Washington, DC by Amerind Publishing, New Delhi, 1979).

Pelier, Louis C. and G. Etzel Pearcy, *Military Geography* (Princeton, NJ: D. Van Nostrand, 1966).

Playfair, I. S. O., F. C. Flynn, C. J. C. Molony and S. E. Toomer, *The Mediterranean and Middle East*:

Vol. I: *The Early Successes against Italy (to May 1941)* (London: Her Majesty's Stationery Office [HMSO], 1954).

Vol. II: *The Germans Come to the Help of their Ally (1941)* (London: HMSO, 1956).

Vol. III: *British Fortunes reach their Lowest Ebb* (London: HMSO, 1960).

Vol. IV: *The Destruction of the Axis Forces in Africa* (London: HMSO, 1966).

Potter, Elmer B., *The Great Sea War: The Story of Naval Action in World War II* (Englewood Cliffs, NJ: Prentice-Hall, 1960).

Potter, Elmer B. and Chester W. Nimitz, eds, *Seemacht: Eine Seekriegsgeschichte von der Antike bis zur Gegenwart* (Herrsching: Manfred Pawlak, rev. edn, 1986).

——, eds, *Sea Power: A Naval History* (Englewood Cliffs, NJ: Prentice-Hall, 1960).

Richmond, Herbert, *National Policy and Naval Strength and Other Essays* (London: Longmans, Green, 1928).

——, *Naval Warfare* (London: Ernest Benn, 1930).

―, *Sea Power in the Modern World* (London: G. Bill & Sons, 1934; repr. New York: Arno Press, 1972).

―, *Statesman and Sea Power* (Oxford: Clarendon Press, 1st edn, 1946; repr. 1947).

Robinson, Oliver P., *The Fundamentals of Military Strategy* (Washington, DC: US Infantry Association, 1928).

Robertson, Horace B., Jr, ed., *The Law of Naval Operations* (Newport, RI: Naval War College, 1991).

Rodgers, William L., *Naval Warfare under Oars 4th to 16th Century: A Study of Strategy, Tactics and Ship Design* (Annapolis, MD: Naval Institute Press, 1967).

Rohden, Peter R., *Seemacht und Landmacht: Die Gesetze ihrer Politik und ihrer Kriegfuehrung* (Leipzig: Wilhelm Goldmann Verlag, 1942).

Roskill, Stephen, W., *The War At Sea 1939–1945*, 3 vols (London: Her Majesty's Stationery Office, 1954–61).

Vol. I: *The Defensive* (London: HMSO, 1954).

Vol. II: *The Period of Balance* (London: HMSO, 1956).

Vol. III, *The Offensive*, pt. 1, 1st June 1943–31st May 1944 (London: HMSO, 1960).

Vol. III: *The Offensive*, pt. 2: 1st June 1944–14th August 1945 (London: HMSO, 1961).

Ruge, Friedrich, *Sekerieg: The Germany Navy's Story 1939–1945*, trans. M. G. Saunders (Annapolis, MD: Naval Institute Press, 1957).

―, *The Soviets as Naval Opponents 1941–1945* (Annapolis, MD: Naval Institute Press, 1979).

Salewski, Michael, *Die deutsche Seekriegsleitung 1935–1945*, 2 vols (Frankfurt a.M: Bernard & Graefe, 1970 and 1975).

Vol. 1: *1935–1941* (Frankfurt a.M: Bernard & Graefe, 1970).

Vol. 2: *1942–1945* (Frankfurt a.M: Bernard & Graefe, 1975).

Sallagar, Frederick M., *Lessons from an Aerial Mining Campaign* (Santa Monica, CA: Rand Corporation, April 1974).

Schoefield, B. B., *The Russian Convoys* (Philadelphia, PA: Dufour Editions, 1964).

Selburn, P. A., *The Evolution of Sea Power* (London: Longmans, Green, 1919).

Sheppard, Gilbert A., *The Italian Campaign 1943–45* (New York: Frederick A. Praeger, 1968).

Simpson, Mitchell B. III, ed., *The Development of Naval Thought: Essays by Herbert Rosinski* (Newport, RI: Naval War College Press, 1977).

Smith, Peter C., *Hold the Narrow Sea: Naval Warfare in the English Channel 1939–1945* (Ashbourne, Derbyshire: Moorland Publishing, and Annapolis, MD: Naval Institute Press, 1984).

Sokol, Anthony, E., *Seapower in the Nuclear Age* (Washington, DC: Public Affairs Press, 1961).

Sondhaus, Lawrence, *The Habsburg Empire and the Sea: Austrian Naval Policy 1797–1866* (West Lafayette, IN; Purdue University Press, 1989).

Sprout, Harold M. and Margaret T. Sprout, *The Rise of American Naval Power 1776–1918* (Princeton, NJ: Princeton University Press, 1939).
Stenzel, A. and Hermann Kirchoff, *Seekriegsgeschichte in ihrem wichtigste Abschnitte mit Beruecksichtigung der Seetaktik*, 6 vols (Hannover: Hahn, 1907–11).
——, *Kriegfuehrung zur See: Lehre vom Seekriege* (Hannover/Leipzig: Hahn, 1913).
Stevens, William, O. and Allan Westcott, *A History of Sea Power* (Garden City, NY: Doubleday, 1929).
Till, Geoffrey, ed., *Maritime Strategy and the Nuclear Age* (New York, NY: St Martin's Press, 2nd edn, 1984).
Tracy, Nicholas, *Attack on Maritime Trade* (Toronto: University of Toronto Press, 1991).
Tschischwitz, Erich von, *The Army and Navy during the Conquest of the Baltic Islands in October 1917* (Berlin: Eisenschmidt, 1931).
Uhlig, Frank, Jr, *How Navies Fight: The US Navy and Its Allies* (Annapolis, MD: Naval Institute Press, 1994).
Wegener, Wolfgang, *The Naval Strategy of the World War*, trans. Holger Herwig (Annapolis, MD: Naval Institute Press, 1989).
Westcott, Allan, *Mahan on Naval Warfare: Selections from the Writings of Rear-Admiral A. T. Mahan* (Boston, MA: Little, Brown, 1918).
——, ed., *American Sea Power since 1775* (Chicago, IL: J. B. Lippincott, 1947).
Whetton, Frederick, E., *Moltke* (London: Sampson, Low, Marston, 1936).
Wiese, Andreas, *Kleine Seemachtstheorie* (Herford: Koehlers Verlagsgesellschaft, 1968).
Woodward, David, *The Russians at Sea: A History of the Russian Navy* (New York: Frederick A. Praeger, 1966).
Wuest, Georg, *Kleine Wehrgeographie des Weltmeeres* (Munich: Verlag E. S. Mittler & Sohn, 1938).
Wylie, J. C., *Military Strategy: A General Theory of Power Control* (New York: Rutgers University Press, 1967).
Zehrer, Hartmut, *Der Golfkonflikt: Dokumentation, Analyse und Bewertung aus militaerischer Sicht* (Herford/Bonn: Verlag E. S. Mittler & Sohn, 1992).

Monographs

Behrendt, Hartmut, *Die Handlungsfreiheit der militäerischen Führung– Möeglichkeiten und Grenzen aufgrund des heutigen Kriegsbildes* (Hamburg: Fuehrungsakademie der Bundeswehr, January 1968).
Bowden, Scott, *Forward Presence, Power Projection, and the Navy's Littoral Strategy* (Arlington, VA: IRIS Corporation, July 1997).
Hanschermann, Volk, *Die seestrategische Bedeutung der Ostseeausgaenge in*

der Geschichte (Hamburg: Fuehrungsakademie der Bundeswehr, March 1960).

Heinstein, Rudolf, *Zur Strategie des Mehrfrontenkrieges: Das Problem der 'inneren und äusseren Linie' dargestellt am Beispiel des Erste Weltkrieges* (Hamburg: Fuehrungsakademie der Bundeswehr, November 1975).

Hollifield, Carlos R., *Littoral Warfare: The Transformation of US Naval Strategy To Meet the 21st Century* (Carlisle Barracks, PA: US Army War College, 1996).

Noeske, Rolf, *Die Bedeutung des Ostseekriegschauplaetzes fuer den Ausgang des Ersten Weltkrieges* (Hamburg: Fuehrungsakademie der Bundeswehr, October 1974).

O'Brien, John, *Coup d'Oeil: Military Geography and the Operational Level of War* (Ft Leavenworth, KS: School of Advanced Military Studies, US Army Command and General Staff College, 1991).

Robbins, Thomas H., *The Employment of Naval Mines* (Newport, RI: Naval War College), 20 March 1939.

Rodriguez, Joseph, Jr, *How to Maximize the Advantages of Interior Lines at the Operational Level* (Ft Leavenworth, KS: School of Advanced Military Studies, US Army Command and General Staff College, 1987).

Seebens, Dieter, *Grundlagen: Auffassungen und Plaene fuer eine Kriegfuehrung in der Ostsee 1935–1939* (Hamburg: Fuehrungsakademie der Bundeswehr, August 1971).

Vanderbeek, Walter A., *The Decisive Point: The Key to Victory* (Ft Leavenworth, KS: School of Advanced Military Studies, US Army Command and General Staff College, 1991).

Articles

Arthur, Stanley R., and Martin Pokrant, 'Desert Storm At Sea', *Proceedings/Naval Review 1991*.

Almog, Ze'ev, 'Israel's Navy Beat the Odds', *Proceedings*, 3 (March 1997), pp. 106–7.

Ambrosius, H., 'Das britische Kuestenvorfeld als deutsche Blokadezone', *Wissen und Wehr*, 9 (September 1941), pp. 330–4.

Blumenson, Martin, 'Beyond the beaches', *Military Review*, 9 (September 1962).

Breuning, 'Die englische Blockadestellung in der noerdlichen Nordsee im Kriege 1914 bis 1918', *Marine Rundschau*, 9 (September 1939).

Brown, C. R., 'The Principles of War', *Proceedings*, 6 (June 1949).

Cagle, Malcolm W., 'Wonsan: The Battle of the Mines', *Proceedings*, 6 (June 1957).

Connolly, Richard L., 'The Principles of War', *Proceedings*, 1 (January 1953).

Conrady, Dietrich von, 'Die Kriegsmarine und der letzte Kampf um

Sewastopol im Mai 1944', *Wehrwissenschaftliche Rundschau*, 6 (June 1961).

Ditten, Meuffling von, 'Kriegerische Landungen', *Wissen und Wehr*, 3 (March 1929), pp. 167–82.

Donner, P., 'Ueber Kuestenkrieg', *Marine Rundschau*, 8 (August 1928), pp. 338–45.

——, 'Die Seekriegfuehrung im begrenzten Raum', *Nauticus 1939*, pp. 84–98.

Dowdy, William L., 'Naval Warfare in the Gulf: Iraq versus Iran', *Proceedings*, 6 (June 1981).

Ebert, Paul, ' Natuerliche Abwehrkraefte der Kueste', *Marine Rundschau*, 6 (June 1928).

Editors, 'Die Sperre der Otranto-Strasse 1917–18', *Marine Rundschau*, 10 (October 1933) pp. 449–58.

Engelmann, H., 'Die Sicherstellung von Seeoperationen', *Militaerwesen* (East Berlin), 3 (March 1980).

Filonov, S., 'Morskaya Operatsiya' (A Naval Operation), *Morskoy Sbornik* (Moscow), 10 (October 1977).

Fock, Harald, 'Seestreitkraefte in Kuesten- und Randmeergebieten (I), *Wehrkunde*, 11 (November 1974), pp. 579–85; (II) *Wehrkunde*, 12 (December 1974), pp. 630–4.

Frei, Daniel, 'Neutralisierte Zonen: Versuch einer strategischemacht-politischen Theorie', *Wehrwissenschaftliche Rundschau*, 12 (December 1969).

Fuchs, Hans, 'Die Diversion als strategisches Mittel zur Erzielung eines Kraefteausgleiches, dargelegt an geschichtlichen Beispielen', *Marine Rundschau*, 4 (April 1938), pp. 235–51.

Gadow, 'Strategische Verteidigung', *Wissen und Wehr*, 2 (February 1936).

——, 'Die seestrategischen Leitgedanken der schwachen Seemaechte', *Marine Rundschau*, 7 (July 1926).

——, 'Flottenstuetzpunkte', *Militaerwissenschaftliche Rundschau*, 4 (April 1936), pp. 511–32.

Groos, Otto, 'Die strategische Bedeutung des Aermelkanals in der Geschichte und Gegenwart', *Marine Rundschau*, 7 (July 1940).

Groote, Wolfgang von, 'Land- und Seemacht an den Meerengen: Ein militaerischer Ueberblick', *Wehrkunde*, 1 (January 1964), pp. 20–5.

Handel-Mazzetti, Peter, 'Kuestenangriff und Kuestenverteidigung', *Marine Rundschau*, 5 and 6 (May and June 1936), pp. 213–19.

Hansen, 'Von Seeschlachten und einigen ihrer Lehren', *Marine Rundschau*, 9 (September 1934).

Hoffman, Roy F., 'Offensive Mine Warfare: A Forgotten Strategy', *Naval Review*, 1977.

Huan, Claude, 'Die sowjetische Landungs-Operationen auf der Krim 1941–42', *Marine Rundschau*, 6 (June 1962).

Huemmelchen, Gerhard, 'Unternehmen "Eisstoss": Der Angriff der Luftflotte

1 gegen die russische Ostseeflotte im April 1942', *Marine Rundschau*, 4 (April 1959), pp. 226–32.
Jablonsky, Walter, 'Die Seekriegfuehrung im vierten Nahostkrieg', *Marine Rundschau*, 11 (November 1974), pp. 645–65.
——, 'Die Seekriegfuehrung im Vierten Nahostkrieg aus aegyptischer Sicht', *Marine Rundschau*, 7 (July 1978), pp. 441–7.
Kempf, Thomas, 'Der britische Rueckzug aus Kreta im Mai 1941', *Truppenpraxis*, 7 (July 1981).
Kupfer, 'Die strategische Verteilung der Hauptflotten im Hinblick auf ihre Friedens- und Kriegsaufgaben', Marine Rundschau, 6 (June 1936), pp. 289–301.
LaRoque, Alexander, 'The Role of Geography in Military Planning', *Canadian Army Journal* (July 1955).
Liddell Hart, B. H., 'The Value of Amphibious Flexibility and Forces', *Journal of the Royal Services Institution*, Vol. CV, 620 (November 1960).
Lorey, 'Der Minenkrieg im Schwarzen Meer: Eine kritische Betrachtung', *Marine Rundschau*, 5 (May 1933), pp. 210–19.
Maerz, Josef, 'Betrachtungen zur Stuetzpunktpolitik der Seemaechte', pt. 1, *Marine Rundschau*, 4 and 7 (April and July 1923).
Maltzahn, 'Das Dardanellenunternehmen: politisch, strategisch und taktisch', *Wissen und Wehr*, 1 (January 1921).
Monitor, Hans-Joerg, 'Neuzeitliche Kuestenverteidigung', *Wehrkunde*, 2 (February 1965), pp. 89–92.
Mundy, Carl E., 'Thunder and Lightning: Joint Littoral Warfare', *Joint Force Quarterly* (Spring 1994), pp. 45–50.
Murphy, Frank J., Jr, 'Littoral Warfare: Adapting to Brown-Water Operations', *Marine Corps Gazette* (September 1993), pp. 65–73.
Navoytsev, P., 'Zakonomernosti, soderzhaniye I Kharakternye Cherty Sovremennykh Morskikh Operatsii' (Regularities, Content, and Characteristics of Modern Naval Operations) *Morskoy Sbornik* (Moscow), 7 (July 1986).
'Newell, Clayton R., 'Levels of War', *Army*, 6 (June 1988).
——, 'What Is Operational Art?' *Military Review*, 9 (September 1990).
Olson, William J., 'The Iran–Iraq War and the Future of the Persian Gulf', *Military Review*, 3 (March 1984).
Palmer, Michael A., 'The Navy Did Its Job', *Proceedings/Naval Review 1991*.
Refaat, Ashraf M., 'How the Egyptian Navy Fought the October War', *Proceedings*, 3 (March 1995), pp. 94–7.
Rohwer, Juergen, 'Die sowjetische U-Bootwaffe in der Ostsee 1939–1945', *Wehrwissenschaftliche Rundschau*, 10 (October 1956), pp. 547–68.
——, 'Der Minenkrieg im Finnischen Meerbusen, Part I: June–August 1941', *Marine Rundschau*, 1 (January 1967), pp. 16–25; 'Part II: September–November 1941', *Marine Rundschau*, 2 (February 1967), pp. 94–102.
Ruge, Friedrich, 'Der Anteil der deutschen Seekriegfuehrung an der Niederzwingung Russlands', *Marine Rundschau*, 6 (November–December 1924).

——, 'Die Verwendung der Mine im Seekriege 1914 bis 1918. Ihre Erfolge und Misserfolge', *Marine Rundschau*, 6 (June 1927); *Marine Rundschau*, 7 (July 1927).

Schellmann, Holm, 'Die Aufgaben von Luftstreitkraeften im Kuestenvorfeld', *Wehrkunde*, 5 (May 1956), pp. 237–40.

Schulz, Otto, 'Das militaerische Mittelmeerproblem vom italienischen Standpunkt', *Wissen und Wehr*, 10 (October 1941).

Schulze, Ernst, 'Der Verlust der Insellage Englands', *Marine Rundschau*, 11 (November 1939).

Sell, Manfred, 'Der nordamerikanische Kampf um die innere Linie', *Marine Rundschau*, 1 (January 1932), pp. 22–6.

Shaw, James C., 'The Rise and Ruin of Rabaul', *Proceedings*, 6 (June 1951), pp. 625–9.

Sokol, Anthony E., 'Naval Strategy in the Adriatic Sea during the World War', *Proceedings*, 8 (August 1937).

——, 'Seapower in the Mediterranean 1940 to 1943', *Military Review*, 8 (August 1960).

Stange, Rudolf, 'Kuestenlaenge und Flottenstaerke', *Marine Rundschau*, 4 (April 1934), pp. 145–50.

Stewart, James, 'The Evolution of Naval Bases in the British Isles', *Proceedings*, 7 (July 1957).

Telem, Benyamin, 'Die israelischen FK-Schnellbooten im Yom-Kippur-Krieg', *Marine Rundschau*, 10 (October 1978), pp. 635–41.

Vego, Milan, 'Seekriegfuehrung in der Adria', *Marine Rundschau*, 8 (August 1979), pp. 485–98.

——, 'The Baltic Naval Operations', *NAVY International*, 2 (February 1983).

——, 'Shallow Water ASW: Sweden's Problems', *Jane's Defence Weekly*, 18 January 1986.

Vilovic', Tihomir, 'Specifičnosti priprema i izvodjenja operacija na uskim morima' (Peculiarities of Preparing and Conducting Operations in Narrow Seas), *Mornarički Glasnik* (Belgrade), 2 (March–April 1976).

Welsch, D., 'Die Sperrung der daenischen Ostsee–Zugaenge: Eine wenig bekannte Episode aus den ersten Augusttagen 1914', *Marine Rundschau*, 7 (July 1934).

Wolfram, Eberhard, 'Minen im Kattegat', *Marine Rundschau*, 11 (November 1934).

Zohar, Abraham, 'Die israelische Marine im Libanonkrieg 1982', *Marine Rundschau*, 2 (February 1985), pp. 97–101.

Index

Aaland (Åaland) Islands, 43, 76, 172, 186
ABDA (Australian–British–Dutch–American), 132 88, 132, 189
Aberdeen, 62
Aboukir (Abūqir), 216
Abū Mūsá, 55, 243
Abu Rueis, 242
action against an enemy fleet, 147–55
Adabia, 154
Aden, 7, 63
Adramyti (Edremit), Gulf of, 166
Adriatic Sea: and bases, 62, 65, 68, 69; and contesting command, 206, 213, 216, 217; and exercising command, 186, 187, 194, 198, 199, 200; and factors of space, 18, 19, 22, 23, 30, 31; and positions, 51; as scene of conflict, 9; and sea control, 112, 114, 125; and securing command, 149, 159, 164, 165, 168; and support of army flank, 285, 286; mentioned, 103, 295
advanced naval base, 65–6
Aegean islands, 124, 174; *see also* names of islands
Aegean Sea, 22, 31, 48, 50, 115, 122, 125, 166, 173, 174–5, 229, 234
Africa, 105, 113, 286; *see also* North Africa; names of places and countries
Afrika Korps, 48, 69, 228, 270
AFSOUTH (Allied Force South), 200
air power, 12, 26, 31, 32, 64, 71, 72, 122–7, 152–3, 154, 173, 188–9, 227–8, 230, 231, 233, 249, 258–9, 261–2, 263, 279, 293, 296, 298–9

Ajaccio, 26, 45
Al Fāw, 172, 232, 244, 245
Al Fāw Peninsula, 172, 244
Al Kalia, 153
Albania, 66, 69, 172, 277, 285
Albion, Operation, 187
Alborán, 51, 54
Alexandria, 45, 48, 63, 68, 69, 91, 105, 115, 118, 153–4, 195, 230, 234, 246, 255, 275
Algeria, 216
Algiers, 26, 45, 106
Allied Tactical Air Force (ATAF), 5th, 200
Allies: World War I, 76, 165, 166, 194–5, 213, 217, 220, 244, 277; World War II, 27, 33, 44, 48, 50, 55, 57, 62, 69, 71, 81, 82, 88, 98, 115, 118, 122, 123, 124, 125, 126, 132, 134, 135, 137, 143, 144, 151, 156, 188, 189–90, 192, 215, 219–20, 228, 229, 230, 233, 235, 239, 240, 247, 248, 255, 256, 262, 263, 268, 271, 272, 273, 277, 278–9, 282, 285, 286, 287
Alsen, 9
American Civil War (1861–65), 238, 267
American War of Independence (1776–83), 103
American–Spanish War (1898), 161, 187
amphibious landings, 32, 33, 141, 185–93
Anapa, 285
Anatolia, 272
Ancona, 186, 285

Anglo-Dutch Wars, 148
angular or salient base of operations, 77
Antwerp, 54, 158, 171, 235
Apennine Peninsula, 19
Apolda barrier, 170
Apulia, 159
Aqaba (Al 'Aqaba), 244, 248, 261
Aqaba, Gulf of, 119, 201, 242
Arab–Israeli conflicts, 11, 144, 196, 245, 269, 276; *see also* Six-Day War; Yom Kippur War
Arabian (Persian) Gulf: and attacks on maritime trade, 227, 234, 237, 243, 244, 248; and bases, 66, 68; and contesting command, 221; and exercising command, 196, 197, 201; and factors of space, 18, 22, 23, 29, 30, 31, 34, 38; and positions, 55; and protection of maritime trade, 255, 264; and sea control, 110–11; and securing command, 144, 154, 156, 172; and strategic objectives, 98, 102; and support of army flank, 269, 287; mentioned, 10, 53, 74, 297
Arctic, 5, 93, 255, 268
Argentinian navy, 201
Ark Royal, HMS, 195
army flank, support of: defensive major operations, 274–82; difficulty of, 296; importance of, 267–70; offensive defense, 282–7; offensive major operations, 271–3; prerequisites, 270–1
Army of the Nile, 275
Asalūyeh, 264
Asia, 17, 57, 247; *see also* South East Asia; West Asia; names of countries
Athenians, 208, 255–6
Athens, 48
Atlantic, 6, 16, 18, 19, 27, 43, 44, 47, 51, 53, 62, 67, 76, 77, 103, 105, 106, 150, 157, 163, 168, 169, 171, 177, 187, 209, 214, 217, 218, 248, 256, 257
attacks, 129; *see also* attacks on coast; attacks on maritime trade
attacks on coast, 193–6
attacks on maritime trade: combined arms, 233–4; commerce raiding, 248–9; commercial blockade, 237–46; commercial counter-blockade, 246–8; methods, forces and assets, 229–33; objectives and prerequisites, 225–9; offensive mining, 234–7
attenuated shape, 21
Attica, 277
attrition, 155–6
Australia, 105, 201, 287; Royal Australian Navy, 287
Austria, 8, 9, 17, 50, 65, 99, 100, 149, 172, 186, 187; Army, 186; Navy, 100
Austria-Hungary, 62, 66, 68, 69, 102, 103, 105, 165, 168, 172, 188, 194, 206, 213, 217, 239, 277
Austro-Italian War (1866), 186–7
Axis, 26, 47, 48, 69, 91, 113, 114, 115, 118, 119, 151, 172, 177, 178, 190, 219, 227, 229, 233, 234, 246, 247, 255, 261, 262, 263, 270, 271, 272–3, 275, 278, 279, 280, 284, 286
Azores, the, 54
Azov, Sea of, 5, 34, 53, 113, 177, 178, 278, 284, 285

Bab el-Mandeb Strait, 53, 55, 89, 119, 166, 201, 237, 242, 245
background noise, 37
Bacon, Admiral Reginald, 117
Baghdad, 196, 244
Bahrain, 66, 201
Balearic Islands, 51, 54, 158, 215
Balikpapan, Battle of (1942), 131–2
Balkan wars (1912–13), 105
Balkans, 192, 285
Baltic Fleet: Russian, 101, 103, 106, 107, 209; Soviet, 10, 47, 56, 62, 66, 100, 101, 102, 107, 113, 151, 152, 153, 169, 170, 176, 268
Baltic Islands, 176, 177, 187, 268, 277
Baltic Sea: as area of conflict, 8, 9; and attacks on maritime trade, 231, 234, 235, 236, 239, 246; and bases, 62, 63, 66; and contesting command, 206, 208, 211, 214, 220; and exercising command, 185, 187, 191; and factors of space, 17, 18, 19, 22, 23, 27, 28, 29, 30, 31, 33, 38, 39; and fleet distribution, 104, 106; and positions, 43, 47, 50, 51, 53, 57; and

312

INDEX

protection of maritime trade, 254, 256, 258, 259; and sea control, 112, 113, 115, 118, 120, 121, 122; and securing command, 150, 160, 162, 163, 166, 172, 175, 176, 177; and strategic objectives, 100, 101, 102; and support of army flank, 267, 268, 271, 274, 275, 276, 280, 283, 284, 286; and theater and its elements, 76, 77, 91, 92; mentioned, 5, 10, 98, 133, 144, 297
Baltic states, 47, 101
Baltic Straits, 43
Baltic–White Sea Canal, 121
Balṭīm, Battle of (1973), 151
Bandar Abbas (Bandar-e 'Abbas), 23, 243
Bandar Khomeini (Bandar-e Khomeyni), 23
Banias (Bāniyās), 195
Bantry Bay, Battle of (1689), 8
Bar, 198
Barbarossa, Plan, 10, 172
Barents Sea, 6, 18, 27, 29, 44, 107, 120
Bari, 198
base of operations, 75–80
bases, naval: attacks on, 153–4; characteristics, 63–5; defense of, 274–6; importance of, 61–2, 292; location, 67–72; prerequisite to protection of maritime trade, 256; purpose, 62–3; seizing, 165, 295; seizing from the land, 175–9; type, 65–7
Basra (Al Baṣrah), 23, 102, 154, 196, 232, 244, 245
Bastia, 279
battles, naval, 131–2, 147–50; *see also* names of battles
Batumi, 62, 231, 276
Baytown, Operation, 273
beach defenses, 221, 222
Beachy Head, Battle of (1690), 117, 148, 207
Bear Island, 29
Beirut, 26, 45, 69, 234
Belgium, 21, 23, 80, 118, 171, 201, 235, 245, 267
Belomorsk, 121
Bengal, Bay of, 54
Benghazi (Banghāzī), 91, 272
Bergen, 27, 44

Bering Sea, 5
Biscay, Bay of, 44, 76, 91, 230–1
Bismarck Archipelago, 125, 156
Bismarck Sea, 70
Bizerte (Bizerta), 45, 55, 68, 106
Black Sea: and attacks on maritime trade, 231; and factors of space, 17–18, 22, 34; and positions, 47, 53–4; as scene of conflict, 9; and sea control, 112, 114, 118, 120; and securing command, 170, 175, 177, 178; and strategic objectives, 103, 106; and support of army flank, 267, 268, 272, 274, 275, 278, 279, 280, 284, 287; mentioned, 133, 144, 185, 211, 219
Black Sea Fleet, 62, 100, 107, 151, 274, 275, 276, 281, 282
blockades, 28–9, 156–67, 213–15, 237–48, 272–3, 295
'Bomb Alley', 124, 230
bombardments of the coast, 194–6, 210
Bon, Cape, 48, 53
Bône (Annaba), 45, 55, 216, 272
Bonifacio Strait, 63, 279
Borneo, 88, 122, 132, 239
Bornholm, 21
Bosnia and Herzegovina, 199, 200, 201
Bosnian Muslims, 198
Bosnian Serbs, 200
Bosporus, 7, 103, 206, 219, 234; *see also* Turkish Straits
Bothnia, Gulf of, 23, 29, 30, 91
Bougainville, 81, 281
Bougie (Bejaïa), 55
Boulogne, 82, 118, 123, 172
Breslau, 216, 217
Brest, 19, 44, 86, 106, 158, 159, 160, 171, 214, 246
Brindisi, 65, 69
Bristol Channel, 44, 260
Britain: 8th Army, 69, 229, 263, 271, 273, 282; 10th Squadron, 245; Admiralty, 91, 104, 105, 123, 216, 217, 260; Army, 91–2; Anglo-Persian pipeline, 9; and attacks on maritime trade, 227–8, 229, 230, 231, 232, 233–4, 235, 236, 238, 239, 243, 244–5, 246, 247, 248; and bases, 61–2, 63, 66, 67–8, 69–70, 71; Bomber Command, 235; Channel

313

Fleet, 92; Chiefs of Staff, 48; Coastal Air Command, 169; Coastal Command, 227, 258; Coastal Convoy System, 258; Coastal Forces, 118, 232; combining diplomacy and naval strength, 8; and contesting command, 207, 208, 209, 210, 211, 212–13, 214, 215, 216, 217, 219, 220; Dover Command, 235; East Coast Convoy System, 258; and exercising command, 186, 188, 189, 191, 192, 194, 195–6, 200, 201; (British) Expeditionary Force (BEF), 100–1, 171; and factors of space, 16, 19, 21, 22, 23, 26, 27, 28, 29; and fleet distribution, 104–5, 106; Force H, 195; Grand Fleet, 56–7, 66, 101, 115, 123, 150, 162, 163, 171, 192, 209, 210, 217, 238, 246, 257; Home Fleet, 67, 105, 210, 258; and methods, 137; Mediterranean Fleet, 282; Naval Staff, 275; Navy *see* Royal Navy; Nore Command, 235; Northern Patrol, 67, 166, 238; Ocean Convoy System, 258; and positions, 43, 44, 45, 47–8, 50, 53, 54, 56, 59; and protection of maritime trade, 255, 256, 257, 258, 259, 262, 263, 264; RAF, 115, 124; Scandinavian Convoy System, 258; Scottish Convoy System, 258; and sea control, 6–7, 113, 115, 116, 117, 118, 119, 121–2, 123, 124, 125; and securing command, 147, 149, 150, 151, 152, 154, 157–8, 159–60, 161, 162, 163, 166, 168, 169, 172, 173, 175; South Coast Convoy System, 258; Southern Patrol, 166; and strategic objectives, 98, 100–1, 102; and support of army flank, 267, 268, 270, 271, 272, 275, 277, 282–3; and theater and its elements, 76, 77, 86, 90, 91–2; War Office, 123; *see also* England; Scotland
Brittany, 215, 228
broken base of operations, 77
Bruix, Admiral Eustache, 158
Brunsbuettel, 80
Būbiyān, 68, 152
Būbiyān Channel, 153
Bud, 236
Bugaz, 219
Buka, 281
Bulgaria, 113, 178, 191, 199, 217, 219, 231, 234, 284
Bungo Suidō, 82, 171
Burgas, 219
Burma, 239
Būshehr, 243, 264
Byzantium, 17

Cable Light, 243
Cairo, 247
Calabria, 273
Calais, 19, 118, 123, 168, 172
Cambodia, 241, 242
campaigns, 144–5, 294
Canada, 149, 188, 201
canals, 56, 89, 121; *see also* names of canals
Cape (of Good Hope), the, 26, 27, 105, 247, 255
Carentan Plain, 33
Caribbean Sea, 5, 7, 10, 34, 50, 70, 187, 196, 260, 295
Cartagena, 51
Casablanca, 106
Casos (Kásos), 50
Caspian Sea, 5
Castex, Raoul, 211
castling (leap-frogging), 31
Catania, 261, 271
Catherine, Operation, 53, 98
Cattaro, 114
Cattaro, Bay of, 63, 64, 66, 68, 165, 168, 213
Caucasus, 9, 62, 176, 178, 231, 270, 272, 284, 285
Celebes, 122
central position (country), 15
central *vs* exterior position (forces), 56–9
Central Powers, 29, 165, 191, 217
Cephalonia (Keffalinīa), 68
Cerberus, Operation, 214
Cervera, Admiral Pasquale, 161
Ceuta, 51
Champlain, Lake, Battle of, 149
Charles, Archduke of Austria, 55
Cherbourg, 123, 171
Chersonnes Peninsula, 280
Chesapeake Bay, 238, 267

314

INDEX

China, 10, 21, 54, 64, 114, 149, 241, 273, 281, 283
Chinnanpo (Namp'o), 212, 220, 283
choke point: control, 120–3, 164–7, 245; seizing, 171–3
Chromite, Operation, 192
Churchill, Winston, 53, 98, 191, 210
climate, 29, 38–40
close blockades, 157–61, 162, 295
Clyde, River, 258, 263
Coalition (Gulf War), 110, 117, 137–8, 151–2, 153, 201, 221, 244
coalition or alliance strategy, 1
coast: attack on, 193–6; base on, 70; character, 30–2; configuration, 29–30; control of, 43; defense of, 221–2; strategic point on, 56
coastal defense forces, 221
coastal gun batteries, 232–3
coastal seaborne trade, 256–7
coastal waters, 6
Cold War, 120, 171
Colomb, Admiral Phillip, 147
Colón, 51
combat craft, 298, 299
combined arms, 233–4
combined major naval operation, 137–8
Combined Task Force 440 (CTF 440), 199
Comiso, 261
commerce raiding (*guerre de course*), 248–9
commercial blockade, 237–46
commercial counter-blockade, 246–8
communications: land, 32, 66, 93–4, 273; land lines of, 88–9; sea lines of, 88–93
compact physical concentration, 21
compact shape, 21
Communists, 93, 94, 241, 273, 281, 283, 285; *see also* names of Communist countries
concentration, 140–1, 215
Confederate states, 238
configuration, 29–30
conflict prevention, 196
Constanta, 219, 231, 280
Constantinople, 9, 106, 216, 217, 231, 270, 272
contesting control of the sea: counter-blockade, 213–15; defense of the sea coast, 221–3; defensive mining, 218–21; fleet in being, 207–13; general aspects, 206, 295–6; strategic diversion, 215–18
continental (or land) theater, definition of, 73
continental shelf, 6
control of the air, 124–7
control of the sea: contesting *see* contesting control of the sea; exercising *see* exercising control of the sea; as prerequisite to protection of maritime trade, 255–6; as prerequisite to support of army flank, 267; securing *see* securing control of the sea; *see also* sea control
convoys, 258, 259–61, 264
Copenhagen, 23, 56, 160
Coral Sea, 6, 10; Battle of the (1942), 132, 141
Corbetha barrier, 170
Corbett, Sir Julian, 110
Corfu, 50, 66, 68, 165
Corinth (Korinthia), Gulf of, 256
Cornwallis, Admiral William, 158
Corsica, 19, 26, 45, 159, 279
corvettes, 298
Cos (Kos), 48, 50, 125, 174
Cotentin Peninsula, 123
counter-blockades, 28–9, 213–15, 246–8
country, position of, 15–19
Courland, 280, 283
Crete, 16, 47, 48, 68, 70, 102, 114, 115, 122, 124, 125, 150, 153, 173–4, 175, 217, 230, 247, 277
Crimea/Crimean Peninsula, 53, 77, 114, 177, 178, 190, 219, 231, 276, 278, 280, 281, 284, 285
Crimean War (1853–56), 9, 160, 192, 211, 272
Croatia, 21, 30, 198
Cromarty Firth, 65
Cuba, 47, 50, 51, 70, 114, 187
Cuban Missile Crisis (1962), 10–11, 117, 164
curved base of operations, 77, 80
Curzola (Korčula), 114
Cuxhaven, 23, 80
cybernetic decisive points, 82

315

Cyprus, 16, 45, 68, 70, 174, 175
Cyrenaica, 48, 50, 70, 230

Dago (Hiiumaa), 175, 187
Dairen, 149
Dalmatia, 31, 50, 64, 69, 100, 114, 165, 186, 194, 285
Damietta, 195; Battle of (1973), 151
Dampier Strait, 122
Danish Straits, 7, 18, 23, 51, 56, 120, 166, 171, 172, 273; *see also* Kattegat; Skagerrak
Danube, River, 9, 199, 270
Danube Estuary, 54, 103, 151, 167, 219, 231, 272
Danzig (Gdańsk), 76, 235, 280
Danzig (Gdańsk), Gulf of, 62, 235
Dardanelles, 7, 68, 159, 166, 194, 195, 206, 216, 217, 219, 220, 272; *see also* Turkish Straits
Dayton Peace Accords (1995), 199, 201
Decisive Endeavour, Operation, 200
decisive points, 81–2
defense of the sea coast, 221–2
defensive major naval operations, 137, 274–82
defensive mining, 218–21
defensive tactical actions, 132–3
Deliberate Force, Operation, 200
Demon, Operation, 277
Denmark, 9, 17, 18, 19, 21, 23, 27, 28, 43, 51, 68, 122, 149, 160, 166, 171, 236; Navy, 206
Denmark Strait, 169
Deny Flight, Operation, 200
deployment, 139
Desert Storm, Operation, 196
Determined Guard, Operation, 200
Dieppe, 188
distances, 23, 26–7
distant blockades, 29, 161–7, 295
diversion, strategic, 215–18
Dnieper (Dnepr), River, 276
Dniester, River, 178
Dodecanese Islands, 47, 48, 103, 125, 173, 174
Doenitz, Grand Admiral Karl, 113
domestic policy, 2
Don, River, 219, 285
Doorman, Admiral Karel, 88
Dover, 63, 115, 168, 231

Dover, Strait of, 7, 17, 22, 47, 56, 76, 123, 168, 169, 214, 232, 238, 258
Dover Barrage, 168, 213
Drake, Admiral Sir Francis, 147–8
ducting, 40
Duilio, 195
Dunkirk (Dunkerque), 123, 171, 267, 277
Durazzo (Durrës), 66, 172, 277
Dutch, 18, 104, 114, 148, 157, 160, 287
Dutch coast, 30, 118, 232, 235, 236
Dutch ports, 171, 238
Dutch Provinces, 16, 51
Dutch Republic, 7
Dutch Revolt (1576–85), 186
Dvina (Daugava) estuary, 220
Dynamo, Operation, 277

Earnest Will, Operation, 197
East Africa, 247
East China Sea, 240
East Coast Barrage, 169
East Germany, 47, 66, 68; *see also* Germany
East European Revolution (1989), 47
East Indies, 105; *see also* Netherlands East Indies
Eastern Solomons, Battle of the (1942), 119
economic exclusion zone (EEZ), 11; *see also* exclusion zones
Edinburgh, 67
Egypt, 45, 48, 69, 70, 113, 119, 151, 154, 166, 191, 195, 215, 216, 242, 247, 263, 270, 275, 276, 277, 283; 3rd Army, 276
Eilat, 242, 247
Eisstoss, Operation, 153
El Alamein, 69, 229, 262
El Dorado Canyon, Operation, 197
Elba, 157
Elbe, River, 246
Elbe estuary, 149
Elbing (Elbląg), 283
Elliot Islands (Changsan, Qundao, Korea Bay), 65
elongated shape, 21
El'tigen, 190, 191
Emden, 23, 80, 235
Ems, River, 155, 236

INDEX

enclosed sea, definition of, 5
engagements, 131
Engano, Cape, battle off (1944), 131
England, 7, 16, 17, 19, 22, 30, 55, 65, 66, 76, 82, 86, 103, 104, 114, 115, 117, 147–8, 149, 157, 158, 171, 186, 215, 216, 230, 235; see also Britain
English Channel (La Manche): and attacks on maritime trade, 228, 230, 231, 232, 235, 247; and contesting control, 208, 211, 214, 215, 216, 217; and exercising control, 186, 188; and factors of space, 16, 17, 19, 29, 34; and positions, 44, 47, 56, 59; and protection of maritime trade, 257, 258, 260, 263; and sea control, 114, 115, 116, 117, 118, 120, 123; and securing control, 148, 149, 158, 159, 160, 163, 169, 171; and support of army flank, 267, 273, 277; and theater and its elements, 76, 86, 89, 91; mentioned, 7, 9, 105
Entente agreement (1912), 105
Entente powers, 29, 50, 68, 77, 102, 112, 163, 165, 166, 168, 187, 191, 194, 206, 209, 217, 239, 257
Erie, Lake, 5, 149
Esperance, Cape, Battle of (1942), 132
Estonia, 175, 176, 177, 187, 214, 284
evacuation, 276–81
Even Keni Bay, 220
Ewe, Loch, 67
exclusion zones, 11, 29, 164, 199–201, 243
exercising control of the sea: amphibious landings, 185–93; attack on the coast, 193–6; operations in peacetime, 196–202; projecting power ashore, 184–5, 295
exterior position, 59

Faeroes (Faroes), 44, 169
Famagusta, 68
Fano, 165
Far East, 9, 18, 64, 105, 106, 209, 248, 255; see also names of countries
Fārsī Island, 172, 243
Fāw Peninsula, 172, 244
Feodosiya, 190, 219
Ferdinand Max, Archduke, 100

Filey, 210
Finisterre (Finistère), Cape, 89
Finland, 8, 30, 31, 50, 53, 56, 91, 100, 101, 112, 170, 172, 175, 177, 187, 213, 214, 219, 220, 275, 277
Finland, Gulf of, 27, 29, 34, 47, 56, 76, 101, 103, 112, 113, 118, 153, 170, 172, 175, 176, 186, 213, 214, 219, 268, 277, 284
fire support, 282–5
Fisher, Admiral John, 191
Flamborough Head, 264
Flanders, 69, 168, 171–2, 230
Flanders Flotilla, 168
fleets: actions between, 147–55; distribution, 104–8; fleet in being, 207–13
Fletcher, Admiral Frank J., 188
Florida, Strait of, 51
focal area, 89–90
force-oriented decisive points, 82
foreign policy, 2
Formosa (Taiwan), 57
Formosa (Taiwan) Strait, 273
Forth, Firth of, 23, 56, 67
Forth, River, 260
'Four Day Battle' (1666), 148
France: and attacks on maritime trade, 230, 246; and bases, 68; and contesting control, 207, 211, 212, 215–16, 217; and exercising control, 186, 188, 194, 200, 201; and factors of space, 15, 16, 17, 18, 19, 22, 29, 33; and fleet distribution, 104, 105–6; Navy, 106, 149, 166; and positions, 44, 45, 47, 55, 57; and sea control, 112, 114, 117; and securing control, 148–9, 156, 157, 158, 159, 160, 161, 165, 166, 172; and strategic objectives, 100, 103, 104, 134; and support of army flank, 267, 272, 279; and theater and its elements, 76, 86, 91
Franco-Prussian War (1870–71), 161
Frederick the Great, 194
freedom of navigation (FON) and overflight, 197–8
French, General Sir John, 91–2
French Revolutionary Wars (1792–1800), 158, 215–16
frigates, 298

317

Frisian Islands, 208
Funen, 9

gaining/losing space, 27–9
Galati, 270
Gallipoli, 187, 191, 194, 272
Ganāveh, 264
Gävle, 92
Gela, 282
general fleet action, 147
Genichevs'k (Genichesk), 285
Genoa, 195
Genoa, Gulf of, 45
geographically oriented decisive points, 82
Germany: 1st Air Fleet, 152, 153; 5th Mountain Division, 173–4; 6th Mountain Division, 173; 7th Parachute Division, 173; VIII Air Corps, 276; X Air Corps, 150–1, 173, 261; XI Air Corps, 173; 17th Army, 276, 278; 18th Army, 175, 176; 42nd Army Corps, 175; Admiralty, 166; Army, 100, 176, 268, 284; Army Group A, 278; Army Group North, 10, 112, 152, 175, 176, 214, 268, 271, 276, 282; Army Group South, 177, 189–90, 268; and attacks on maritime trade, 227, 228–9, 230–1, 231–2, 233, 234, 235, 236, 238–9, 244, 245–6, 246–7, 248; and bases, 61, 62, 63, 67, 68, 69, 70; and contesting control, 206, 209, 210, 211, 212–13, 214–15, 216–18, 219–20; and exercising control, 187–8, 190–1, 192, 194, 201; and factors of space, 17, 19, 21, 23, 27, 28, 29, 30, 33; and fleet distribution, 104, 105, 106; General Staff, 100; Hermann Goering armored division, 282; High Seas Fleet, 47, 56, 76–7, 100, 101, 115, 118, 150, 155, 162, 168, 192, 208, 209, 210, 231, 246; *Kriegsmarine* (1934–45), 10, 76, 100, 102, 103, 112, 116, 124, 176, 247, 268, 271, 276, 284, 286; and methods, 137, 144; Naval Staff (High Command), 48, 62, 100, 170, 176, 259, 280, 285; Navy, 10, 67, 76, 80, 100, 101, 102–3, 104, 106, 120, 121, 144, 162, 163, 171, 209, 247, 248, 280, 284, 297; North Sea Command, 235; and positions, 43–4, 45, 47, 48, 50, 51, 53, 54, 55, 56, 57; and protection of maritime trade, 254–5, 256, 257, 258, 259, 260, 261–2; *Reichsmarine* (1919–33), 100; and sea control, 9–10, 112, 113, 114, 115, 116, 117, 118, 119, 120,121, 122, 123, 124–5, 126; and securing control, 149, 150, 151, 152–3, 155, 157, 161, 162, 163, 165, 166, 168, 169, 170, 171–2, 173–5, 175–8; and strategic objectives, 100–1, 102–3; and support of army flank, 268, 270, 271, 272, 273, 274, 275, 276, 277, 278, 279–80, 281–2, 283, 284, 285, 286; Supreme Command of the *Wehrmacht* (OKW), 10, 124, 176, 259, 268, 274, 284; and theater and its elements, 74, 76, 77, 80, 82–3, 91, 92, 93; *Wehrmacht*, 100, 259, 268, 284, 286; *see also Luftwaffe*
Ghent (Gent), 149
Gibraltar, 7, 26, 48, 63, 68, 69, 70, 91, 103, 113, 118, 158, 159, 195, 216, 228, 246, 255, 256
Gibraltar, Strait of, 6, 18, 26, 27, 45, 51, 54, 89, 120, 121, 169, 172, 273
GIUK (Greenland–Iceland–United Kingdom) Gap, 121–2
Glasgow, 67
Gloucester, Cape, 122
Gneisenau, 214
Goeben, 216, 217
Goetz von Berlichingen, Operation, 153
Good Hope, Cape of, 26, 27, 105, 247, 255
Goodwin Sands, 238
Gorleston, 210
Gotenhafen (Gdynia), 280
Gotland, 34, 186
Granital, Cape, 53
Grant, General Ulysses S., 267
Great Belt (Store Bælt), 68, 166, 236
Great Lakes, 86
Great Tueters, 170
Greater Tunb Island (Tunb al Kubrá), 55, 172
Greece, 21, 34, 47, 48, 68, 98, 115,

122, 124, 125, 150, 166, 173, 187, 189, 201, 217, 277, 286
Greenland, 18, 121, 238
Grimsby, 62
Gris-Nez, Cap, 232
Guadalcanal, 71, 81, 119, 125, 132, 135, 137, 155–6, 188, 189, 190, 281
Guantanamo, 261
Gulf crisis (Feb.–March, Nov.–Dec. 1998), 10
Gulf states, 244, 245; *see also* name of country
Gulf War (1990–91), 11, 26, 66, 74, 110–11, 116, 117, 137, 151–2, 153, 154, 193, 196, 244, 269
gun batteries, coastal, 232–3
Gūreh, 264
Guzmán, Alonzo Pérez de, Duke of Medina Sidonia, 208

Haifa, 69, 234, 275
Haiphong (Hâi Phòng), 242
Hamburg, 9, 121
Hangö, 50, 101, 170, 172, 175, 176, 220, 275, 277–8
Hardanger Fiord, 168
Hartlepool, 62, 210
Harwich, 62, 66, 115
Haugesund, 23
Haus, Admiral Anton, 194
Hawke, Admiral Edward, 160
Hel Peninsula, 62
Heligoland, Battle of (1864), 9, 23, 80, 149
Heligoland Bight, 43, 76, 101, 115, 162, 166, 168, 210, 231
Helsinki, 23
Henderson Field, 189
Hengion, 172
Herbert, Arthur, Earl of Torrington, 207
Herkules, Operation, 174
Hideyoshi, Toyotomi, 273
hinterland, physical character of, 29–30
Hiroshima, 170, 240
Hitler, Adolf, 174, 214, 247, 258, 259, 272, 276, 279, 280
Hoche, Lazare Louis, 158
Hochland, 170
Holland, 16, 19, 69, 149, 156; *see also* Dutch; Dutch coast; Dutch ports; Netherlands

Honshū, 240
Hormuz, Strait of, 19, 22, 23, 53, 55, 68, 89, 120, 172, 197, 201, 243, 245, 248, 263, 264
Howard of Effingham, Lord, 208
Hudson Bay, 5
Hull, 23
Humber, River, 264
Hundred Years War, 186
Hungary, 199
Hüngnam, 241, 281, 283
Huon Gulf, 122
Husky, Operation, 137, 282
Hussein, Saddam, 193

Iberian Peninsula, 8
ice conditions, 39
Iceberg, Operation, 170
Iceland, 44, 67, 121, 169, 238
IFOR (multinational peace Implementation Force), 199, 200
Ijmuiden, 232
Inchon (Inch'ŏn), 192, 283
independent major naval operation, 137
India, 215
Indian Ocean, 17, 53, 105, 244
Indonesia, 16, 21
Indonesian straits (Selat Sunda and Selat Lombok), 89–90
Inland Sea, 16, 82, 120, 170, 171, 240
Inner Leads, 92, 236, 254, 255; *see also* Norwegian Leads
inshore waters, definition of, 6
insular position, 16, 17
interdiction, 273
Invergordon, 67
Ionian Islands, 215
Ionian Sea, 22, 68
Iran, 19, 23, 30, 31, 34, 55, 68, 98, 113, 119, 156, 172, 197, 227, 228, 232–3, 234, 237, 242, 243–4, 245, 248, 261, 263–4; Revolutionary Guard, 172; Navy, 102
Iran–Iraq War (1980–88), 11, 29, 74, 102, 119, 172, 227, 228, 232–3, 243–4, 245, 248, 261, 263–4
Iraq, 18, 23, 69, 110–11, 117, 119, 137, 151, 152, 153, 154, 172, 193, 196, 201, 221, 227, 228, 232, 234, 243–4, 245, 248, 261, 263, 264; Navy, 102, 153

Irben Passage, 170, 220
Ireland, 8, 18–19, 22, 44, 47, 86, 158, 159, 186, 245
Irish Channel, 219
Irish Sea, 16, 19, 22, 230
islands, 16, 17, 31, 32, 47–8, 50–1, 54, 55–6, 70–1; *see also* names of islands
Isle of Wight, 207
Isonzo, River, 149
Israel, 119, 151, 154, 167, 195, 242, 245, 248, 261, 276, 283; Navy, 151, 154, 196, 242, 245, 269
Istria, 149, 187
Italy: Air Force (*Regia Aeronautica*), 230; and attacks on maritime trade, 228, 230, 233, 242, 246, 247; and bases, 63, 65–6, 68, 69, 70; and contesting control, 210, 212, 216; and exercising control, 186–7, 191, 194, 195–6, 198, 200, 201; and factors of space, 19, 29, 30, 33; Naval Staff (*Supermarina*), 102; Navy, 68, 101–2, 103, 125, 187, 210, 212, 228, 256, 262; and positions, 45, 47, 48, 50; and protection of maritime trade, 256, 261, 262; and sea control, 118, 123, 125, 126; and securing control, 149, 150, 151, 152, 153, 154, 163, 171, 172, 174; and strategic objectives, 102, 103–4; and support of army flank, 268, 272, 276, 277, 278, 279, 282, 285, 286; mentioned, 91, 105, 137, 144
Ito, Admiral Sukenori, 149
Iwo Jima, 239

Jacobites, 8
Jade, River, 155, 236
Jamaica, 51
James II, King, 8
Jan Mayen, 29
Japan: and attacks on maritime trade, 234, 236–7, 239–40, 243; and bases, 61, 64, 65, 70, 71; Central Invasion Force, 189; Combined Fleet, 141; and contesting control, 208, 209, 211–12; Eastern Invasion Force, 189; and exercising control, 186, 188, 189, 190, 192; and factors of space, 16, 17, 27; First Diversionary Attack Force, 135, 143; General Staff, 212; and methods, 132, 135, 141, 143; Navy, 84, 120, 148, 206, 297; and positions, 45, 57; and protection of maritime trade, 255; and sea control, 114, 119, 120, 121–2, 122–3, 125; and securing control, 149, 150, 154, 156, 161, 167–8, 171; and strategic objectives, 99–100, 103; and support of army flank, 273, 281; and theater and its elements, 81, 82, 84, 85, 86, 88; mentioned, 9, 10
Japan, Sea of, 9, 10, 16, 18, 23, 45, 61, 86, 120, 150, 196, 240
Jāsk, 264
Java, 57, 88, 121, 122
Java Sea, 54, 88
Jellicoe, Admiral Sir John, 123, 194
Joint Endeavour, Operation, 199, 200, 201
Jordan, 244, 248, 261
Jubilee, Operation, 188
Judgment, Operation, 137, 152
Julian Alps, 186
Juminda barrier, 170
Juminda Cape, 175
Jutland, Battle of (1916), 150, 155
Jutland Peninsula, 17, 21, 32, 38, 51, 76, 77, 228

Kaliningrad (Koenigsberg), 27, 47
Kallbadargrund, 170
Kamchatka, Sea of, 18, 120
Kamchatka Peninsula, 18
Kanga, 264
Karlskrona, 30
Karmö Island, 236
Kattegat, 7, 9, 17, 43, 51, 115, 206, 235, 236; *see also* Danish Straits
Kedah, 54
Kentish Knock, 114
Kerch, 282
Kerch Peninsula, 177, 282
Kerch Strait, 34, 53, 113, 170, 177, 178, 190, 191, 278, 284, 285
Key West, 51, 260
Khafji (Al Khafjī), 152
Kham Sound, 236
Kharg Island, 227, 243, 261, 263, 264
Kherson, 178, 282
Khiós, 166

INDEX

Khor Abdullah lagoon, 68
Khor Abdullah waterway, 102, 233
Khor al-Amaya, 245
Kiel, 23, 63
Kiel Canal (Nordsee–Ostsee Kanal), 17, 23, 63, 89, 121, 236
Kipinola minefield, 170
Kirkenes, 254, 259
Kiruna-Gallivare district, 92
Knight's Move (*Roesselsprung*), Operation, 137
Kōbe, 240
Koenigsberg (Kaliningrad), 10, 283
Kolberg, 280
Kolombangara, 71, 156, 281
Korea, 45, 114, 121, 186, 212, 220, 239, 241, 242, 268, 269, 273, 281; see also Korea, Republic of; Korean War; North Korea; South Korea
Korea, Republic of, 27
Korea Bay, 212
Korea Strait, 45, 64, 121, 122, 171, 186, 206, 212, 273
Korean War (1950–53), 10, 26, 93, 119, 192–3, 220–1, 268, 269, 281, 283, 287
Krasnodar, 177
Kronstadt (Kronshtadt), 112, 152, 176, 274
Kronstadt (Kronshtadt), Bay of, 112
Kuban', 178, 276, 278, 285
Kuban', River, 177
Kunda, 175
Kuolemanjarvi minefield, 170
Kure, 170, 240
Kurita, Vice-Admiral Takeo, 143
Kuwait, 27, 68, 111, 153, 172, 193, 196, 197, 221, 243, 244
Kuwait City, 172, 193
Kyūshū, 240

La Hague (Barfleur), Battle of (1692), 149
La Maddalena, 63, 158
La Perouse Strait, 86, 121, 171, 206
La Valletta, 26, 153
Lae, 125
Lampedusa, 189
land communications, 32, 66, 93–4, 273
land lines of communication, 88–9
landform, 32–3

landings, amphibious, 32, 33, 141, 185–93
Lapeyrère, Admiral Boué de, 216
Lārak Island, 53, 172, 264
Lastovo, 64
Latakia (Al Lādhiqīyah), 195; battle off (1973), 151
Lāvān island, 264
Lavansaari (Lavansari), 56, 176
Le Havre, 123
Lebanon, 45, 269
Leghorn (Livorno), 45, 157, 195
Lemonos (Límnos), 115
Lengeh, 264
Leningrad, 10, 47, 56, 62, 100, 112, 121, 152, 153, 176, 177, 274, 278, 284
Léros, 50, 125, 174
Lesina (Hvar), 114
Lesser Tunb Island, 55
Levant, 159
Leyte, 82, 84, 85, 88, 131–2, 137, 141, 143
Liaotung Peninsula, 45, 161, 212
Libau, 43, 62, 170, 175, 220, 227, 228, 283, 286
Libya, 45, 48, 68, 69, 102, 103, 113, 118, 173, 237, 247, 261, 262, 263
Liepāya, 23, 68
Ligurian coast, 45
limited conflict, 11
Lincoln, President Abraham, 238
lines of communications, 88–93
lines of operations, 85–8, 139–40, 292–3
Linguetta, Cape, 168
Lissa, 50, 99, 149, 165, 186; Battle of (1866), 50, 99, 149, 187
Lithuania, 169
Little Belt (Lillebælt), 166, 236
littoral, 7, 184–5
Liverpool, 258
Livonia, 118
local conflict, 11
logistic vessels, 299
Loire, River, 91
Lombok (Selat Lombok), Strait of, 89–90, 120
London, 121, 263
losing/gaining space, 27–9
Louis XIV, King of France, 149

Low Countries, 8, 158
Lowestoft, 194
Luebeck, 10, 236
Luftwaffe, 44, 56, 90, 115, 116, 124, 125, 173, 174, 175, 176, 188, 191, 227, 230, 247, 263, 268, 276, 277, 284
Luleå, 23, 28, 50, 92
Lunga Point, 132
Lussin (Lošinj), 100
Lussin Piccolo (Mali Lošinj), 65
Luzon, 57, 141, 239

MacArthur, General Douglas, 122, 281, 286
McClellan, General George B., 267
Madang, 122, 125
Maddalena Bay, 160
Madeira, 54
Magenta, 100
Mahan, Admiral Alfred T., 2–3, 42, 47, 55
main point of defense, 142
major naval base, 65
major naval operations, 133–44, 150, 152, 229–30, 249, 269–70, 271–82, 294
Makarov, Admiral S. O., 167–8
Makassar Strait, 89
Malacca, Strait of, 19, 53–4, 89–90, 120, 121, 122, 273
Malaya, 10, 19, 54, 88, 122, 124, 189, 239
Malaya, HMS, 195
Mallorca, 158
Malta, 7, 19, 48, 50, 62, 68, 69, 70, 71, 91, 102, 105, 118, 124, 153, 159, 172, 174, 189, 212, 217, 228, 230, 246, 248, 256, 261, 262, 263, 270, 272
Manchuria, 93, 211, 212, 239, 241
Mannerheim Line, 177
Maradin Island, 153
marginal sea, definition of, 6
maritime combat zones, 74
maritime exclusion zone, 164; *see also* exclusion zones
Maritime Guard, Operation, 198, 199
maritime intercept operation (MIO), 201
maritime theater of war, definition of, 73; *see also* theater and its elements

Market Time, Operation, 242
Marmarica, 166
Marseilles, 26, 45, 106
Marshall Islands, 71
MCM (mine countermeasures) ships, 299
Medina Sidonia, Alonzo Pérez de Guzhán, Duke of, 208
Mediterranean Sea: and attacks on maritime trade, 228, 229, 230, 233, 242, 244, 246, 247, 248; and bases, 63, 68, 69, 70; and contesting control, 210, 212, 215, 216, 217; and exercising control, 188, 196, 201; and factors of space, 17, 18, 19, 22, 23, 26, 34; and fleet distribution, 104, 105, 106; and positions, 45, 47, 48, 50, 51, 53, 54; and protection of maritime trade, 255, 256, 258, 262, 263; and sea control, 112, 113, 114, 115, 117, 118–19, 122, 124–5, 126; and securing control, 151, 152, 154, 157, 158, 159, 160, 164, 165, 167, 168, 169, 172, 173, 174; and strategic objectives, 102, 103, 104; and support of army flank, 270, 277, 278, 282, 285, 287; mentioned, 5, 6, 7, 9, 133, 294, 295
Mekong Delta, 242
Melilla, 51
Melos (Mílos), 115
Memel, 169
Merkur (Mercury), Operation, 124
Mersa Matrûh, 275
Mersa T'lemet, 283
Mersey, River, 263
Mesopotamian campaign, 10
Messina, 216, 217, 271, 278
Messina, Strait of, 19, 53, 113, 123, 217, 271, 273, 278, 282
methods of combat force employment: attacking maritime trade, 229–30; campaigns, 144–5, 294; differences between the services, 3; major naval operations, 133–4, 294; naval tactical actions, 129–33, 293–4
Mexican border, 238
Mexico, Gulf of, 47, 51
Middle East, 26, 113, 120; *see also* names of countries
Mikawa, Gunichi, 132

INDEX

Miletus, 208
military objectives *see* strategic objectives
military strategy, definition of, 1–2
Milne, Admiral Archibald Berkeley, 216, 217
Mina al-Bakr, 245
mines, 35, 37, 167–71, 218–21, 234–7
minor naval base, 65
Minorca, 103, 159
Mississippi River, 238
Missouri, 196
Mius, River, 177
MO, Operation, 132, 141
Moltke, Field Marshal Helmuth von, Sr, 9, 86–7
Moluccas archipelago, 54
Mona Passage, 51
Mons, 200; Battle of, 123
Monte Gargano peninsula, 50, 194
Montenegro, 199
Montgomery, General Bernard, 33
Moon (Muhu), 175, 187
Moonsund, 170, 220
Moray Firth, 67
Morgan oil fields, 247
Mudros, 68, 166
multinational intercept force (MIF), 201
Munda, 71, 156
Murmansk, 268
Muskö, 64

Nagoya, 240
Naples, 26, 159, 286
Napoleon I, Bonaparte, 82, 158–9, 211, 215, 216, 246
Napoleonic Wars, 112, 158–9
narrow sea: features of combat in, 11–12; use of terminology, 5–7
Narva, 175, 176, 177, 254
Narvik, 92, 236, 254, 255, 259
Nāsariÿa, 227
Nasser, Gamal Abdel, 167
national (or national security) strategy, definition of, 1
NATO (North Atlantic Treaty Organization), 27, 171, 198, 199, 200, 201
Naval Act (Germany, 1900), 106, 212
naval bases *see* bases, naval

naval combat zones, 74
naval control of shipping, 257
naval policy, definition of, 2
naval strategy, 2, 4, 292
naval tactical actions *see* tactical actions
NAVSOUTH (Allied Naval Forces Southern Europe), 199
navy, building a balanced, 297–9
Near East, 113; *see also* names of countries
Nelson, Admiral Horatio, 159–60, 208, 211, 216
Neptune, Operation, 33, 134
Netherlands, 8, 186, 244; *see also* Dutch; Dutch coast; Dutch ports; Holland
Netherlands East Indies, 10, 57, 70, 71, 84, 88, 122–3, 141, 189, 239
neutral territories, 28, 166, 244–5
neutrality zone, 28
neutralization, 156–7; *see also* blockades
New Britain, 70, 81, 125
New Georgia, 70, 81, 125, 156
New Guinea, 71, 81, 122, 125, 156, 268, 286, 287
New York, 260–1
Niemen, River, 54
Nieuport (Nieuwport), 267
nights, length of, 38
Niigata, 240
Nikolayev (Mykolayiv), 178
Nile Delta, 50
Nile Estuary, 270
Nile Valley, 45
Nimitz, Admiral Chester, 240
Norfolk, VA, 27
Normandy, 33, 188, 273
North Africa, 26, 45, 48, 50, 69, 91, 104, 105, 113, 115, 120, 174, 212, 230, 246, 247, 261, 262, 268, 270, 282
North Cape, 76
North Channel, 19, 258
North Foreland, 264
North Korea, 10, 23, 93–4, 192, 193, 220–1, 240–1, 268, 281, 287; Army, 192–3; Navy, 269
North Passage, 47, 67
North Sea: as area of conflict, 8–9; and attacks on maritime trade, 228, 231, 234, 235, 236, 238, 239, 247; and bases, 61, 62, 63, 65, 66, 67, 68, 69;

323

and contesting control, 206, 209, 210, 214, 218; and factors of space, 16, 17, 18, 19, 22, 23, 27, 28, 34, 38; and fleet distribution, 104, 105, 106; and positions, 43, 44, 47, 48, 51; and protection of maritime trade, 256, 257, 258, 263; and sea control, 112, 114, 115, 117, 120, 121; and securing control, 149, 150, 155, 159, 162, 163, 166, 168, 169, 171; and strategic objectives, 100, 101, 102; and theater and its elements, 76, 80; mentioned, 5, 7, 133, 188, 295
North Vietnam, 241, 242
Northern Barrage, 168, 169
Norway, 8, 18, 23, 27, 28, 29, 43, 44, 51, 67, 92–3, 98, 114, 115, 122, 124, 168, 169, 171, 173, 188, 189, 210, 214, 232, 236, 238, 245, 255, 258, 259
Norwegian Leads, 27, 44, 232; *see also* Inner Leads
Norwegian Sea, 120
Novorossiysk, 62, 177, 178, 217, 219, 276, 285
NSFS (naval surface fire support) 32
Nyköping, 34

oceanography, 34–7
Odensholm, 101, 220, 278
Oder, River, 283
Odessa, 177, 178, 217, 231, 275, 278, 279–80, 284
Oeland, 169
Oesel (Saaremaa), 175, 187
offense–defense, 282–7
offensive major naval operation, 137, 271–3
offensive mining, 167–71, 234–7
offshore patrol vehicles (OPVs), 299
Okhotsk, Sea of, 5, 18, 29, 39, 107, 120
Okinawa, 170, 171, 240
Oman, 19, 53
Oman, Gulf of, 53
Onega, Lake, 121
Ontario, Lake, 5
operational art, 2–4, 5
operational commander, 135, 294
operational concentration, 140–1

operational intelligence, 135
operational maneuver, 141–2
operational objectives, 80–1, 134, 150
operational pauses, 142–3
operational protection, 136–7
operational realities, 4
operational space, 134
operational success, 143
operational-tactical (joint task force) commander, 136
Oran, 45
Oranienbaum (Lomonosov), 153, 176
Orfordness, 264
Orkneys, 23, 66, 168, 169
Oro, 170
Ōsaka, 240
Oslo, 236
Oslo fiord, 235
Ostend (Oostende), 23, 80, 123, 171
Otranto, 63, 66, 165, 166, 169
Otranto, Strait of, 19, 23, 50, 53, 56, 63, 66, 68, 69, 102, 165, 168–9, 171, 213, 217
Otranto Barrage, 168–9
Overlord, Operation, 134
Oxelösund, 92

Pacific, 6, 10, 16, 18, 53, 57, 81, 86, 120, 170–1, 196, 234, 239, 240
Pacific Ocean Area, 81
Pakerort, 170
Palembang, 54
Palermo, 261
Palestine, 45, 69
Panama Canal, 51, 89
Pantelleria, 189, 233
Paris, 198
peace operations (PO), 197
peace-enforcement operations (PEO), 198, 202; expanded, 201
peacekeeping operations (PKO), 198, 202; expanded, 201
peacetime operations, 196–202
Pedestal, Operation, 228
Peloponnesian War (412–11 BC), 208
Peloponnesus, the, 256, 277
Peninsula Campaign (1862), 267
peninsulas, 16, 55, 70; *see also* names of peninsulas
People's Republic of China *see* China
Perekop, 281

Perim, 53, 55
peripheral sea, definition of, 6
Pernau, 175
Persano, Admiral Carlo di, 99, 186
Peter the Great, 8
Petropavlovsk, 167
Philippines, 16, 21, 57, 81, 84, 85, 88, 122, 141, 189
Phillippeville (Skikda), 45, 216
physical objectives, 80–1
physically fragmented country, 21
Piedmont, 100
Pillau (Baltiysk), 76, 283
Piombino, 286
Piraeus (Pireás), 256, 277
Pitt, William, the Elder, 157
Plevna (Pleven), 9, 270
Ploče, Cape, 50
Ploče-Ston area, 64
Plymouth, 44, 260
Po, River, 33
Pohang (P'ohang), 193, 268
point of main attack, 142
points, 55–6
Pola (Pula), 62, 65, 66, 68, 69, 100, 165, 231
Poland, 17, 21, 47, 66, 99
policy, 1, 2
political objective, 97–8
Polyarnoye (Polyarnyy), 268
Pomerania, 191, 192, 208, 213, 280
Pontus, 255–6
Port Arthur (Lüshun), 45, 62, 64, 65, 104, 115, 149, 150, 161, 167, 188, 208, 209, 211, 212
Port Mahón, 215
Port Moresby, 132
Port Said, 69, 154, 195
Port Sudan, 242
Port of Hurgada, 154
Portland, 238
Portsmouth, 44, 115, 160, 260
Portugal, 246
position of country, 15–19
positions, strategic, 42–59, 292
Potenza, River, 194
Poti (P'ot'i), 62, 177, 276
primary theater of war, definition of, 73–4
Prinz Eugen, 214
projecting power ashore, 184–5, 295;
see also amphibious landings; attacks on coast
prorupt shape, 21
protection of maritime trade: coastal seaborne trade, 255–6; importance of, 197, 254–5; methods, 259–65; naval control of shipping, 257–9; prerequisites, 256–7
protection of one's basing and deployment areas, 132–3
Provence, 45
Prussia, 8, 9, 149; Eastern, 283
Prusso-Italian war (1866), 50
Pusan, 23, 121, 192, 193, 273, 281, 283

Qatar, 201
Qeshm, 172, 264
Quiberon Bay, Battle of (1759), 157, 160
Quoins, 53

Rabaul, 70, 71, 81, 125, 156, 189
radio waves, 40
Raeder, Admiral Erich, 115, 125, 210, 259
RAF (Royal Air Force), 115; Fighter Command, 124
Ragusa (Dubrovnik), 114
raids, naval, 130–1, 194; *see also* commerce raiding
railway communications, 32, 66, 93
Ras al-Kuh, 243
Ras Al Qualayah, 196
Ras Sadaat, 247
reconstitution, 143
Red Army, 101, 274
Red Sea, 7, 38, 53, 55, 74, 119, 154, 164, 167, 196, 201, 242, 244, 245, 248, 295
redeployment, 143
re-entrant base of operations, 77, 80
regeneration of combat power, 143
Reggio Calabria, 261
regional conflict, 11
relief, 32
Renown, HMS, 195
reorganization, 143
restricted waters, definition of, 6–7
Reval (Tallinn), 101, 175, 177, 275
Revolutionary Guards (*Pasdaran*), 173
Rhodes, 48, 50, 125, 175

Rhône, River, 33
Rhône estuary, 45
Ricasoli, Baron Bettino, 149
Richmond, 267
Richmond, Admiral Herbert, 147, 156–7
Riga, 175, 176
Riga, Bay of, 101, 169, 220
Riga, Gulf of, 175, 176, 187, 268, 283
Rimini, 194
risk theory, 212–13
rivers, 33–4, 54–5; *see also* names of rivers
road communications, 32, 66, 93, 94
Robeck, Admiral John de, 220
Roesselsprung (*Knight's Move*), Operation, 137
Romania, 50, 114, 177, 178, 191, 199, 214, 219, 231, 275, 276, 278, 280, 281–2, 284; Navy, 284
Rome, 165, 199
Rommel, Field Marshall Erwin, 48, 228, 262
Rosetta, 195
Rostock, 23, 68
Rosyth, 62, 67–8
Rota, 51
Rotterdam, 235
Royal Navy, 6–7, 8, 16, 22, 43, 47, 48, 53, 56, 62, 65, 66, 67, 86, 91, 103, 104–5, 112, 117, 118, 150, 155, 157–8, 159, 162, 166, 169, 172, 174, 191, 194, 206, 207, 208, 209, 210, 215, 238, 239, 245, 256, 258, 271, 272
Rozhestvenskiy, Admiral Zinovi P., 86
Ruegen, 47, 66, 68
Ruhr, 10
Russell Islands, 71
Russia: and attacks on maritime trade, 231, 234, 246; Baltic Fleet, 101, 103, 106, 107, 209; Baltic Sea and Black Sea as scene of naval actions of, 8, 9; bases, 62, 63, 64–5, 68; and contesting control, 209, 211–12, 213, 217; and exercising control, 186, 187–8, 192; and factors of space, 15, 17, 18, 26, 28; Far Eastern Squadron, 103; fleet distribution, 104, 106–7; Navy, 101, 103, 104, 106, 150, 206, 211, 267; and positions, 44, 45, 47, 53; and protection of maritime trade, 255, 256, 261; and sea control, 110, 114, 115, 118, 122; Second Pacific Squadron, 103, 106; and securing control, 148, 149, 150, 160, 161, 166, 167, 168, 174; and strategic objectives, 100, 101, 103; and support of army flank, 270, 272; and theater and its elements, 86; Twelfth Army, 187; mentioned, 8–9, 10; *see also* Russian Federation; Soviet Union
Russian Federation, 18
Russo-Japanese War (1904–5), 86, 99, 103, 106, 114, 149–50, 161, 162, 167–8, 188, 209, 211–12
Russo-Turkish War (1877–78), 9, 167, 270
Ruyter, Admiral Michael A. De, 148
Ryūkyūs, the (Nansei Shotō), 240

S-boats, 231–2, 235
St George's Channel, 19, 169
St Nazaire, 91
Saint Vincent, Cape of, 51
Salamaua, 125
Samar, battle off (1944), 131, 143
Samland, 283
Sámos (Sámi), 166, 174, 208
San Bernardino Strait, 82
San Giovanni di Medua, 277
sanction enforcement, 201
Santiago, 161
Santiago Bay, 161
Santiago de Cuba, 209
Sarajevo, 200, 201
Sardinia, 9, 45, 63, 68, 100, 124, 158, 160, 172, 189, 195, 215, 230, 233, 263, 279
Saronic (Saronikós) Gulf, 256
Sasebo, 170, 240
Sassnitz, 68
Saudi Arabia, 89, 193, 201, 244, 247, 248, 261
Savo Island, Battle of, 132
Scandinavia, 29, 53, 56, 122, 238, 244, 264; *see also* names of countries
Scapa Flow, 23, 27, 44, 63, 65, 66, 67, 76, 163, 238, 258
Scarborough, 210
Scarpanto, 48, 50

326

INDEX

Scharnhorst, 214
Scheer, Admiral Reinhard, 194, 210
Scheldt (Schelde), River, 54, 236
Schepel, 176
Schleswig-Holstein, 191
Schooneveldt Channel, 148
Schwarzkopf, General Norman, 193
Scotland, 43, 47, 56, 67, 76, 169, 231, 238, 258, 263
Scott, General Winfield, 238
sea area, 22–3
sea-based strategic nuclear forces (SSBNs), 154–5
sea control: choke point control, 120–3; control of the air, 124–7; control of the sea, 110–16; degree of, 116–19; duration of, 116; sea denial, 119–20
sea denial, 119–20
sea lines of communication (SLOC), 88–93
Sebenico (Šibenik), 66, 69
secondary theater of war, 73, 74
sector of main effort, 142
sector of secondary effort, 142
securing control of the sea: actions between fleets, 147–55; attrition, 155–6; by land-based aircraft, 173–5; close blockade, 157–62, 295; distant blockade, 162–7, 295; neutralization, 156–7, 294–5; offensive mining, 167–71; seizing basing areas from the land, 175–9; seizing choke points, 171–3
Seeloewe, Plan, 115
Seiskari, 176
semi-central position, 15, 16, 17
semi-enclosed sea, definition of, 5
semi-isolated position, 16
Senyavin, Admiral Dmitry, 114
Seoul, 192, 273
Serbia, 198, 199; Army, 277
Sevastopol', 9, 62, 68, 103, 114, 177, 211, 217, 231, 276, 278, 280, 281, 282, 284
Seven Years War (1756–63), 112, 157, 194, 215
shallow waters, definition of, 6
shape, 21–2
SHAPE (Supreme Headquarters, Allied Powers Europe), 200

Sharm el-Sheikh, 242, 247, 248, 276
Sharp Fence, Operation, 198, 199
Sharp Guard, Operation, 199
Shatt el-Arab, 10, 22, 68, 232, 244, 245
Sheffield, HMS, 195
Shetlands, 18, 22, 23, 27, 43, 44, 67, 171, 232
Shikoku, 240
Shimonoseki Strait, 170, 171, 236–7, 240
Shortlands, the, 281
Sibuyan Sea, Battle of the (1944), 131, 135
Sicilian Narrows, 7, 53, 55, 189, 230, 233, 247, 272
Sicily, 19, 45, 48, 91, 123, 124, 138, 159, 172, 173, 189, 216, 230, 233, 247, 261, 263, 271, 278–9, 282, 286
Sidra, Gulf of, 197
Sihanoukville (Kâmpóng Saô), 241, 242
Simpson, Admiral William T., 161
Sinai, 167, 195, 245, 247, 248, 276
Singapore, 55, 121, 239
Sino-Japanese War (1894), 9, 114, 149
Sinope (Sinop), 160
Sirrī Island, 172, 243, 261, 263–4
Six-Day War (1967), 119, 167; see also Arab–Israeli conflicts
Sjaelland (Zealand), 9, 21, 32, 161
Skaens Odde, 68
Skagerrak, 7, 17, 43, 51, 93, 114, 115, 121, 166, 168, 188, 206, 231, 236, 254, 255; see also Danish Straits
Skåne, 30
Skaw, 23
Sky Monitor, Operation, 199
'Slot', the ('Middle Course'), 70
Smyrna Patrol, 166
Soelosund, 170
Sole Bay, Battle of (1672), 148
Solferino, 100
Sollum, 275
Solomon Sea, 6
Solomons, 10, 71, 81, 125, 132, 144, 155, 156, 189, 190, 281
Somerville, Admiral James, 195
sonar conditions, 35–7, 39
Songjin, 241
sortie control, 164
Sorve Peninsula, 177
Souchon, Admiral Wilhelm von, 216, 217

Soúdha Bay, 68, 153
Sound, the, 68, 160, 166, 236
South China Sea, 5, 57, 89, 239
South East Asia, 70; *see also* Asia
South Korea, 23, 193, 221; First Corps, 281; Navy, 287; Third Division, 281
South Tirol, 149
South Vietnam, 241, 242
Southwest Pacific Area (SOWESPAC), 81
Southwestern Approaches, 47, 230
Soviet Union: and attacks on maritime trade, 231, 241, 242; Baltic Fleet, 10, 47, 56, 62, 66, 100, 101, 102, 107, 113, 151, 152, 153, 169, 170, 176, 268; and bases, 61, 62, 66; Black Sea Fleet, 62, 100, 107, 151, 274, 275, 276, 281, 282; and contesting control, 213–14, 219, 220; and exercising control, 190–1; and factors of space, 17, 27, 28; failure to build balanced navy, 297; and fleet distribution, 106, 107; ground forces, 176; and methods, 144; Navy, 10, 106, 120, 144; Northern Fleet, 107; Pacific Fleet, 107, 171; and positions, 47, 50–1, 53–4, 56; and protection of maritime trade, 255; Red Army, 101; and sea control, 10, 110, 112, 113, 114, 119, 120, 121; and securing control, 151, 152, 153, 169, 170–1, 172, 175, 176, 177, 178; and strategic objectives, 100, 101, 102; and support of army flank, 268, 274, 275–6, 277, 278, 279, 280, 281–2, 283, 284, 285, 286; and theatre and its elements, 76, 77, 92; *see also* Russia
space: character of coast, 30–2; climate, 38–40; configuration, 29–30; country's position, 15–19; distances, 23, 26–7; gaining/losing space, 27–9; landform, 32–3; oceanography, 34–7; rivers, 33–4; sea area, 22–3; shape, 21–2
Spain, 7, 16, 114, 147, 148, 156, 158, 160, 161, 186, 187, 209, 246; Armada, 148, 208; Navy, 114
Spanish–American War (1898), 161, 187

special forces, 153–4
Spithead, 157
Srivayayan Empire, 54
Stabilization Force (SFOR), 200, 201
Stadtlander, 236
Stalin, Joseph, 282
Stalingrad, Battle of (1943), 100, 177
Standing Naval Force Atlantic (STANAVFORLANT), 198
Standing Naval Force Mediterranean (STANAVFORMED), 198, 201
Starvation, Operation, 240
Stavanger, 27, 44, 236
Stettin (Szczecin), 10
Stockholm, 23, 92, 186
Stockholm archipelago, 64
Store Bælt (Great Belt), 68, 166, 236
storms, 38
straight base of operations, 77
straits, 51–4, 56, 89–90, 120–3, 272–3; *see also* names of straits
strategic diversion, 214–18
strategic objectives, 80, 81, 97–108, 134, 293, 294
strategic points, 55–6
strategic position, 42–59
strategy, 1–2; *see also* naval strategy
strikes, 129–30, 152, 293
submarines, 34–5, 37, 230–1, 297; *see also* U-boats
Suez, 63, 115, 154, 172
Suez, Gulf of, 154, 242, 247, 248, 276
Suez Canal, 7, 27, 45, 48, 69, 89, 90, 121, 167, 174, 237, 247, 273, 275, 283
Sulina, 219
Sumatra, 54, 57, 122
Sunda Strait (Selat Sunda), 53–4, 89–90, 120, 121, 122
Sunderland, 194
super-refraction, 40
surface combatants, 297–8
surging actions, naval, 131, 194
Surigao Strait, 82; Battle of (1944), 131, 143
Suursaari, 278
Svalbards (Spitsbergen), 29
Sweden, 8, 10, 28, 30–1, 43, 51, 64, 91, 92, 102, 113, 160, 169, 176, 186, 213, 214, 235, 236, 239, 254, 259

INDEX

Swinioujscie (Świnioujście), 47, 66
Switzerland, 244
synchronization, 143
Syria, 45, 69, 151, 195, 244, 248, 261

tactical actions, 129–33, 229, 230, 249, 269, 293–4
tactical objectives, 81
tactics, 4–5
Taganrog, 284
Takhona, 170
Tallin, 23
Taman Peninsula, 53, 284–5
Taranto (Tarent), 50, 68, 137, 152
Tarifa, 51
Tartar Strait, 122
Tartūs, 195
Task Force 38, 141
Task Force 95, 283
Task Force 436, 200
Tegetthoff, Admiral Wilhelm von, 9, 50, 99, 149, 187
terminal area, 89
Termoli, 65
Terschelling, 232
Texel, 158, 232
Texel–Yarmouth line, 56
Thailand, 21
Thames, River, 69, 148
Thames estuary, 171, 245, 264
theater and its elements: base of operations, 75–80; decisive points, 81–2; definitions, 73–4; interior *vs* exterior lines, 85–8; land communications, 93–4; line of operations, 82–5; physical objectives, 80–1; sea lines of communication, 88–93
theater strategy, 1
thermal gradient, 36–7
Thursfield, J. R., 207
Tilsit, Peace of (1807), 8
Timen, Riever, 93
Timor, 57
Tirpitz, 210
Tirpitz, Admiral Alfred von, 171, 212
Tirso (Omodeo), Lake, 195
Tobruk, 45, 50, 272
Togo, Admiral Heichichiro, 86, 161, 209
Tokyo, 240

Tokyo Express, 70, 119, 156
Tomahawk land-attack missiles (TLAMs), 196
Torbay, 157
torpedoes, 35
Torrington, Arthur Herbert, Earl of, 207
total exclusion zone, 64; *see also* exclusion zones
Toulon, 45, 68, 106, 158, 159, 215
Tourville, Admiral (Comte) de, 149, 207
trade, maritime *see* attacks on maritime trade; protection of maritime trade; sea lines of communication
Trafalgar, Battle of (1805), 158, 211
Transcaucasus, 113
Trapani, 261, 286
Trappenjagd, Operation, 282
Trasimeno, Lake, 286
triangle-like base of operations, 80
Trieste, 69, 186
Tripoli, 91
Tromp, Admiral Maarten H. van, 157
Trondheim, 44
Troubridge, Admiral E. C. T., 217
Tsugaru Strait, 86, 121
Tsushima, Battle of (1905), 9, 86, 103, 106, 148, 209
Tsushima Strait (Korea Strait), 121, 240, 273
Tuapse, 62, 177, 231, 276
Tunis, 26, 68, 104
Tunis, Strait of, 273
Tunisia, 48, 55, 91, 189–90, 233, 262, 263, 272, 286
Turkey, 9, 17–18, 19, 106, 113, 151, 159, 160, 166, 167, 171, 178, 187, 191, 194–5, 217, 220, 228, 244, 248, 261, 270, 272, 274; Navy, 103, 206
Turkish Narrows, 194, 195, 220
Turkish Straits, 7, 17–18, 23, 48, 68, 103, 120, 121, 171, 178, 195, 219, 231; *see also* Bosporus; Dardanelles
Turner, Admiral R. K., 188
Tyrrhenian Sea, 22, 45, 103; Upper, 195

U-boats, 43–4, 45, 62, 76, 90, 101, 102, 113, 150, 163, 165, 168, 169, 172, 177, 209, 210, 214, 219, 230–1, 235, 247, 257, 264, 297

329

Ukraine, 151, 177, 178, 256
Ulsan, 281
Umm Qasr, 68, 102, 153, 196, 245
underground shelters, 64
Union Army, 267
Union Navy, 238, 267
United Arab Emirates (UAE), 201
United Nations (UN): Charter, 196, 198, 201; Eighth Army, 283; and Korean War, 10, 94, 192, 241, 268, 269, 273, 281, 283, 287; and peace operations, 196, 198, 199, 200; Protection Forces (UNPROFOR), 199; Security Council, 199, 244; Security Council Resolutions (UNSCR), 199, 200, 201; Tenth Corps, 283
United States: 1st Cavalry division, 193; 1st Marine Division, 122, 221; 1st Provisional Marine Brigade, 268; 3rd Division task force, 281; 6th Fleet, 201; 7th Army, 282; 7th Fleet, 269; X Army Corps, 281; 13th Marine Expeditionary Unit, 193; XXI Bomber Command, 240; 24th Division, 121, 193; 25th Division, 121; Air Force, 197; and attacks on maritime trade, 220, 221, 236, 237, 239–40, 241, 242, 244, 245; and bases, 71; continental shelf, 6; and exercising command, 187, 188, 189, 192, 193, 196, 201; and factors of space, 27, 33; and methods, 132; Navy, 3, 7, 66–7, 93, 98, 114, 117, 139, 151–2, 154, 156, 164, 184, 187, 196, 197, 198, 200, 201, 221, 234, 237, 241–2, 244, 260–1, 268, 269, 281, 283, 297; and protection of maritime trade, 255, 258, 260–1; and sea control, 119; and securing control, 149, 152, 153, 161, 166, 168, 171; and support of army flank, 268, 269, 281, 282, 287; Task Force 90, 281; and theater and its elements, 71, 74, 94; mentioned, 10, 51
Ushant, 91
Ust'-Dvinsk, 175

Valona (Vlorë), 55, 66, 68, 69, 172
Varna, 219, 231, 234, 272
VDS (variable depth sonar), 37

Vella Lavella, 281
Venetia, 149
Venetia Giulia, 186
Veneto, Republic of, 9
Venice, 30, 50, 65, 66, 100, 149, 159, 215, 285, 286
Vestfiord (Vestfjorden), 236
Vian, Admiral Philip, 263
Vicenza, 200
Vienna, 186; treaty of (1864), 9
Vietcong, 241, 242
Vietnam War (1965–75), 11, 26, 119, 241–2, 269
Vila, 156
Virginia, 267
Vis, 65
Visby, 186
Vitgeft, Admiral Vilgelm K., 212
Vitiaz Strait, 122
Vladivostok, 23, 45, 86, 114, 188, 209, 211, 212, 255
Vyborg, 177
Vyborg Bay, 160

War of 1689, 8
War of 1812, 86, 149
War of the Grand Alliance, 148
Warba, 68
Warsaw Pact, 77, 171
Wartburg barrier, 169, 170
Watchtower, Operation, 137, 188–9
Wegener, Admiral Wolfgang, 43, 44, 101
Wei-hai-wei, 149
Wellington, Duke of, 149
Weser, River, 155, 236
Weseruebung, Operation, 188
West Asia, 105; *see also* Asia
West Indies, 103
West European Union (WEU), 198, 199, 201
Western Approaches, 258
Western Dvina (Daugava), River, 54
Western Front, 268
Westphalia, 194
Wewak, 122, 125
White Sea, 32, 107
Wiking, Operation, 278
Wilhelmshaven, 62, 80
William, King, 8
wind noise, 39

INDEX

Windau (Ventspils), 170, 175, 286
Windward Passage, 51
Wisconsin, 196
Wonsan (Wŏnsan), 23, 220, 221, 241, 281
World War I (1914–18): attacks on maritime trade, 230–1, 234, 238–9, 244–5, 247; bases, 62, 63, 64, 66, 67, 68, 69; contesting control, 206, 208, 209, 210, 213, 216–18, 219, 220; exercising control, 187–8, 191, 192, 194, 195; factors of space, 27, 28, 29; fleet distribution, 106; importance of narrow seas in, 9–10; positions, 43–4, 47, 54, 56; protection of maritime trade, 257, 258, 264; sea control, 112, 113, 115, 117, 118, 123; securing control, 150, 155, 162, 163, 165–6, 168–9, 171–2; strategic objectives, 100–1, 102–3; support of army flank, 267, 268, 272, 277; theater and its elements, 76–7, 80, 91–2; mentioned, 7
World War II (1939–45): attacks on maritime trade, 227–8, 228–9, 230, 231–2, 233–4, 235, 236–7, 239–40, 247, 248; bases, 62, 63, 66, 67, 68, 69, 70, 71; contesting control, 210, 212, 213–15, 219–20; exercising control, 188, 189–91, 192, 195–6; factors of space, 19, 26–7, 28, 29; and fleet distribution, 107; importance of narrow seas, 10; methods, 131, 132, 133–4, 135, 137, 138, 141, 144; positions, 44, 47, 48, 50–1, 53–4, 55, 56, 57, 59; protection of maritime trade, 254–5, 256, 257, 258, 259, 260–1, 261–2, 262–3; sea control, 112–13, 114, 115–16, 118–19, 122–3, 123–5; securing control, 150, 151, 152–3, 153–4, 155–6, 163, 169–71, 172, 173–8; strategic objectives, 98, 100, 101, 102, 103; support of army flank, 268, 270, 271, 272–3, 274, 275–6, 277–81, 281–2, 282–3, 284–5, 285–6, 286–7; theater and its elements, 76, 77, 81, 82, 84, 85, 88, 90, 91, 92; use of term *narrow seas*, 7
Wrangel, Count Karl Gustav, 160

Yalta, 219
Yalta hills, 280
Yalu, River, 93; Battle of (1894), 149
Yamato task force, 171
Yellow Sea, 9, 10, 45, 114, 120, 149, 150, 168, 188, 212, 241, 295
Yentoa Bay, 161
Yevpatoriya, 219, 284
Yokohama, 240
Yom Kippur (Ramadan) War (1973), 90, 119, 151, 154, 166, 242, 247, 283; *see also* Arab–Israeli conflicts
Yucatan Channel, 51
Yugoslavia, 11, 18, 64, 125, 198–201, 285, 286
Yuya Bay, 240

Zagreb, 200
Zeebrugge, 23, 69, 80, 171
Zonguldak, 272